The Origin and Evolution of Galaxies

NATO ADVANCED STUDY INSTITUTES SERIES

Proceedings of the Advanced Study Institute Programme, which aims at the dissemination of advanced knowledge and the formation of contacts among scientists from different countries

The series is published by an international board of publishers in conjunction with NATO Scientific Affairs Division

A	Life Sciences	Plenum Publishing Corporation
B	Physics	London and New York
C	Mathematical and Physical Sciences	D. Reidel Publishing Company Dordrecht, Boston and London
D	Behavioural and Social Sciences	Martinus Nijhoff Publishers
E	Engineering and Materials Sciences	The Hague, London and Boston
F	Computer and Systems Sciences	Springer Verlag
G	Ecological Sciences	Heidelberg

Series C – Mathematical and Physical Sciences

Volume 97 – The Origin and Evolution of Galaxies

The Origin and Evolution of Galaxies

Proceedings of the NATO Advanced Study Institute held at Erice, Italy, May 11-23, 1981

VIIth Course of the International School of Cosmology and Gravitation

edited by

B. J. T. JONES

Observatoire de Meudon, France and Institute of Astronomy, Cambridge, U.K.

and

J. E. JONES

Institut d'Astrophysique, Paris, France

D. Reidel Publishing Company

Dordrecht : Holland / Boston : U.S.A. / London : England

Published in cooperation with NATO Scientific Affairs Division

Library of Congress Cataloging in Publication Data

CIP

NATO Advanced Study Institute (1981 : Erice, Italy)
The origin and evolution of galaxies.

(NATO advanced study institutes series. Series C, Mathematical and physical sciences ; v. 97)
Includes index.
1. Galaxies–Congresses. 2. Cosmology–Congresses.
I. Jones, B. J. T. (Bernard Jean Trefor), 1946- . II. Jones, J. E. (Janet Elizabeth), 1944- . III. International School of Cosmology and Gravitation. IV. Title. V. Series.
QB857.N37 1981 521'.582 82-18563
ISBN 90-277-1507-6
ISBN-13: 978-90-277-1517-3 e-ISBN-13: 978-94-009-7939-0
DOI:10.1007/978-94-009-7939-0

Published by D. Reidel Publishing Company
P.O. Box 17, 3300 AA Dordrecht, Holland

Sold and distributed in the U.S.A. and Canada
by Kluwer Boston Inc.,
190 Old Derby Street, Hingham, MA 02043, U.S.A.

In all other countries, sold and distributed
by Kluwer Academic Publishers Group,
P.O. Box 322, 3300 AH Dordrecht, Holland

D. Reidel Publishing Company is a member of the Kluwer Group

Softcover reprint of the hardcover 1st Edition 1985

CONTENTS

PREFACE

The Origin and Evolution of Galaxies is the outstanding problem of modern cosmology. Fortunately, we have a firm cosmological framework on which to base our theories (the hot big bang) and recently there has been substantial progress in providing observations which potentially can constrain these theories. The problem of galaxy formation is, as a consequence, one involving many diverse branches of physics and astrophysics. It has been the aim of the school, and this compendium of lectures and seminars, to bring together these diverse aspects at a level enabling research workers to understand what is going on in other corners of the subject and to see how progress in each area impinges on the others.

We are grateful to the contributors to this volume for allowing us considerable editorial license with their articles. We have attempted to provide a representative sample of the talks that were given at the school besides the texts of the invited lecturers. It is regrettable that for reasons of space we have had to leave out a number of other contributions.

The school was supported by NATO as part of their 'Advanced Study Institutes' program: we would like to thank them for their support. We also gratefully acknowledge financial support from the Italian Ministry of Public Education, the Italian Ministry of Scientific and Technological Research, the Italian National Research Council and the Sicilian Regional Government. The school was held in the incomparable environment of the Ettore Majorana Center for Scientific Culture at Erice: we thank the director of the center, Prof. A. Zichichi, for his hospitality and Dr Alberto Gabriele and his staff for the excellent local organisation. The directors of the International School of Cosmology and Gravitation, Professors N. Dallaporta and V. de Sabbata conceived the subject of this school and were responsible for much of the scientific organisation which resulted in so successful a meeting.

It is a great pleasure to acknowledge the help of Rosemary Wyse in the proof-reading and criticism of the manuscripts which make up this volume, and Sally Roberts for typing the whole book.

Bernard Jones,
Observatoire de Meudon,
and Institute of Astronomy
Cambridge.

Janet Jones,
Institut d'Astrophysique,
Paris.

WORDS OF WISDOM

'We do not understand what we are doing - a common experience in cosmology'

Peebles

'A Beethoven symphony is like 1/f noise'

Peebles

'The easiest way to go forward is to jump up straight in the air'

Centrella

'Isothermal fluctuations are not really isothermal'

Hogan

'Adiabatic fluctuations are not really adiabatic'

Bonometto

'You choose a cosmological model carefully - but that doesn't make any difference'

Wall.

PARTICIPANTS AND OBSERVERS

ADAMUTI, I.J., Bucharest, Romania
ALLEN, B., Cambridge, UK
ANILE, M., Catania, Sicily
APPLETON, P., Jodrell Bank, UK
BARBUY, B., Meudon, France
BAZANSKI, S., Warsaw, Poland
BEAN, A.J., Durham, UK
BHAVSAR, S.P., Sussex, UK
BONOMETTO, S., Padova, Italy
BRANCEWICZ, H., Krakow, Poland
CARLBERG, R., Cambridge, UK
CARR, B., Cambridge, UK
CAVALIERE, A., Roma, Italy
CECCARELLI, C., Roma, Italy
CENTRELLA, J., Austin, Texas
CHAUBEY, N., Sultanpurcavadhj, India
CHIOSI, C., Padova, Italy
CORKELL, R.W., Cambridge, UK
CORONA-GALINDO, M.G., Konstanz, West Germany
D'ALESSANDRO, F., Lecce, Italy
DALLAPORTA, N., Padova, Italy
DANESE, L., Padova, Italy
DE SABBATA, V., Bologna, Italy
DE ZOTTI, G., Padova, Italy
DI FAZIO, A., Frascati, Italy
DOMINGUEZ, R., Meudon, France
DWIGHT, V., Toronto, Canada
ELLIS, R., Durham, UK
FASANO, G., Padova, Italy
FINDLAY, D.A., Cambridge, UK
FLIN, P., Krakow, Poland
FONTI, S., Lecce, Italy
FRANCESCHINI, A., Trieste, Italy
FRANKEL, N.E., Melbourne, Australia
FRY, J.N., Chicago, Illinois
FRY, W., Madison, Wisconsin
FUSCO FEMIANO, R., Frascati, Italy
GASPERINI, M., Bologna, Italy
GERBAL, D., Meudon, France
GERHARD, O., Cambridge, UK
GODART, O., Louvain, Belgium
GRABINSKA, T., Wroclaw, Poland
GRIMLEY, P., Cardiff, UK
GURBUZ, F.G., Cambridge, UK
HICKSON, P., Vancouver, Canada
HOFFMAN, Y., Tel-Aviv, Israel
HOGAN, C.J., Chicago, Illinois
INGLIS, I.A.A., Durham, UK

JONES, B.J.T., Cambridge, UK
JONES, J.E., Cambridge, UK
JUSZKIEWICZ, R., Warsaw, Poland
KAISER, N., Cambridge, UK
KALINKOV, M., Sofia, Bulgaria
KASHLINSKY, A., Cambridge, UK
KRAWCZYK, S., Torun, Poland
KREMPEC-KRYGIER, J., Torun, Poland
LACEY, C.G., Cambridge, UK
LACHIEZE-REY, M., Saclay, France
LAKE, G., Cambridge, UK
LUMINET, J-P., Meudon, France
MACCALLUM, M.A.H., London, UK
MATTEUCCI, F., Padova, Italy
MILLER, J., Oxford, UK
MOFFAT, J.U., Toronto, Canada
MONTI, R., Bologna, Italy
MOTTA, S., Catania, Sicily
OCCHIONERO, F., Frascati, Italy
OORT, M., Leiden, Netherlands
PALMER, P.L., Cambridge, UK
PANDEY, S.N., Gorakhpur, India
PARTRIDGE, B., Haverford, Pennsylvania
PEEBLES, P.J.E., Princeton, New Jersey
PETROU, M., Cambridge, UK
PIETRANERA, L., Roma, Italy
PINEAU DES FORETS, G., Meudon, France
POLLOCK, M.D., Padova, Italy
PUGET, J-L., Paris, France
RECILLAS-CRUZ, E., Mexico City, Mexico
RIZZATI, P., Ferrara, Italy
SELLWOOD, J., Cambridge, UK
SHANDARIN, S., Moscow, USSR
SIROUSSE-ZIA, H., Paris, France
SOKOLOWSKI, L., Krakow, Poland
SREDNIAWA, B., Krakow, Poland
SUNDT, G., Trondheim, Norway
TARRAB, I., Paris, France
TENORIO-TAGLE, G., Munchen, West Germany
TERLETSKII, Y., Moscow, USSR
TYTLER, D.R, London, UK
URBANIK, M., Krakow, Poland
VAGNETTI, F., Frascati, Italy
VITTORIO, N., Frascati, Italy
DE WAARD, G.J., Leiden, Netherlands
WALL, J.V., RGO, UK
WHITE, S.D.M., Berkeley, California
WILKINS, D., Frascati, Italy
WINDHORST, R., Leiden, Netherlands
WYSE, R.F.G., Cambridge, UK
YUNQUIANG, Y., Trieste, Italy
ZABIEROWSKI, M., Wroclaw, Poland

INAUGURAL LECTURE

N. Dallaporta

Istituto de Fisica, Galileo Galilei, Padova, Italy

Among the several problems still unsolved in the field of cosmic physics, one of the most important - if not the most important - is certainly the question of the origin and evolution of galaxies. Perhaps the main reason for our still somewhat rudimentary understanding is the fact that galaxies are by far the most complicated configurations of celestial bodies met up to now, requiring the knowledge of almost every field of physics to cover their interpretation. Moreover, galaxies are in a sense situated midway between the domains of universal cosmology and the astrophysics of individual bodies. This leads to a bifurcated view of the subject so that research on galaxies is divided into different lines which are sometimes only loosely bound together. Theories concerning one or the other of these domains frequently focus on a simple viewpoint, without much regard to other aspects which are by no means less important or fundamental than the one considered. The most frequent polarization we meet is, on one hand, considering galaxies just as density fluctuations of the primeval medium, and, on the other hand, attributing the rather complex phenomenology of the whole body of observation on individual objects to the variation of one single physical parameter. Therefore, a comprehensive and impartial overlook over the whole field is a somewhat difficult undertaking and in fact there are not many up-to-date extended summaries of the kind presented in this School.

This is the reason why, several years ago, it was felt that a whole Course devoted to the origin and evolution of galaxies would be of the utmost interest. The Erice International School of Cosmology and Gravitation, planned and realized by Prof. De Sabbata in 1972 and running uninterrupted since then, provides

B. J. T. Jones and J. E. Jones (eds.), The Origin and Evolution of Galaxies, 1–7.

the ideal framework for such a Course. At the time when this was first considered, the situation was less involved than it is now. The main field of research was then focussed on two main items rather loosely related to each other: the problem of the amplification of primeval fluctuations as they evolve through the classical cosmic eras, and the main collapse phases of proto-galaxies.

Since then however, many new data and new points of view have appeared concerning both the intrinsic properties of galaxies and clusters and their connection with general cosmology. As a consequence, there is little doubt that a Course held this year has to offer a general overview on a field with a rather different emphasis than it would have had some five or six years ago, when the first idea of the Course was conceived.

In order to stress this programme and owing to the fact that I personally belong to a generation which has been more involved with the past problems than with the present ones, I have thought it worthwhile as an introduction to look back on the subject as it was some time ago, and leave it to the different speakers to stress in their lectures the main changes that have taken place in the meanwhile.

The main streams of research in the past concerning galaxies could be summarised by posing two questions: 1) Why do galaxies exist, and 2) why, given that they exist, are they as we observe them? The first is related to the early cosmological phase, and the second to the later astrophysical one. The apparently philosophical nature of the first question has its experimental basis in the obvious contrast between our direct observation of the cosmos, divided into lumps of matter and separated by large volumes of space filled with matter at much lower densities, as opposed to the cosmological theory which assumes the universe to be everywhere homogeneous and isotropic.

It was shown that spontaneous thermodynamical fluctuations of galactic mass are much too weak to be able to grow to the observed density contrasts in so short a period as the lifetime of the universe. On the other hand, attempts to generate the necessary fluctuations by invoking cosmic turbulence failed. In order that such turbulence should survive to the epoch of recombination it would have to have been so strong as to perturb the isotropy of the background black body radiation by an amount not compatible with observations. The question: 'why do galaxies exist?' was replaced by less fundamental questions, e.g. 'given the present observed fluctuations in density, what was their amplitude at recombination and how far in the early phases of the Big Bang can we push back their origin?'

An answer to this new view of the problem has been given by the gravitational instability theory. We need, at the epoch of recombination, relative fluctuations $\delta\rho/\rho$ of the order of one thousandth to one hundredth (depending on the choice of the present cosmic density parameter Ω) in order to yield the present-day observed density contrasts. It is adequate to postulate primeval fluctuations of constant amplitude of the order of 10^{-4} for initially physically disconnected portions of the cosmos entering the horizon during the radiation era. The fate of these fluctuations and their complicated behaviour when the ionization varies around the time of recombination are rather different for the two components, adiabatic and isothermal, into which the perturbation can be decomposed.

For adiabatic fluctuations, involving both matter and radiation, the damping turns out to be quite strong for smaller masses during the ionized matter era, and the outcoming fluctuation spectrum at recombination follows a power law in mass, $M^{-1/3}$, with a lower boundary around 10^{12}–10^{14} solar masses. This result imposes a strong constraint on the type of theory to be chosen for galaxy formation: it yields only large mass fluctuations (corresponding to clusters of galaxies) and hence indicates that individual galaxies form later from these larger clouds by further fragmentation. A theory of hierarchical derivation of galaxies from clusters has to be looked for.

If one considers instead isothermal fluctuations, which involve matter only, the damping during the ionized era is negligible and a general spectrum of the same power law type but extending down to the Jeans mass at recombination, (around 10^6 $M_\odot$, is expected to emerge. In this case, most of the matter turns out to be condensed into lumps with mass of the order of globular clusters, and galaxies should be then considered as being formed from the merging of such smaller clouds into larger entities.

Arguments in favour of both theories have been advanced by different authors, (observation of pancake like shaped clusters of galaxies for the first, and absence of any favoured galactic scale resulting from the correlation functions for the second), so that the question of whether the fluctuations are to be considered as prevalently of the adiabatic or of the isothermal type has been one of the most debated problems: no doubt it will be abundantly discussed in different courses. In any case the unsettled state of this question is largely responsible for the amount of uncertainty concerning the initial conditions defining a protogalaxy.

Turning to the second main question, whatever the mechanism of its formation, it is currently assumed that a given mass corresponding to a single galaxy starts its individual evolution by first following the general cosmic expansion. Owing to its higher average density with respect to the surrounding medium, it lags behind the general expansion, reaches a maximum extension and collapses back owing to its self-gravitational attraction. The bulk of the work concerning this collapse phase is based on some common assumptions which may be summarized as follows:

a) The initial geometrical shape has almost always, in the older models, being rather arbitrarily assumed to be of constant density with spherical symmetry or at most flattened rotational symmetry. In order to collapse, the mass of the protogalaxy has to be larger than the Jeans mass. It follows that the virial condition is not satisfied, temperature and gas pressure effects are irrelevant and the contraction will be of the free fall type.

b) Star formation is assumed to have been going on since the expansion phase. Thus that at maximum expansion two components, a gaseous one and a stellar one, will be present, which will undergo two different kinds of collapses: a violent relaxation non-dissipative collapse for the stellar component, and an energy non-conserving dissipative collapse for the gaseous component. These two different collapse processes will thus give rise to the stellar halo and to the gaseous disk respectively, in both cases the final state being a rotationally supported virial equilibrium configuration.

c) The different values for the mass ratio of disk to halo is considered in these approaches to be responsible for the different morphological types of the Hubble sequence. The variation of disk to halo ratio is generally attributed to the variation of a single parameter in the initial conditions. Different choices have been proposed by various authors for this fundamental differentiating factor, the main proposals essentially being: A) a variation with time (of the stellar birthrate function) during the phases of the collapse: or B) a different mass ratio at maximum expansion of the stellar and gas components. This latter variation is due in turn either (i) to a variation of the redshift value z_m at maximum expansion, because of some time spread in the formation of the fluctuations at recombination (early z_m leading to ellipticals, late z_m to spirals), or (ii) to different values of the total angular momentum, (large momenta yielding spirals, low momenta ellipticals). The angular momentum has been thought to arise from torques produced by tidal interactions among galaxies, or as a statistical mean value resulting from the sum of the component momenta of the single units, as in the cloud merging picture for the original formation of the protogalaxy. These different assumptions appeared to yield comparable results in the interpretation of the Hubble sequence, so some combination of the different factors invoked could yield a better result.

As is now generally known, this investigation of highly symmetric models has been truncated by the observation, in the seventies, that the rotation velocity of ellipticals is generally too low to support rotational equilibrium of the structure. Several ideas have been put forward in order to try to explain these surprising data.

One such theory is based on a low z_m redshift value for ellipticals, corresponding to a totally gaseous initial state. For non-virial initial configurations in pressure free gravitational collapse, the possible initial spherical symmetry, even if postulated, is unstable to the growth of both non-homology and non-sphericity. Hence the first phase consists of a dissipative collapse of the primeval gas, leading to a sheet-like, very flattened pancake, from which, after stars have formed, a violent relaxation of these stars yields a non-rotationally supported equilibrium state possibly corresponding to what is observed. Such a result is thus mainly a consequence of an inversion in time of the dissipative and non-dissipative collapse phases. Unfortunately, the one-parameter hypothesis for the Hubble sequence is rather spoiled in this view.

This new viewpoint concerning the origin of ellipticals has been the starting point for investigations aiming mostly to stress the fundamental importance of asymmetry in the process of galaxy formation. On the observational side, much stress has been given to the apparent twisting of the axes of the isophotes in ellipticals which could correspond to triaxial equilibrium configurations. On the theoretical side, it was surmised that spherically symmetrical initial fluctuations are quite unlikely to occur. Thus a much more realistic starting point consists of considering irregular and random motions in different directions of irregularly shaped fluctuations, colliding and thus forming shock waves which will propagate and increase the temperature up to virial values, so that gas pressure will become of importance. One then has to consider as a new alternative a pressure supported collapse of the primeval gas of the protogalaxy after its maximum expansion has been reached. The fate of the gravitating cloud will then be ruled by energy loss due to radiative cooling, and the relative ratios of the cooling time, the free fall time and the Hubble expansion time. The different possibilities lead to quite new views of galactic evolution, with some remarkable coincidences with the observed mass range and dimension of galaxies.

The picture I have drawn has been for some years the basic line of thought concerning galactic formation and evolution, notwithstanding several caveats. Its apparent simplicity is due to the assumption that a galaxy evolves in isolation under the influence of its own self gravity alone. There is however increasing evidence that a variety of other factors, entirely neglected in what I have presented above, may play a dominant role. With no aim of completeness I will just mention some of them.

First, the whole phenomenology concerning radio galaxies and quasars, although developed since the sixties, has been only recently linked to phenomena occurring in the nuclei of large galaxies. Whichever will be the adopted assumptions concerning the source of the enormous amounts of energy developed in these events, it will certainly modify the early phases foreseen for galactic evolution.

New information has been obtained from the chemical composition analysis of different galaxies and parts of galaxies, so that now the chemical evolution strongly complements the dynamical information.

An entirely new phenomenological picture is developing owing to the realisation that most galaxies are not isolated but part of clusters. Comparisons of galaxies which are members of clusters with those in the field indicate that morphology is in part determined by interactions between galaxies and their neighbours, or an external gaseous medium. There are further indications of the merging of smaller galaxies to form larger ones. This could produce the ionized medium in which the cluster galaxies are embedded, whose properties are mostly revealed through X-ray emission. In a more general sense, the nature of the clustering has been uncovered by the studies of correlation functions and by numerical simulations of the gravitational processes. These studies seem to indicate that the masses implied for cluster galaxies should be much larger than generally assumed, which underlines in a more acute way the old problem of the missing mass of the clusters.

It is currently thought that a possible solution to this dilemma may be provided by a non-zero mass for the neutrino, as has been surmised from some recent experiments of particle physics. If this were so, and the neutrino mass were as large as nowadays proposed, the neutrinos present in the cosmos would be enough to close the universe and thus considerably change the whole frame of cosmic evolution.

However, the most important viewpoints on the field are probably those concerning the effects of galaxies on their cosmological surroundings at different cosmic epochs. From a cosmological point of view galaxies are just a consequence of perturbations of the cosmic medium. However their effects on their cosmic surroundings may be tested directly. Thus one is led to study the role of the fluctuations in the very initial phases of the big bang, their bearing on the early models of the universe, whether symmetric or asymmetric, and what effects may be reflected in the possible asymmetries of the microwave radiation.

This brief and incomplete overview may give a rough idea of the large number of new ideas which have now entered the study of the formation and evolution of galaxies. As may be seen, the main stress of research in this field has now moved from the interpretation of the Hubble sequence to the more fundamental questions associated to the interaction of galactic fluctuations with their environment. The goal of the present Course, as organized and directed by Dr Bernard Jones, aims essentially to be as far as possible a first and comprehensive overview of most of these new fundamental data and ideas I have mentioned, which have been the subject of research and discussion mainly in these past five years. The main line of the Course, according to Dr Jones' own indications, will consider the origin of the fluctuations leading to galaxies and the details of the early cosmological models consistent with them, then the gravitational enhancement of the fluctuations to form galaxies and clusters with the resultant star formation. An important aspect of the Course will be a close look at the observational consequences of evolution as relates to chemical changes, to the spectrum and distribution of the microwave radiation and to deep observations of galaxies, quasars and radio sources.

My last task in this introduction is to thank all the staff of the Ettore Majorana Center for Scientific Culture, and especially Dr Gabriele and Miss Savall for their outstanding work concerning the organization of the Course, and to give a hearty welcome to all participants, both teachers and students. We have among us Professors Bonometto, Centrella, Chiosi, Ellis, Flin, Hogan, MacCallum, Partridge, Peebles, Wall and White, who have kindly agreed to give a number of lectures, and who are all very well known for their work in different domains of cosmology and galactic evolution. Moreover, several other scientists will give seminars closely related to the main topics. The number of students enrolled is large, thus giving an indirect confirmation of the need that was generally felt for an Course on galaxies. As stressed by the Director of the Course, it is in the spirit of the school and indeed one of its aims to maintain and promote the largest possible personal contacts between teachers and students: this maximises the benefit for every participant. It is with the wish that everyone enjoys a very pleasant and fruitful stay in Erice that I close this introductory speech.

RELATIVISTIC COSMOLOGY FOR ASTROPHYSICISTS

M.A.H. MacCallum

Department of Applied Mathematics, Queen Mary College,
Mile End Road, London E1 4NS

1. ANISOTROPIC AND INHOMOGENEOUS COSMOLOGIES

1.1 Introduction

My aim in these lectures is to provide that background information about modern mathematical work in relativistic cosmology which I feel astrophysicsts should know. It is thus a personal selection (though it omits certain things I might have said had I not known that Dr Centrella would ably say them).

The really basic things, the theory of relativity and of the isotropic spatially-homogeneous models, were omitted on the grounds that astrophysicists already knew them. Quantum field theory and quantum gravity were omitted because I do not properly know them, and also for lack of time. Other lectures at the school discuss these matters.

The style is not that of a formal review: indeed some friends and colleagues may be surprised by my lack of rigour. The idea was to present some folklore, which, like most astronomical folklore, is full of errors of detail and ignores technical pitfalls, but will still (I hope) give a reasonable overall picture.

For the same reasons, the bibliography is far from complete. I have reviewed the material of section 1 (and parts of sections 2 and 3) more formally elsewhere (MacCallum 1979a). Other similar reviews are Ryan and Shepley (1975), Criss et al. (1975) and MacCallum (1973, 1979b). The discussion of horizon definitions in section 2 uses MacCallum (1971), but such topics are much more

B. J. T. Jones and J. E. Jones (eds.), The Origin and Evolution of Galaxies, 9–39.

fully covered in Hawking and Ellis (1973). The latter, together with an excellent more recent review (Tipler, Clarke and Ellis 1980), gives a full account of the singularity theorems and related work which is discussed in section 3. To these references, I added only a few more recent papers, and some whose importance is perhaps not sufficiently stressed in the literature already quoted. Otherwise I relied on these references to supply the bibliography, with the exception of section 3 where I took the opportunity to comment on recent controversies, which necessitated a fuller list of references.

1.2 Why consider non-standard models?

The standard models are those with metric (in my favourite coordinates)

$$ds^2 = -dt^2 + R^2(t)(dr^2 + f^2(r)[d\theta^2 + \sin^2\theta \, d\phi^2]), \tag{1}$$

where f = sin r, r or sinh r when the curvature parameter k is 1, 0 or −1. They fit the observations remarkably well. In what follows, I call these FRW models: F for Friedman, the first to study their dynamics in a general way, and RW for Robertson and Walker, who proved some geometrical properties of (1). They provide a comfortable mattress on which astrophysical cosmology rests: however, relativists, like the princess in the fairy story, feel some peas under this mattress.

The first P is Physics. Equation (1) cannot be exact. Perturbations of it include some modes which grow in each direction of t. Why then should (1) be thought a good approximation at all t? Moreover, it would be nice to know if certain features of (1) are general: for example, the existence of a big bang, the irrelevance of spatial curvature at early times, the effects of horizons, or the difficulties in explaining galaxy formation.

A second P is Prudence. In case anisotropy or inhomogeneity show up in the data, we should be prepared.

The third is Philosophy. It seems unsatisfactory to rely on initial conditions, rather than physical processes, to explain observed uniformity, yet we 'see', by the microwave radiation, regions which in FRW models were not, at decoupling, causally connected but had (almost) identical properties (see Prof. Partridge's lectures).

1.3 What models are there?

One can arrange the possibilities in ascending order of mathematical difficulty. (For exact solutions of each type see Kramer et al. 1980.)

1. Homogeneous spacetimes. No physical variable varies in space or time. These are 'steady-state' universes. The equations become algebraic. Such models are mathematically simple, but physically (probably) irrelevant.

2. Spatially-homogeneous models. Here the equations involve a single independent variable, time, so they are ordinary differential equations (ODEs). The isotropic cases are FRW. The anisotropic cases are the Kantowski-Sachs and Bianchi models (see below).

3. Temporally-inhomogeneous models (e.g. static spherically-symmetric models). These obey ODE in one space variable. They have been used in cosmology (see e.g. MacCallum 1979a), but are not generally believed to be correct. (Some mathematical aspects are considered in Harness 1982).

3. Spacetimes with two-dimensional spatial symmetry. The equations become partial differential equations (PDE) in one space and one time variable. These include spatially self-similar models (in which the space dependence is a simple scaling), nonstatic spherically-symmetric models with

$$ds^2 = -A^2\, dt^2 + B^2\, dr^2 + C^2\, (d\theta^2 + \sin^2\theta\, d\phi^2), \qquad (2)$$

where A, B and C are functions of r and t, and spaces with metrics dependent on only two variables. Note that this last category does not include (2), since that metric has $\sin^2\theta$ in it: it does include Gowdy metrics, cylindrically-symmetric metrics, and the metrics considered by Dr Centrella.

5. The rest. In general the equations are PDE in four variables though there are special cases (other than 4 above), dependent on only 2 or 3 variables (see e.g. Kramer <u>et al.</u> 1980). One interesting class is the Szekeres metrics

$$ds^2 = -dt^2 + e^{2A}\, dx^2 + e^{2B}\, (dy^2 + dz^2), \qquad (3)$$

where A and B depend on all four coordinates, and fluid flows on lines x^i = constant. (My conventions are that Latin indices run from 1-3, Greek from 1-4, and the time coordinate is x^4. I strongly urge this last convention on relativistic cosmologists so that confusion with subscript zero, denoting initial conditions is avoided.)

All the above classes have given useful examples in cosmology (see MacCallum 1979a).

1.4 Spatially-homogeneous metrics

Physically, these correspond to universes in which all comoving observers record the same history of the universe. Mathematically, this implies that for any two observers A and B there is a symmetry mapping the universe, considered as centred on A, to the same universe, considered as centred on B. Without proof, I state that the only possible cases are the Kantowski-Sachs metric,

$$ds^2 = -dt^2 + A^2\, dx^2 + B^2\, (d\theta^2 + \sin^2\theta\, d\phi^2), \tag{4}$$

where A and B depend only on t (for which reason it is misleading to compare (4) with (2): $x = 0$ is not a centre), or Bianchi models, which have a three-dimensional group of translation symmetries. Lie group theory tells us such groups are characterized (up to global topology) by the commutators of infinitesimal generators (in this case, infinitesimal translations): the idea is familiar to all physicists in the particular case of the rotation group and the angular momentum operators. The most general form of the commutators, for three generators $\underline{\xi}_a$, is

$$\begin{aligned}
[\underline{\xi}_2, \underline{\xi}_3] &= N_{11}\underline{\xi}_1 + (N_{12}-A_3)\underline{\xi}_2 + (N_{13}+A_2)\underline{\xi}_3,\\
[\underline{\xi}_3, \underline{\xi}_1] &= (N_{12}+A_3)\underline{\xi}_1 + N_{22}\underline{\xi}_2 + (N_{23}-A_1)\underline{\xi}_3,\\
[\underline{\xi}_1, \underline{\xi}_2] &= (N_{13}-A_2)\underline{\xi}_1 + (N_{23}+A_1)\underline{\xi}_2 + N_{33}\underline{\xi}_3,
\end{aligned} \tag{5}$$

This has obvious cyclic symmetries, N_{ab} is the symmetric part, and A_b represents the antisymmetric part of the array of coefficients (called structure constants) of the right hand sides of equations (5). The consistency condition for (5), i.e. the Jacobi equation,is

$$N_{ab}\, A^b = 0\,, \tag{6}$$

By linear transformations of the basis vectors $\underline{\xi}_a$ one can achieve $N_{ab} = \mathrm{diag}\,(N_1, N_2, N_3)$, where the N_i are 1,0 or −1, and $A_b = (A,0,0)$. (By (6) A_b must be an eigenvector of N_{ab}.) The distinct cases (allowing re-numbering of the N_i) are the Bianchi types given in the table below.

Bianchi's type III is the same as VI_h, $h = -1$. The FRW universes are special cases of types I (or VII_o) for $k = 0$, V (or VII_h) for $k = -1$, and IX for $k = 1$.

Type	I	II	VI_o	VII_o	VIII	IX	V	IV	VI_h $h=-A^2$	VII_h $h=A^2$
A	0	0	0	0	0	0	1	1	A	A
N_1	0	1	0	0	-1	1	0	0	0	0
N_2	0	0	-1	1	1	1	0	0	-1	1
N_3	0	0	1	1	1	1	0	1	1	1

Without proof, I state that one can choose a frame of vectors $\underline{e}_a$ with components $e_a{}^i$ dependent only on spatial variables, the surfaces of symmetry being t = constant, so that the $\underline{e}_a$ have commutators of the same form as (5) and the spacetime metric is

$$ds^2 = -dt^2 + g_{ab}\, e^a{}_i\, e^b{}_j\, dx^i\, dx^j, \qquad (7)$$

where $e^a{}_i$ is the matrix inverse of $e_a{}^i$. The evolution of the universe is then represented by the time-dependence of the six components of g_{ab}.

I shall not write out the Einstein equations or give a full discussion of all properties of (7). I mention only a few points of later relevance.

(a) Choice of variables

A very useful way to parametrise the matrix $g_{ab}(t)$ is to find all transformations of the $\underline{e}_a$ which preserve the structure constants. One then has

$$\underline{e}'_a = \bar{M}_a{}^b \underline{e}_b \qquad (8)$$

for some matrix $\bar{M}_a{}^b$. One can take $\bar{M}_a{}^b$ to be time-dependent, and choose it in such a way that

$$g_{ab} = D_{cd}\, M^c{}_a M^d{}_b \qquad (9)$$

where $M^c{}_a$ is the inverse of $\bar{M}_a{}^c$, and D_{cd} is specially simple (diagonal, usually).

An example is given by type VII metrics. Here the appropriate $M^c{}_a$ is given by

$$M^c{}_a = \begin{bmatrix} 1 & b_2 & b_3 \\ 0 & e^{\lambda}\cos\phi & e^{\lambda}\sin\phi \\ 0 & -e^{\lambda}\sin\phi & e^{\lambda}\cos\phi \end{bmatrix} , \qquad (10)$$

and one can then make

$$D_{cd} = \text{diag}\ (a^2,\ e^{\mu},\ e^{-\mu}).$$

The function $\phi(t)$ can be interpreted in terms of vortical motion or cylindrically polarized waves.

The idea appears in Doroshkevich, Lukash and Novikov (1973), Collins and Hawking (1973) and Bogoyavlenskii and Novikov (1975) and is stated most clearly in Jantzen (1979) and Siklos (1980). The forms of $M^a{}_b$ were also found in Harvey (1980). For Bianchi IX, the first case intensively studied, it happens that $M^a{}_b$ is an ordinary rotation matrix, which misled some of us into thinking that this was always a good choice.

(b) Generality

The generality of a Bianchi metric is measured by the number of free initial values that give physically distinct solutions. Since the allowed reparametrisations are given by the possible (time-independent) $M^a{}_b$ described in the previous paragraph, the most general types are those in which $M^a{}_b$ has the fewest parameters, which are essentially those with the largest number of non-zero entries in the table (because normalization of a non-zero entry reduces the re-scalings allowed). The most general are types VIII, IX, VI_h and VII_h.

(c) Diagonalization

One can always restrict g_{ab} to be diagonal. However, this may imply restrictions on the $T_{\mu\nu}$. Only in certain cases can $T_{\mu\nu}$ also be diagonal, and diagonal g_{ab} is only compatible with the canonical choice of N_{ab} if T_{4a} is parallel to A_a. Conversely, if T_{4a} is zero, restrictions are imposed on g_{ab}.

If the energy-momentum tensor is that of a perfect fluid, the fluid flow need not be along the x^i = constant lines of (7). If not, the model is called 'tilted'. In this case T_{4a} is non-zero, and the fluid flow has an (antisymmetric) vorticity tensor, whose components (for four-velocity $u_{\mu} = (u_i, u_4)$ where $u_i = U_a\, e^a{}_i$) are

$$w_{a4} = 1/2\ U_{a,4}, \qquad w_{23} = N_1\, U_1\ ,$$

$$w_{31} = N_2\, U_2 - A\, U_3, \quad w_{12} = N_3\, U_3 + A\, U_2\ ,$$

in the basis of (7).

1.5 Why one might expect anisotropy

For a metric of the form

$$ds^2 = -dt^2 + g_{ij}\, dx^i\, dx^j \tag{11}$$

(of which (7) is a special case), one obtains Einstein equations

$$\ell^{-3}(\ell^3\theta^a{}_b)^\cdot + {}^3R^a{}_b = \text{matter terms}, \tag{12}$$

$$2\theta^2/3 = \sigma^a{}_b\sigma^b{}_a - {}^3R + \text{matter terms}, \tag{13}$$

where: $\theta_{ab} = 1/2\ d(g_{ab})/dt$ (Centrella's $K_{ab} \equiv -\theta_{ab}$),

$\theta = \theta^a{}_a$, $\sigma^a{}_b = \theta^a{}_b - \theta/3\ \delta^a{}_b$:

${}^3R^a{}_b$ is the three-dimensional curvature of t = constant, and ${}^3R = {}^3R^a{}_a$; and $\ell^6 = \det g_{ab}$.

As $\ell \to \infty$, one expects ${}^3R^a{}_b \sim \ell^{-2}$ and the matter terms $\sim \ell^{-3}$. Thus unless ${}^3R^a{}_b$ is isotropic, $\theta^a{}_b$ cannot be (by (12)). Therefore universes with anisotropic ${}^3R^a{}_b$ cannot approach FRW.

In fact, even in those universes with isotropic ${}^3R^a{}_b$ the FRW cases may be unstable. For VII_h ($k = -1$) this was shown by Collins and Hawking (1973), and for VII_0 an instability exciting the vortex mode (ϕ in (10)) is predicted to arise from Hawking radiation from black holes by Barrow and Tipler (1978). Thus FRW universes could be unstable to perturbations not preserving the I, V or IX symmetry.

As $\ell \to 0$, if ${}^3R^a{}_b$ and matter terms are negligible in (12), $\theta^a{}_b \sim \ell^{-3}$ and this dominates the right side of (13). An obvious exception is when $p = \rho$ (note that to convert metrics above to normal units with time τ, one replaces t by $c\tau$ and p by p/c^2), which gives $\rho \sim \ell^{-6}$.

These arguments are far from rigorous, but they suggest that most relativistic cosmologies start with anisotropies measured in terms of, say,

$$\varepsilon = |3\sigma^a{}_b\sigma^b{}_a/2\theta^2|^{1/2},$$

around 1 and finish with $\varepsilon \sim O(1)$, maybe $\varepsilon \sim 1/4$.

in the basis of (7).

1.5 Why one might expect anisotropy

For a metric of the form

$$ds^2 = -dt^2 + g_{ij}\, dx^i\, dx^j \tag{11}$$

(of which (7) is a special case), one obtains Einstein equations

$$\ell^{-3}(\ell^3 \theta^a{}_b)^{\cdot} + {}^3R^a{}_b = \text{matter terms}, \tag{12}$$

$$2\theta^2/3 = \sigma^a{}_b\sigma^b{}_a - {}^3R + \text{matter terms}, \tag{13}$$

where: $\theta_{ab} = 1/2\ d(g_{ab})/dt$ (Centrella's $K_{ab} == -\theta_{ab}$),

$\theta = \theta^a{}_a$, $\sigma^a{}_b = \theta^a{}_b - \theta/3\ \ \delta^a{}_b$:

${}^3R^a{}_b$ is the three-dimensional curvature of t = constant, and ${}^3R = {}^3R^a{}_a$; and $\ell^6 = \det g_{ab}$.

As $\ell \to \infty$, one expects ${}^3R^a{}_b \sim \ell^{-2}$ and the matter terms $\sim \ell^{-3}$. Thus unless ${}^3R^a{}_b$ is isotropic, $\theta^a{}_b$ cannot be (by (12)). Therefore universes with anisotropic ${}^3R^a{}_b$ cannot approach FRW.

In fact, even in those universes with isotropic ${}^3R^a{}_b$ the FRW cases may be unstable. For VII_h ($k = -1$) this was shown by Collins and Hawking (1973), and for VII_0 an instability exciting the vortex mode (ϕ in (10)) is predicted to arise from Hawking radiation from black holes by Barrow and Tipler (1978). Thus FRW universes could be unstable to perturbations not preserving the I, V or IX symmetry.

As $\ell \to 0$, if ${}^3R^a{}_b$ and matter terms are negligible in (12), $\theta^a{}_b \sim \ell^{-3}$ and this dominates the right side of (13). An obvious exception is when $p = \rho$ (note that to convert metrics above to normal units with time τ, one replaces t by $c\tau$ and p by p/c^2), which gives $\rho \sim \ell^{-6}$.

These arguments are far from rigorous, but they suggest that most relativistic cosmologies start with anisotropies measured in terms of, say,

$$\varepsilon = |3\sigma^a{}_b\sigma^b{}_a/2\theta^2|^{1/2}\ ,$$

around 1 and finish with $\varepsilon \sim O(1)$, maybe $\varepsilon \sim 1/4$.

1.6 The actual isotropy, and how to achieve it

(a) Local measurements, e.g. of the Hubble flow, set limits of the order $\varepsilon < 1/4$, and

$$\eta = |w_{\mu\nu}w^{\mu\nu}/\theta^2|^{1/2} < 1/4.$$

(Kristian and Sachs, 1966: a modern re-analysis could be done.)

(b) Probing further back in time, the microwave background, which measures the integrated effects since the last scattering, gives (somewhat model-dependent) limits of order 10^{-3} (Collins and Hawking 1972).

(c) Assuming $\theta^a{}_b \sim \ell^{-3}$, the need for θ to approximate its FRW value during the nucleosynthesis era implies (see (13)) a present-day value of $\varepsilon < 10^{-6}$ (Barrow 1976).

(d) Barrow and Matzner (1977) suggested that on grounds of total entropy ε must always have been small: assuming a large ε, later reduced by dissipation, leads to too many photons per baryon, if the energy goes into photons. The creation of baryons by GUTS (see Dr Fry's contribution) may limit the values of this argument by restoring the photon/baryon ratio

In any case it seems clear that now, and in the recent past, anisotropy has been small (unless one takes some special anisotropic model with $\sigma^a{}_b$ oscillating and by chance small enough during nucleosynthesis).

There are very general reasons for supposing that no ordinary (classical) physics can explain the isotropy. Essentially this is because any process governed by regular differential equations can be started from observationally forbidden values and integrated backwards in time from now (t_0) to any given earlier time t_1, so giving an open set of conditions at t_1 which lead to impossible models now. To get round this argument one needs processes working all the way back to the singularity at $\ell = 0$ ($t = 0$), or processes with discontinuities (e.g. quantum jumps).

The conclusion is that either one must believe in special initial conditions, or appeal to the 'magic' of early universe quantum processes to impose low anisotropy. (One interesting reason for expecting FRW initial conditions is given by Penrose (1979).)

I do not personally feel at ease with this conclusion (see Section 3.6 for some further comments).

2. HORIZONS

2.1 Definitions

[Technical note: I am assuming various nice properties, e.g. 'hole-free-ness', wherever necessary.]

Our past null cone is the set of points on light rays reaching us now (at $\tau = \tau_0$ on our proper-time clocks).

Our particle horizon at time τ_0 is the (boundary of) the set S of particles which could have causally influenced us at times $\tau' \leq \tau_0$ (i.e. whose worldlines intersect our past null cone at some time $\tau_1 \leq \tau_0$, and hence at all $\tau_1 \leq \tau \leq \tau_0$). A spacelike surface $\tau = \tau_0$ can be defined in all 'stably causal' spacetimes (Hawking and Ellis 1973), though not just by naive generalization of the FRW t coordinate. One can then take the boundary B(S), a set of worldlines, or the set $B_0(S)$ where B(S) meets $\tau = \tau_0$, to be the particle horizon, cf. MacCallum (1971).

Our event horizon is the set of events that can causally influence us at some time $\tau \leq \infty$.

Our creation light cone (once called the 'particle horizon', for reasons related to note 5 below) is the set of points on light rays emitted by us at our creation, $\tau = 0$ (more formally, the boundary of the point set of light rays emitted at $\tau > 0$.)

Now see Fig. 1. The singularity is, as is conventional, represented by a wiggly line: I assume it is a spacelike big-bang (see section 3); U labels our worldline, A,C,D,F those of other particles. τ is our proper time since the big bang. Its value at certain points on U is indicated. A surface $\tau = \tau_0$ is drawn: it meets U at our present position P. PX and PY represent our past null cone, OM and ON our creation null cone: the worldlines of D and F represent B(S), and the points V and W represent $B_0(S)$.

Notes on definitions

1. A is inside our particle horizon at τ' but C is not. Both are at τ_0.

2. A is inside our particle horizon at all time $\tau \geq \tau_A$. A particle cannot leave our particle horizon, unless it leaves spacetime altogether, i.e. is destroyed (assuming spacetime has no holes).

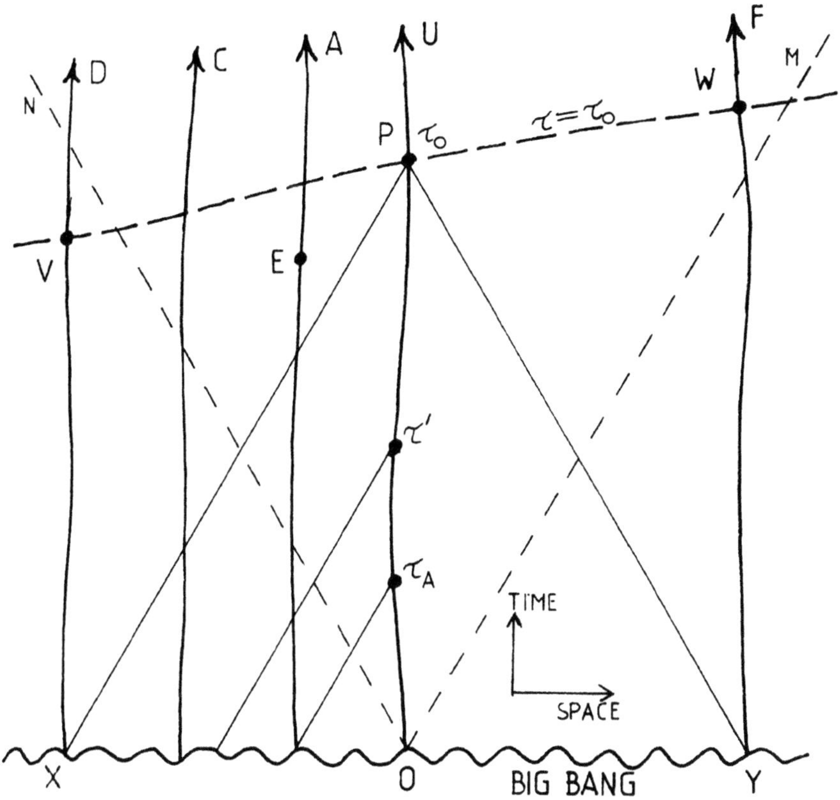

Figure 1. Our past null cone, our particle and event horizons, and our creation light cone.

3. Although A is in our particle horizon at τ_0, we have not yet recorded the event E. Signals take time, even within particle horizons, and one should avoid speaking as if communication within particle horizons were instantaneous.

4. A may experience events E outside our event horizon, although A is in our particle horizon. For example, A may fall into a black hole: if we do not, we never see the events A experiences inside the black hole. Nevertheless, we continue to receive light from A, emitted just as A fell over our event horizon, but with redshift $z \to \infty$ as $\tau \to \infty$.

5. In an FRW universe, ON and OM meet $\tau = \tau_0$ at W and V. In general this need not be so. In Figure 1, D is outside our creation light cone at $\tau = \tau_0$, or, equivalently, we are outside D's particle horizon at τ_0. It is possible for us to be outside D's horizon, while D is inside ours (this depends on the definition of 'while' ,i.e. of $\tau = \tau_0$).

6. In Fig. 2, at $\tau = \tau_0$, C is in D's particle horizon and F is in G's. No particle is in both D's and G's particle horizon. However, H's horizon includes both C and F. Thus one cannot consider space as partitioned into distinct particle horizons.

7. The definitions do not consider opacity, which could prevent actual photons going all the way along the light rays (Sato 1968).

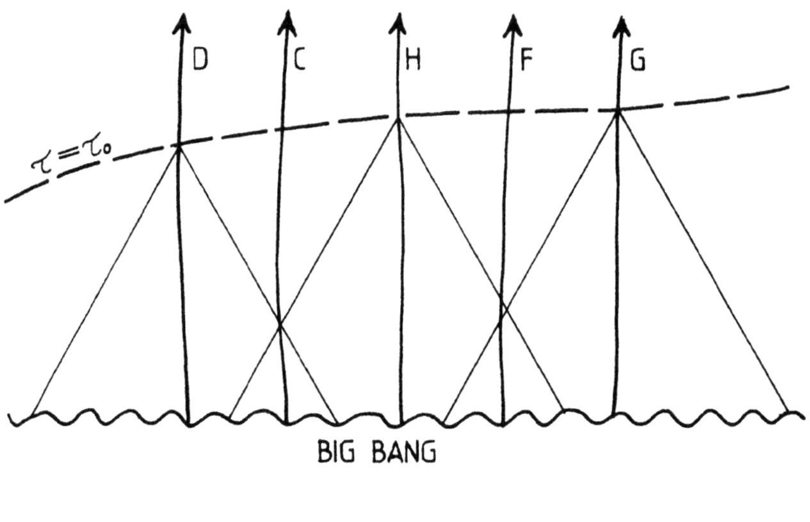

Figure 2. Space cannot be partitioned into distinct particle horizons.

2.2 The meaning of horizons

1. A physical process depending on signals from one particle A to another, C, can only work if A is in C's particle horizon.

2. Monopole (inverse square law, 'Coulomb') forces are not signals. Such gravitational, electromagnetic, or other forces are fixed by initial conditions: in relativity, four of Einstein's equations are essentially consistency conditions between these forces and the initial energy-momentum distribution (cf. Centrella's lectures).

Therefore such forces can work on scales greater than the particle horizon. For example, perturbations greater than the horizon size may begin gravitational collapse (though they may only represent gauge terms, see Sect. 2.3).

3. Perturbations on scales greater than the particle horizon can be damped. The reason is that damping processes respond to local gradients and do not first do a Fourier transform over all space.

To illustrate this, suppose (wrongly) that only Fourier components on scales smaller than the horizon can damp, and all such components do. Take two 2π-periodic functions in one dimension, both equal to sin x in $[0,\pi]$ (so the 'Coulomb' fields, see 2 above, are the same in this range). Let one be sin x in $[-\pi,0]$, with Fourier series sin x. This will not damp until the horizon size reaches π. Let the other be $-\sin x$ in $[-\pi,0]$, with Fourier series

$$f(x) = \frac{2}{\pi} - \frac{4}{\pi}\sum_{1}^{\infty} \frac{\cos 2kx}{4k^2-1} \qquad (14)$$

When the horizon size is $\pi/4$, all components with $k > 4$ damp. But a particle at $x = \pi/2$ cannot know which function is correct. Should it take part in damping or not?

In reality one must consider the actual mechanism and effect of a specific process before saying whether it does significant damping on scales greater than the particle horizon or not.

2.3 A digression: gauge effects and perturbations greater than the horizon size

Press and Vishniac (1980) and Bardeen (1980) have pointed out problems in the usual simple discussions of perturbations of FRW models on scales greater than the horizon size. To illustrate one of these remarks in a crude approximation, let A be a point in a perturbed FRW universe. Try two identifications with the FRW model, one putting A at time t, the other at time t' (if perturbation is an applicable concept, we must have $\Delta t = t-t'$ small). If ρ_A is the density at A, and $\rho_F(t)$ is the Friedman density, we get

$$\delta\rho = \rho_A - \rho_F(t),$$
$$\delta\rho' = \rho_A - \rho_F(t') \tag{15}$$

in the two cases. Suppose for simplicity that $\rho_A = \rho_F(t)$. Then $\delta\rho'/\rho = \Delta t,(d\rho_F/dt)/\rho \propto \Delta t/t$. Provided the identifications have t' = constant not too 'crinkled' with respect to t = constant (so the wavelength of Δt is much greater than the particle horizon), then at successive points on the worldline through it, Δt is constant. This suggests, as can be precisely checked (Press and Vishniac 1980; Bardeen 1980), that perturbations whose dominant term as $t \to 0$ is t^{-1} can have this term eliminated by a new identification between the perturbed and FRW spaces. Note this is crucially different from Newtonian theory, where t is unambiguous. Relativity is required for correct answers.

2.4 Horizon size in simple models

In FRW models (1) the maximum range from which a light ray ($ds = 0$) can reach us is given by assuming it starts at the big bang ($t = 0$): it is a coordinate distance r_h (horizon size at t)

$$r_h = \int_0^t dt/R(t). \tag{16}$$

For Einstein-de Sitter ($R \sim t^{2/3}$), 'radiation' or Tolman solution ($R \sim t^{1/2}$) and de Sitter ($R \sim e^{at}$), r_h is respectively $\sim t^{1/3}$, $\sim t^{1/2}$, and $\sim (1-e^{-at})/a$. The corresponding physical sizes $R(t)r_h$ are $3t(= 3c\tau)$, $2c\tau$, and $(e^{ac\tau}-1)/a$. Note that the last is $\sim t$ at small t but grows exponentially (cf. Fry's review of Guth's ideas).

A simple anisotropic model is given by the Bianchi I vacuum, or Kasner, model

$$ds^2 = -dt^2 + t^{2p_1}\,dx^2 + t^{2p_2}\,dy^2 + t^{2p_3}\,dz^2, \tag{17}$$

$$p_1{}^2 + p_2{}^2 + p_3{}^2 = p_1 + p_2 + p_3 = 1. \tag{18}$$

(18) implies the p_i contain only one free parameter. The Lifschitz-Khalatnikov parametrization is that the p_i are some permutation of $q_1 \leq q_2 \leq q_3$ where

$$q_1 = \frac{-u}{1+u+u^2},\quad q_2 = \frac{u+1}{1+u+u^2},\quad q_3 = \frac{u(u+1)}{1+u+u^2}, \tag{19}$$

and $u > 1$.

A light ray along the x-axis of (17) can travel a coordinate distance, starting at x_1 at time t_1, of

$$x - x_1 = \int_{t_1}^{t} dt/t^{p_1}$$
$$= [(t^{1-p_1})/(1-p_1)]_{t_1}^{t}, \quad (p_1 < 1). \tag{20}$$

[Note $p_1 \leq q_3 \leq 1$: if $p_1 = 1$, $x-x_1 = \ln (t/t_1)$, but this case is actually part of flat space in odd coordinates.] As $t_1 \to 0$, we get the horizon size, which is $\sim \varepsilon^{-1}$ if $\varepsilon = 1-p_1 \ll 1$.

The point of trying to get a big horizon is that all smoothing processes can work over larger scales (assuming that they depend on signals, see section 2.2 above). The answer just obtained looks promising, but if $p_1 \sim 1$, then p_2, $p_3 \sim 0$ and horizons in those directions are still only $\sim t$.

So to use large horizons to resolve the philosophical problems (see section 1.2), one wants each of the p_i in turn to approximate 1. To achieve this, a more complex model was tried by Misner (1969), the Bianchi model (independently, and slightly earlier, examined by Belinskii, Lifschitz and Khalatnikov (1969) for another purpose: see section 3).

2.5 The Bianchi IX model and its horizons

Take a Bianchi IX model with g_{ab} diagonal. Write

$$g_{ab} = e^{2\alpha} (e^{2\beta})_{ab}, \tag{21}$$

where α is a number and β is a traceless matrix with

$$\beta_{ab} = \text{diag}\,(2\beta_+, -\beta_+ + \sqrt{3}\beta_-, -\beta_+ - \sqrt{3}\beta_-), \tag{22}$$

Using α as the time variable, the vacuum $R^a{}_b$ equations (12) come from a Hamiltonian given by

$$\mathcal{H}^2 = p_+{}^2 + p_-{}^2 + e^{4\alpha}V, \tag{23}$$

$$V = 1/3[e^{8\beta_+} - 4e^{2\beta_+}\cosh(2\sqrt{3}\beta_-) + 2e^{-4\beta_+}\sinh^2(2\sqrt{3}\beta_-)]. \tag{24}$$

This potential V has the contours shown in Figure 3. For $|\beta_-/\beta_+| < \sqrt{3}$ and large $|\beta|$, $V \sim (\exp 8\beta_+)/3$, so the walls are a triangular exponential potential.

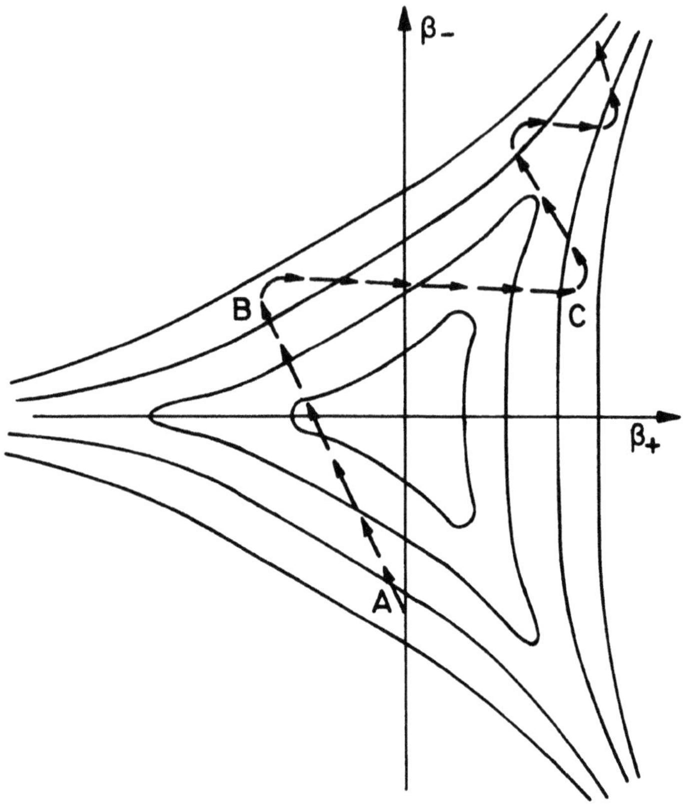

Figure 3. The equipotentials V = const and the path AB depicting motion with initial values $p_1 = q_1$, $p_2 = q_2$. The solution 'bounces' towards a corner of the triangular exponential.

In the early universe, $\alpha \to -\infty$, so a first approximation is to ignore V. Then one gets $g_{ab} = P_{ab}$, where

$$P_{ab} = \mathrm{diag}(t^{2p_1}, t^{2p_2}, t^{2p_3}), \tag{25}$$

with p_i as in (18, 19) and $2\beta_+' = p_1 - 1/3$,

$$\sqrt{3}\beta_-' = p_2 - \frac{1}{2}(p_1 - 1),$$

' denoting $d/d\alpha$. In Fig. 3, AB depicts a motion with $p_1 = q_1$, $p_2 = q_2$. Now what happens when V, as it must, becomes important? Assuming only the simple exponential term is important, an exact result (derived from an exact solution first found by Taub) is obtainable: the change in p_i is given by the change in u and permutation of q_i specified by

$$q_1(u) \rightarrow q_1(u-1), \; q_2(u) \rightarrow q_1(u-1), \; q_3(u) \rightarrow q_3(u-1), \; u \rightarrow u-1. \quad (26)$$

This can be repeated $[u_n]$ times where [ ] denotes the integer part and u_n the starting value. Let $K_n = [u_n]$ and $x_n = u_n - K_n$. The 'bounces' are represented by the diagrammed motion. After K_n bounces, we need a new permutation of q_i and u to make $u > 1$ again. This is given by

$$u \rightarrow 1/u, \; q_1(u) \rightarrow q_1(1/u), \; q_2(u) \rightarrow q_3(1/u), \; q_3(u) \rightarrow q_2(1/u). \quad (27)$$

The picture is thus of 'eras' of K_n bounces, at the end of which the new starting value of u, u_{n+1}, is given by

$$u_n = K_n + 1/(u_{n+1}). \quad (28)$$

Repetition generates the continued fraction

$$u_0 = K_0 + \cfrac{1}{K_1 + \cfrac{1}{K_2 + \cfrac{1}{K_3 + \cfrac{1}{\cdots}}}} \quad (29)$$

The probability distribution of the x_n, for random u_0, was given (by Lifschitz, Lifschitz and Khalatnikov 1970) as

$$p(x) = \frac{1}{(1+x)\ln 2}. \quad (30)$$

Barrow (1981, 1982) has recently re-interpreted the statistical results (Lifschitz et al. 1970) in the language of dynamical systems theory: the system is isomorphic to a Bernoulli shift and hence is ergodic and has 'entropy': initially neighbouring solutions diverge on average like exp(3.4t), cf. (Chitre 1972). (This t is not that of (17).)

It appears that we effectively have each p_i in turn near 1. Hence Misner called this model the 'Mixmaster' because conditions in widely separated regions could be 'mixed' (cf. Sect. 2.4 above). Unfortunately, the work of many authors (those mentioned above plus Novikov, Doroshkevich, Grishchuk, Ryan, Matzner, Chitre and others) showed that in general the effects take place only before the Planck time (when quantum gravity is required) and even then could only arise for a small percentage of the possible initial values. The essential reason is that aiming at $p_i \sim 1$ for some i is equivalent to running at one of the corners in Fig. 3. As $\alpha \rightarrow -\infty$, the potential drops, and the effective corners move out. The required range of x thus diminishes to zero as $\alpha \to -\infty$, and the system cannot hit it.

One might finally add extra complications. [It should be noted, cf. Sect. 3.6, that there is no advantage in using other Bianchi models.] A non-diagonal model of Bianchi type IX containing perfect fluid has 'tilt' and rotation (see Sect. 1.4). We can use (9), $M^a{}_b$ being a rotation matrix parametrized by Euler angles, and D_{ab} diagonal. There are eras in which $M^a{}_b$ is effectively constant, the 'Kasner eras', related by short periods in which $M^a{}_b$ changes abruptly, 'rotation of Kasner axes'. Detailed descriptions were given by Belinskii, Lifschitz, Khalatnikov, and Ryan (1972). Grishchuk, Doroshkevich and Lukash (1971) and Matzner and Chitre (1971) showed that this did not help 'mixing'.

Since it is highly implausible that inhomogeneity would improve matters, it appears that 'mixing' cannot be the answer to the philosophical problem in Sect. 1.2.

3. BIG BANG SINGULARITIES

3.1 Introduction

The FRW models (1), for reasonable physical parameters, all begin at a big bang, R = 0. The natural question, put first by Einstein, was 'is this just a consequence of the symmetry?' Prof. Godart tells me (cf. Heller 1979) Lemaitre answered this in the 1930s by considering (11) with diagonal g_{ij}. Somewhat more generally one has the result found for (11) by Landau, Raychaudhuri and Komar. From the Einstein equation

$$\theta' + \theta^2/3 = 3\,\ell''/\ell = -\sigma^a{}_b\sigma^b{}_a - T_{44} - \tfrac{1}{2}T^\mu{}_\mu + \Lambda\,, \qquad (32)$$

one finds that, for reasonable matter, $\ell'' < 0$ (if $\Lambda \leq 0$) and hence, since $\ell' > 0$ now, ℓ was zero a finite time ago.

This idea is the basis of two modes of approach to the big bang singularity problem, which have now begun to meet. One of these is the work of Belinskii, Lifschitz and Khalatnikov (1970: 1972) and Lifschitz and Khalatnikov (1963), and the other the more rigorous 'global analysis' approach of Hawking and Ellis (1973) and Tipler, Clarke and Ellis (1980).

3.2 The Lifschitz-Khalatnikov paper of 1963

Lifschitz and Khalatnikov (1963: hereafter LK) reported three main calculations on singularities. They first considered the problem of whether $\ell \to 0$ for (11) was a real singularity (cf. Centrella's remarks on 'geodesic slicing').

(a) Fictitious singularities

If one takes a (totally non-singular) point in Minkowski space, O, say, one can draw all future timelike geodesics through it, use proper time on these curves as a coordinate t, and assign space coordinates to each worldline. The metric then has the form (11). However, fields at points outside O's future null cone are not determined by data on t = constant (e.g. P in Fig. 4).

A Cauchy surface is a spacelike surface such that initial data on it determine the whole of spacetime. The surfaces in Fig. 4 are not Cauchy.

LK determined 'generality' of a solution by considering the number of functions (of three variables) that could be specified on t = constant and give physically distinct solutions. The most general cases would have four functions (in vacuum) or eight (with perfect fluid), corresponding physically to the value and derivative of each of the two polarizations of gravitational waves, with, for a fluid, the energy density and initial velocity.

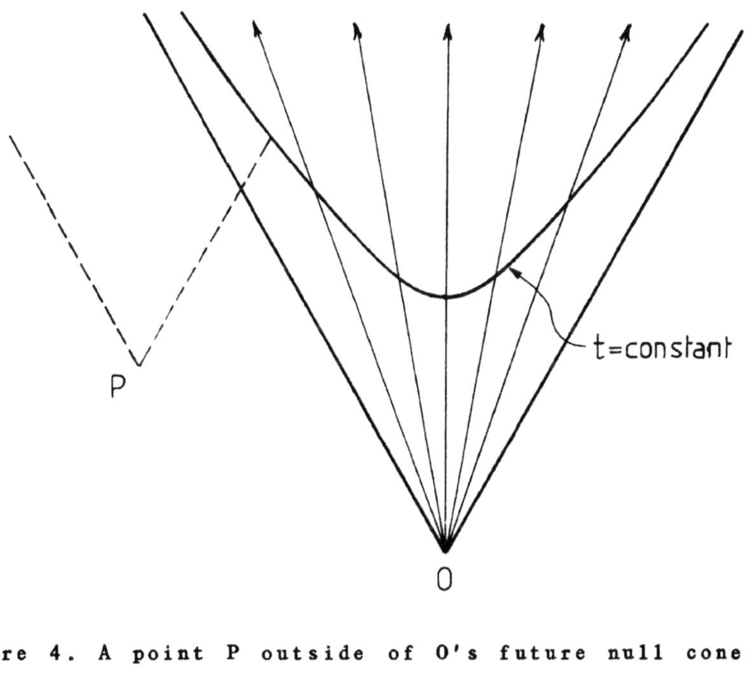

Figure 4. A point P outside of O's future null cone.

LK found that in general the points $\ell = 0$ of (11) formed a caustic, which was timelike since it contained segments of the timelike t-coordinate lines. The caustic was not $t = t_0 =$ constant. To construct such a 'simultaneous singularity' $t = t_0$ had to be a focus. In either case, expansions were used to show that the singularities were 'general' (Lifschitz and Khalatnikov 1963: Lifschitz, Sudakov and Khalatnikov 1961: Belinskii and Khalatnikov 1965).

Comments. With hindsight, the result is obvious. One just applies a construction like Fig. 4 in any spacetime. The t = constant surfaces cannot be Cauchy, however. Barrow and Tipler (1979: hereafter BT) prove this, and take it as a criticism of LK, but since LK say 'such a system cannot contain all of spacetime within itself' and 'these hypersurfaces are positioned entirely within the regions of absolute future or absolute past with respect to the focal surface' they clearly knew. (The term 'Cauchy surface' was not coined until later, see Tipler, Clarke and Ellis 1980.)

It might be fun to find all constructions for 'fictitious singularities'. However, as Belinskii, Lifschitz and Khalatnikov (1980: BLK) remark in their reply to BT, it is now irrelevant (see (d) below).

BT also criticise the function-counting, since it constitutes a proof that the spacetime is general only if the surface is Cauchy: otherwise additional data is required. BLK use this method throughout. My feeling is that, while BT are formally correct, since what BLK claim to have is a description of the most general spacelike singularity, it could still be that the singularity is determined by the t = constant surface even though the whole spacetime is not (see section 3.4).

(b) Kasnerian singularities

LK use a factorization similar to (9), both in 1963 and in later work. They assume that $M^a{}_b$ is (approximately) time-independent, and D_{cd} is (approximately?) diagonal. Taking the form (25) with $p_1 = q_1$, they found the solution self-consistent only if

$$n_{11} = \underline{\ell} \cdot \nabla \times \underline{\ell} = 0 \qquad (32)$$

where $\underline{\ell}$ is treated as a Euclidean vector and $\ell_i = M^1{}_i$. Equation (32) acts as a restriction reducing the number of physical functions to one below the most general. n_{11} is similar to the (constant) N_{11} of the homogeneous models. It seems that LK had examined the effect of a non-zero N_{11} in 1962 (Belinskii and

Khalatnikov 1969), but they then thought the values $q_1 = 1$, $q_2 = 0 = q_3$ were stable and would be approached: since these values give, if exactly true, a fictitious singularity (see Sect. 2.4) the possibility was ignored. Actually, as shown in Misner (1969), Belinskii and Khalatnikov (1969) and Doroshkevich and Novikov (1970), the corner channels in Fig. 3 are unstable (see also Chitre 1972: Ryan 1972: Heller 1979: Khalatnikov and Pokrovskii 1970).

(c) 'Quasi-isotropic' singularities

These approximate FRW in that all length scales go to zero at the singularity. They are not 'general', i.e. they are unstable just as the FRW models are (see Sect. 1.2).

(d) Other results and calculations

A number of detailed computations and special cases are given in Lifschitz and Khalatnikov (1963) and papers listed therein.

Because (c) and (b) above were not general, and general solutions contain fictitious singularities, LK concluded that general solutions did not (also) contain real singularities. It seems to me that this erroneous inference has been the reason for subsequent mistrust of their work, although it did not follow strictly from their calculations, which in themselves may be correct (see Sect. 3.5 below).

3.3 The singularity theorems and their consequences

It is well known that Hawking, Penrose and others proved a series of theorems on the existence of singularities (see Hawking and Ellis 1973: Tipler, Clarke and Ellis 1980). For our purposes the most useful is the following.

Theorem

Provided that every non-spacelike geodesic contains a point at which

$$k_{[a}R_{b]cd[e}k_{f]}k^c k^d \neq 0,$$

where k^a is the geodesic's tangent vector, then the following four conditions are incompatible.

(i) $R_{ab}k^a k^b \geq 0$ for every non-spacelike k^a.

(ii) there are no closed timelike curves

(iii) there is one of
(a) a compact achronal set without edge,
(b) a closed trapped surface,
(c) a point p all of whose past (or future) null geodesics start to refocus,

(iv) all null and timelike geodesics are 'complete'.

The initial technical condition says every geodesic notices some gravity: it eliminates, e.g., flat spacetime. Condition (i) says gravity always attracts: for general relativity, it imposes a condition on the energy-momentum, the 'strong energy condition'. Condition (ii) is a causality condition. Condition (iii) (a) covers closed universes and (iii) (b) black holes; (iii) (c) can be shown (Hawking and Ellis 1968, 1973) to be satisfied in our universe, because of the gravitational effect of the microwave background and matter combined. If one accepts these, (iv) must be false.

The meaning of an incomplete geodesic is that some path of a test massive or massless particle simply stops or starts at a finite (affine parameter) distance. Such a particle must be created or destroyed. The problem is then to decide how this type of 'singularity' (an incomplete geodesic) relates to the 'big bang' type. Where and why do geodesics end?

The first technical problem is to decide whether, if two geodesics both end, their endpoints should be considered the same point or not. Various efforts at solving this problem have all (so far) come to grief (see Tipler, Clarke and Ellis 1980). (However, it seems reasonable, and is often assumed, that if A and B are always in each other's particle horizon and both are on incomplete geodesics, the endpoints are the same.) Despite this lack of definition, it is clear that the nature of a singular endpoint is defined by the gravity field experienced by a body moving on the incomplete geodesic. There are basically three cases (see Tipler, Clarke and Ellis 1980 for references, etc.).

(i) The field is finite and well-behaved. These are 'quasi-regular' singularities. Such singularities are 'locally extendible', that is, the incomplete geodesic has a neighbourhood which is part of a larger space in which the geodesic could be continued. The singularity arises because of the global structure, and can be thought of as like the singularity at the apex of a cone (which, if cut along a generator, opens out into part of a plane). Locally extendible singularities are 'D-specialized' (a Riemann tensor of Petrov type D and electromagnetic field type near the singularity), or spacetime contains 'holes',or the singularity is 'primordial' (extends into the indefinite future). Such a singularity is implausible in cosmology.

(ii) the field becomes infinite in a frame parallelly-propagated along the geodesic, but not in all frames. These are 'non-scalar-polynomial-curvature' singularities. Examples were found by Collins and Ellis (1979), but they are unstable (see King 1975: Belinskii, private communication) and not generic Siklos 1979, 1980a,b).

(iii) the field becomes infinite in every frame, and therefore some scalar polynomial in the Riemann tensor blows up.

These results strongly suggest that the real universe contains singularities at which density, or another scalar in the curvature, becomes infinite. We would like to know the manner in which this happens.

3.4 Further work on 'big bang' singularities

Following methods similar to those of LK, Grishchuk found, for 'dust' ($p = 0$), that the singularities can be general and real, and explained why such solutions may have been overlooked. He also generalized to non-ideal fluids and proved the formal existence of power series (see Grishchuk 1967: Grishchuk and Ulin 1971 and references therein).

The major revision of LK was done by BLK. They investigated the corner channel region of Fig. 3 and found the reflection, anachronistically mentioned in Sect. 3.2. Various treatments, with increasing detail, of the type IX model were given, rotating matter was added, and the results were extended to type VIII. For references see MacCallum 1973,1979a: Belinskii, Lifschitz and Khalatnikov 1970: Barrow and Tipler 1979, which also cover the next development. This was that, parallelling the papers on homogeneous models, a series of papers treating the inhomogeneous models appeared, relating them to the VIII and IX models just like the inhomogeneous Kasnerian models relate to the Kasner model (see Sect. 3.2 (b) above). The final answer is that near the singularity the model behaves, in its time-evolution, like the Bianchi IX models described in Sect. 2.5, including 'rotation of Kasner axes' (Belinskii, Litschitz and Khalatnikov 1972).

This conclusion, as has been pointed out by several authors, must be modified if the extreme equation of state $p = \rho$ (for which Barrow (1978) strongly argued) holds in the early universe (MacCallum 1973: Barrow 1978: Belinskii and Khalatnikov 1972: Ruban 1972), cf. Sect. 1.5. Further complications in energy-momentum will be ignored here.

In general, the ${}^{3}R^{a}{}_{b}$ term in (12) cannot be ignored. When it can, one has the 'velocity-dominated' solutions and the BKL results can be more rigorously proved for this class. (All are Kasnerian or quasi-isotropic in the sense of Sect. 3.2.)

It has been shown that the t coordinate used, if it exists, is unique (Goorjian 1973: Barrow and Tipler 1979): the possibility of finding such a coordinate has been questioned but more recently rather general conditions for its existence, which seem to cover most reasonable models, have been given (Wald and Yip 1981)

Specific examples of the 'quasi-isotropic' type, in Szekeres solutions (see Sect. 1.3) have recently been studied by Goode and Wainwright (1981).

3.5 Criticisms of the BLK work

A number of comments and criticisms have been made concerning the BLK work.

I expressed a concern (MacCallum 1973, 1979a) about the dominance of time derivatives when one might expect space and time to appear on an equal footing. I now feel this overlooks the point that a particular physical situation, in which time had a special role, was under discussion. However, the condition for the existence of Kasner eras (essentially that (32) is approximately true) might fail: in this case Belinskii and I have suggested we may have a situation like the onset of turbulence, the Kasner regime being a 'laminar flow', and the oscillations in the type IX solution being the first unstable mode, the full solution being much more chaotic. I understand Belinskii is looking at some solutions without Kasner eras. Incidentally, I also believe some of the comments on BLK are based on the feeling that the solution is not 'chaotic' enough. (See also Barrow 1982).

I also argued that one might expect VI_h and VII_h to appear as paradigms, along with VIII and IX (see MacCallum 1973: Barrow and Tipler 1979 and Sect. 1.4). At the time no reason against this was known. Subsequently, by methods described in Sect. 3.6, it has been shown that VI_h and VII_h with 'tilt' do follow the BLK evolution, essentially because the BLK frame has to be one in which (32) does not hold (although the rank of N_{ab} is still 2, it may have non-zero entries in all positions).

The strongest criticisms have come from Barrow and Tipler 1979). They made a large number of points. I shall not comment on all of them, but pick out what I consider to be the main ones.

(i) Continuing from the remarks about 'generality' (see Sect. 3.2) BT point out the possibility of timelike singularities. Examples of these have been given (Miller 1979, 1981: Tomita 1978) in cosmology. BT say the arguments of BLK could at best prove the BLK solution 'stable' (i.e. an open set in the space of all solutions) but not 'generic' (i.e. an open dense set). Indeed it seems intuitively clear that there is some other open set corresponding to bodies of mass below the gravitational collapse limits in asymptotically flat space. (Closed universes are not possible without singularity (Barrow and Tipler 1979).)

However, it could be (?) that (a) all reasonable cosmological models contain spacelike singularities, (b) all spacelike singularities can be discussed by setting data on adjacent spacelike surfaces, which need not be Cauchy, (c) that the BLK behaviour correctly describes the asymptotic behaviour near such singularities (with the exception of sets of measure zero from (a), (b) and (c)). This needs further research.

(ii) BT criticize the function-counting of BLK on another ground. (Note that function-counting is often a somewhat ambiguous procedure (Siklos 1979).) Rigorous theorems show that the map of initial data space to spaces of functions is not differentiable at spaces with symmetry (this is the 'linearization stability' problem, see e.g. Fischer, Marsden and Moncrief (1980)). BLK certainly include spaces with symmetries (although they argue (BLK (1980) their general solution is far from such spaces) and note no linearization instability. I do not yet fully understand the background to these points, but I would venture two remarks.

(a) Is differentiability essential? Maybe the homeomorphism stated by BT to exist would suffice for BLK.

(b) BLK's method is not the usual one. They first take synchronous geodesic coordinates with the singularity at $t = 0$. The usual variations would alter the time separation from an initial data surface to the singularity. BT argue that the BLK t = constant surfaces, in two spacetimes close in the usual treatment, could globally diverge, e.g. in Fig. 5 data varied on S_1 so that in U it appears like S_2, could lead to the new BLK surface S_2. Here I think BT are right. But it means one may not be able to discuss the BLK function-counting as if it were the usual one. Could it be that the BLK method gives a essentially different differential structure for initial data space?

(iii) A number of remarks are made concerning the possibility of domains of dependence in local studies failing to contain all (or even any) of the singularity. However, no direct check of these possibilities for the BLK metric was attempted.

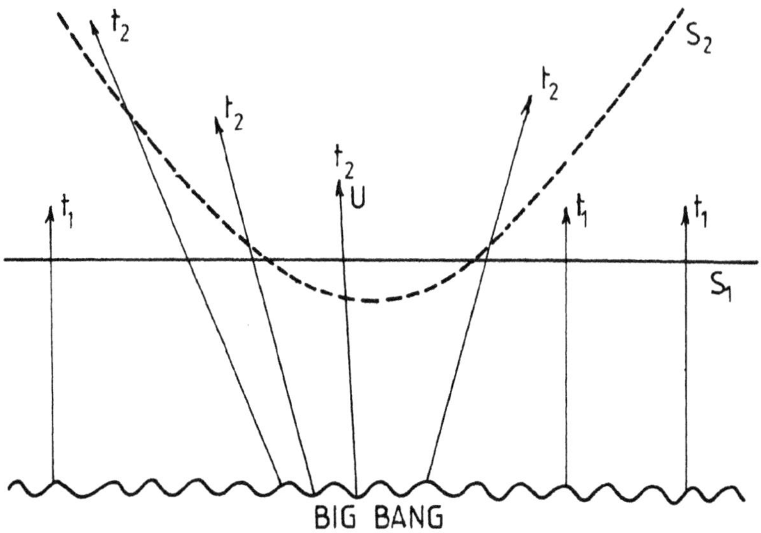

Figure 5. data varied on S_1 so that in U it appears like S_2 could lead to the new BLK surface S_2. Thus the t = constant BLK surfaces, in two spacetimes close in the usual sense, could globally diverge.

Some further BT points suggesting further research are (i) the convergence of the BLK approximations is unproved (ii) there is inadequate understanding of the relations of neighbouring patches approximating different models (of which Centrella's lectures give an example). I also want to understand the relation to Spero and Baierlein (1977, 1978).

My overall impression is that the BT remarks are interesting but inconclusive, and I have the feeling BLK must be right in some domain of validity. The real question is 'What is this domain?'

3.6 Singularities in homogeneous cosmologies

The most detailed studies of the Bianchi models are those by Bogoyavlenskii, Novikov and Peresetskii. Work began with type IX (Bogoyavlenskii and Novikov 1973,1975: Novikov 1973) and other types (Bogoyavlenskii and Novikov 1975) without 'tilt', then type IX with moving matter (Bogoyavlenskii 1976a) and electromagnetic field (Bogoyavlenskii 1975), and then the other Bianchi types again (Bogoyavlenskii 1976a: Peresetskii 1976, 1977). The work has been reviewed and generalised (Bogoyavlenskii and Novikov 1976: Bogoyavlenskii 1977, 1980).

Very briefly the technique is:

(i) to redefine variables so that the equations are on a compact manifold with boundary, all singularities being represented by points $\underline{x}$ on the boundary,

(ii) to find all critical points $\underline{f} = \underline{0}$ of the equations, written as

$$dx/d\tau = \underline{f}(\underline{x}), \tag{33}$$

and find their separatrices. All the critical points, and nearly all the separatrices, lie in the boundary and correspond to singularities,

(iii) use the function ℓ', monotone on the evolution curves (see Sect. 3.1), to show the evolution goes arbitrarily close to the boundary,

(iv) use the analyticity of (33) to argue that once sufficiently near the boundary, the solutions must follow the curves on the boundary. Then the evolution can be stated.

The results are that (for this restricted case) BLK are right, except for certain special cases with different asymptotes. For perfect fluid, $p = (\gamma-1)\rho$, these have power-law form (25) with the values, in the various cases,
(a) T(aub), $p_1 = p_2 = 0$, $p_3 = 1$,
(b) N(ovikov), $p_1 = (2-\gamma)/2\gamma$, $p_2 = p_3 = (2+\gamma)/4\gamma$,
(c) M(acCallum), $p_1 = 1$, $p_2 = p_3 = (2-\gamma)/2\gamma$,
(d) LK (Lifschitz-Khalatnikov), $p_1 = p_2 = p_3 = \gamma/3$ (quasi-isotropic). The names arose from particular exact solutions. (It should be noted that the N-type behaviour was independently found by Lukash and by Ruban in the period 1972-3.)

The most fascinating point is that because only the power-law asymptotes have separatrices going off the boundary, the solutions tend to leave the boundary region near these critical points, although they are unstable (and go to a BLK regime) as 0. Thus they may provide typical states for anisotropic models in expansion, though not at very early times. (The power-law asymptotes were derived in a simple way by other authors also (see Borzeszkowski and Muller 1978: Evans 1978). These last remarks have yet to be extended to types, other than IX, with moving matter, in detail, and I do not myself understand yet the relation to the special non-scalar-polynomial cases (reviewed in Collins and Ellis 1979).

However, it seems to me that to understand the possible

astrophysics of anisotropic models we should next be looking at solutions close to T, N, M and LK types (the last being essentially perturbed FRW). These may even be correct second stages for inhomogeneous models, and also for evolutions after particle creation has reduced initial anisotropy after the Planck time. They certainly should be the first choices for study of such effects as 'hot spots' in the microwave background. Such work has been done by Doroshkevich, Lukash and Novikov (1973), who showed that in Bianchi models the N-type 'damping stage' was the most typical and apparently occurred in all cases except Bianchi types I and V. During this stage, the anisotropy undergoes power-law damping, and in general is followed by a 'Friedman-type' era.

A (very optimistic) scenario - just to prove mathematicians can also play that game - might be a universe, determined by the density and the anisotropy at the Planck time, which would fit all constraints, have significant anisotropies during the hadron and lepton eras, and at later times, and produce a predictable $\Delta T/T$ for the microwave which we actually observe, regardless of whether it follows N, M, T, or LK asymptotes early on.

Acknowledgement

I am grateful to Drs J.D. Barrow and V.N. Lukash for extensive comments from which arose a number of corrections and addenda to my first draft.

REFERENCES

Bardeen, J. 1980. Phys.Rev.D. 22, 1882.
Barrow, J.D. 1978. Nature, 272, 211.
Barrow, J.D. 1981. Phys.Rev.Lett., 46, 963.
Barrow, J.D. 1982. Phys.Reports (in press).
Barrow, J.D. and Tipler, F.J. 1978. Nature 275, 453.
Barrow, J.D. and Tipler, F.J. 1979. Phys.Reports, 56C, 371.
Belinskii, V.A. 1975. Private communication.
Belinskii, V.A. and Khalatnikov, I.M. 1965. Zh.E.T.F. 49, 1000: (Sov.Phys.J.E.T.P. (1966) 22, 694).
Belinskii, V.A. and Khalatnikov, I.M. 1969. Zh.E.T.F. 56, 1700: (Sov.Phys.J.E.T.P. 29, 911).
Belinskii, V.A. and Khalatnikov, I.M. 1972. Zh.E.T.F. 63, 1121. (Sov.Phys.-J.E.T.P. 36, 591).
Belinskii, V.A., Lifschitz, E.M. and Khalatnikov, I.M. 1970. Adv.Phys. 19, 525. (or Usp.Fiz.Nauk. 102, 463: Sov.Phys. Usp. 13, 745).
Belinskii, V.A., Lifschitz, E.M. amd Khalatnikov, I.M. 1972. Zh.E.T.F. 62, 1202: (Sov.Phys.-J.E.T.FP. 35, 838).
Belinskii, V.A., Lifschitz, E.N. and Khalatnikov, I.M. 1980. Phys.Lett.A. 77, 214.

Bogoyavlenskii, O.I. 1975. Teor.mat.fiz. 27, 184: (Theor. Math.Phys. 27, 414).
Bogoyavlenskii, O.I. 1976a. Zh.E.T.F. 70, 361: (Sov.Phys. J.E.T.P. 43, 187).
Bogoyavlenskii, O.I. 1976b. Trudy Seminara I.G. Petrovskii, Moscow State University, 2, 67.
Bogoyavlenskii, O.I. 1977. Dokl.Akad.Nauk. 232, 189: (Sov. Phys.Dokl. 22, 78).
Bogoyavlenskii, O.I. 1980. Methods of the Qualitative Theory of Dynamical Systems in Astrophysics and Gas Dynamics (in Russian), Nauka, Moscow.
Bogoyavlenskii, O.I. and Novikov, S.P. 1973. Zh.E.T.F. 64, 1475: (Sov.Phys.J.E.T.P. (1973) 37, 747).
Bogoyavlenskii, O.I. and Novikov, S.P. 1975. Trudy Seminara I.G. Petrovskii, Moscow State University 1, 7.
Bogoyavlenskii, O.I. and Novikov, S.P. 1976. Usp.Mat.Nauk. 31, 33: (Russian Math.Surveys 31, 31).
Borzeskowski, H.H. and Muller,V. 1978. Ann.d.Phys. 35, 361.
Chitre, D.M. 1972. Ph.D. Thesis. University of Maryland, Tech.Rpt. 72-125.
Collins, C.B. and Ellis, G.F.R. 1979. Phys.Reports 56C, 65.
Collins, C.B. and Hawking, S.W. 1973. Ap.J., 180, 317.
Criss, T.B., Matzner, R.A., Ryan, M.P.,Jr. and Shepley, L.C. 1975. In General Relativity and Gravitation, ed. G.Shaviv and J. Rosen. J.Wiley and Sons, New York, and Israel Universities Press, Jerusalem.
Doroshkevich, A.G., Lukash, V.N. and Novikov, I.D. 1973. Zh.E. T.F. 64, 1457: (Sov.Phys.J.E.T.P. 37, 739).
Doroshkevich, A.G. and Novikov, I.D. 1970. Astr.Zh. 47, 948: (Sov.Astr.A.J. 14, 763).
Evans, A.B. 1978. MNRAS 183, 727.
Fischer, A.E., Marsden, J.E. and Moncrief, V. 1980. In Essays in General Relativity, ed. F.J. Tipler, Academic Press, New York.
Goode, S.M. and Wainwright, J. 1982. MNRAS, 198, 83.
Goorjian, P. 1973. J.Math.Phys. 14, 986.
Grishchuk, L.P. 1967. Zh.E.T.F.Pis'ma 6, 700: (J.E.T.P. Letters 6, 184).
Grishchuk, L.P. and Ulin, V.I. 1971. Proc.State Astron.Inst. Shternberg 173, 32. (In Russian)
Grishchuk, L.P., Doroshkevich, A.G. and Lukash, V.N. 1971. Zh.E.T.F. 61, 3: (Sov.Phys. J.E.T.P. (1972) 34, 1).
Harvey, A.L. 1979. J.Math.Phys., 20, 251.
Hawking, S.W. and Ellis, G.F.R. 1973. The large-scale structure of space-time, Cambridge University Press, Cambridge.
Harness, R.S. 1982. J.Phys., A15, 135.
Heller, M. 1979. In Physics of the Expanding Universe, ed. M. Demianski, Lecture Notes in Physics, vol. 109. Springer-Verlag, Berlin.

Jantzen, R.T. 1979. Comm.math.phys. 64, 24.
Khalatnikov, I.M. and Pokrovskii, V.L. 1970. Zh.E.T.F. 59, 1812: (Sov.Phys.J.E.T.P. (1971) 32, 984).
King, A.R. 1975. Phys.Rev.D, 11, 763.
Kramer, D., Stephani, H., MacCallum, M. and Herlt, E. Exact Solutions of Einstein's Field Equations, Deutscher Verlag der Wissenschaften, Berlin, and Cambridge University Press, Cambridge.
Lifschitz, E.M. and Khalatnikov, I.M. 1963. Adv.Phys. 12,185, (or Usp.Fiz.Nauk. 80, 391: Sov.Phys.Usp. 6, 495)
Lifschitz, E.M., Lifschitz, I.M. and Khalatnikov, I.M. 1970. Zh.E.T.F. 59, 322: (Sov.Phys.J.E.T.P. (1971) 32, 173).
Lifschitz, E.M., Sudakov, V.V. and Khalatnikov, I.M. 1961. Zh.E.T.F. 40, 1847: (Sov.Phys.J.E.T.P. 13, 1298).
MacCallum, M.A.H. 1971. Nature (Phys.Sci.) 230, 112.
MacCallum, M.A.H. 1973. In Cargese Lectures, vol. 6, ed.E. Schatzman, Gordon and Breach, New York.
MacCallum, M.A.H. 1979a. In General Relativity: an Einstein Centenary Survey, ed. S.W. Hawking and W. Israel, Cambridge University Press, Cambridge.
MacCallum, M.A.H. 1979b. In Physics of the Expanding Universe ed. M. Demianski, Lecture Notes in Physics, vol. 109, Springer-Verlag, Berlin.
Matzner, R.A. and Chitre, D.M. 1971. Comm.math.phys. 22, 173.
Miller, B. 1979. J.Math.Phys. 20, 1356.
Miller, B. 1981. J.Math.Phys. 22, 1452.
Misner, C.W. 1969. Phys.Rev. 186, 1319.
Novikov, S.P. 1972. Zh.E.T.F. 62: (Sov.Phys.J.E.T.P. (1973) 35, 1031).
Penrose, R. 1979. In General Relativity: an Einstein Centenary Survey, ed. S.W. Hawking and W. Israel. Cambridge University Press, Cambridge.
Peresetskii, A.A. 1976. Usp.Mat.Nauk. 31, 251.
Peresetskii, A.A. 1977. Mat.Zam. 21, 71: (Russian Math. Notes 21, 39).
Press, W.H. and Vishniac, E.T. 1980. Ap.J., 239, 1.
Ruban, V.A. 1972. Lett.Nuovo Cim. 5, 289: Ap. and Sp.Sci. 46, L.23 (1977).
Ryan, M.P.,Jr. 1972. Ann.Phys. (NY) 70, 301.
Ryan, M.P.,Jr. and Shepley, L.C. 1975. Homogeneous Relativistic Cosmologies, Princeton University Press, Princeton, N.J.
Sato, H. 1968. Proc.Theor.Phys. 40, 781.
Siklos, S.T.C. 1979. Oxford University preprint.
Siklos, S.T.C. 1980a. Phys.Lett.A. 76, 19.
Siklos, S.T.C. 1980b. Comm.math.phys. 58, 255.
Siklos, S.T.C. 1981. Gen.Rel.and Grav. 13, 433.
Spero, A. and Baierlein, R. 1977. J.Math.Phys. 18, 1330.
Spero, A. and Baierlein, R. 1978. J.Math.Phys. 19, 1324.

Tipler, F.J., Clarke, C.J.S. and Ellis, G.F.R. 1980. In General Relativity and Gravitation, ed. A. Held, Plenum Press, New York.
Tomita, K. 1978. Prog.Theor.Phys., 59, 1150.
Wald, R.M. and Yip, P. 1981. University of Chicago Preprint.

NUMERICAL COSMOLOGIES

Joan Centrella

Center for Relativity, University of Texas, Austin, Texas 78712

These lectures are designed for astrophysicists and cosmologists who are interested in probing the physics of the early Universe. The lectures present the formalism and techniques of numerical relativity, using the familiar Friedmann-Robertson-Walker models as examples, and show how to use these methods to construct inhomogeneous cosmological models on a computer.

A great deal of time and effort has been expended over the past decade to develop the mathematical formalism and numerical techniques needed to construct dynamical spacetimes on a computer. This field has produced many fine research articles and a very useful book containing workshop results. The present lectures attempt to collect the basic methods of numerical relativity as of 1981 and to make them accessible to cosmologists who would explore the Universe with a computer. The major texts and papers used in each section are indicated next to the section title. More specialized references to technical points are made in the text itself.

1. INTRODUCTION (Jones 1976: Misner, Thorne and Wheeler 1973: North 1965)

This century cosmology has flowered dramatically as a branch of physical science. On the observational side we have witnessed the recognition of our own Milky Way as one galaxy among many, the discovery of the Hubble flow and Hubble's redshift distance law for galaxies, and the discovery of the 3°K cosmic background radiation. Theorists have made progress as well. Einstein's general relativity predicted the expansion of

B. J. T. Jones and J. E. Jones (eds.), The Origin and Evolution of Galaxies, 41–64.

the Universe. The Friedmann-Robertson-Walker (FRW) models formed the basis for a standard model of the Universe, including matter-radiation decoupling and the synthesis of elements. This standard model is truly a landmark in the application of physics to the study of the Universe, and forms a framework against which we gauge our current theories and observations.

It is necessary to stress the importance of general relativity in modern cosmology. It is of course possible to go back and construct Newtonian cosmological models which can reproduce certain key results derived from Einstein's equations. However, the scientific and philosophical advances due to Einstein's theory, such as the propagation of gravity at the speed of light and the physical effects of spacetime curvature, form the foundations of our modern view of the Universe.

In spite of its great importance in modern cosmology, general relativity suffers from two major misconceptions about its relevance and usefulness. The first of these points is:

Misconception 1: General relativity is not important in astrophysics.

This situation is all for the better since the second point states:

Misconception 2: General relativity is intractable.

We claim that these statements are not true, and will now support this claim with several examples.

General relativity says that gravitational signals travel at the speed of light, in contrast with Newtonian theory, in which gravitational signals travel with infinite speed. This fact gives rise to the existence of horizons (Rindler 1956) in relativistic cosmological models, with a horizon being the distance a light ray has travelled since the Big Bang. Relativity is therefore needed to describe any phenomenon having a scale greater than or of the order of the horizon size. We note that in the standard model the horizon size decreases as we approach the Big Bang. Thus an object which is adequately described by Newtonian physics today required relativity to detail its dynamics at an earlier epoch.

We also need to use Einstein's theory whenever the pressure P is comparable to the rest mass density ρ. The reason for this is that the active gravitational mass density is $\rho + \varepsilon + 3P$ in general relativity, while it is simply ρ in Newtonian physics (Ellis 1971). Here, ε is the internal energy density. Thus, during the radiation era in the early Universe, when the equation of state is $P = 1/3\ \varepsilon$, relativity is essential.

Finally, relativity and Newtonian physics treat anisotropic stresses in the sources differently. Such stress terms appear in the relativistic equations where they are absent from the corresponding Newtonian equations (Ellis 1971). This means that in relativity the anisotropic stresses are more directly coupled to, and therefore have stronger effects on, the tidal part of the gravitational field and the resulting shear of the matter. This difference is significant if the anisotropic stresses are comparable to ρ, as they could be, for example, in periods when there is a strong (neutrino) viscosity in the early Universe or if there are significant cosmic magnetic fields at any time.

With these physical regimes in mind, we now turn our attention to one of the most fundamental cosmological questions, the origin of galaxies. There are two basic categories of theories, smooth and chaotic, that have been advanced to explain galaxy formation. Proponents of the smooth theory take the observation that the Universe is homogeneous and isotropic today to imply that the Universe must always have been uniform, right back to a smooth Big Bang. In order to grow galaxies in this model, however, it is necessary to introduce ad hoc a spectrum of density fluctuations at a certain critical time. Recent work on phase transitions and particle physics, as discussed in Dr Fry's seminar and Dr Hogan's lectures in this volume, may provide mechanisms for generating this structure in a smooth early Universe.

The competing theory of the early Universe, known as chaotic cosmology, originated with Misner (1968). Advocates of this theory believe it is more reasonable to assume the Universe was generally inhomogeneous and anisotropic near the Big Bang. Then, as the Universe expanded, various physical processes conspired to smooth out the chaos and produce the approximately uniform FRW model we observe today.

Until recently, chaotic cosmology in its full generality appeared doomed. Using fairly general arguments (Barrow and Matzner 1977) it is possible to show that if very strong amplitude chaos is dissipated to produce a smooth model by the epoch of nucleosynthesis, then far more entropy per baryon is produced than we observe in the microwave background. This situation has changed with recent progress in particle physics: Grand unified theories show that baryon non-conserving reactions are possible in the early Universe (Weinberg 1979). Thus the subject of very strong initial chaos is again open.

We are still able to use the entropy arguments to limit the amount of dissipation possible once the Universe has cooled below the grand unification temperature. To be conservative, we will also require our cosmologies to be essentially FRW models by the time of nucleosynthesis. In this way, we can use the beautiful prediction of the cosmic helium abundance given by the standard model (Peebles 1971). This region between grand unification and nucleosynthesis is a fascinating and largely unexplored period. This region is in the radiation era, so classical general relativity is valid and important during this time. We can test chaotic cosmology by studying the dynamics and dissipation of inhomogeneities, shock waves, and strong amplitude gravity waves in this era.

But as we venture forth to explore this period we confront the second misconception above, namely general relativity is intractable. Certainly the inherent complexity of the nonlinear, coupled, partial differential field equations has so far permitted only a few perturbative or spatially homogeneous calculations within the chaotic framework. The elegant mathematical work on spatially homogeneous cosmologies is reviewed in Dr MacCallum's lectures in this volume.

During the past decade, however, numerical techniques to solve the Einstein equations have been developed (Smarr 1979a: Wilson 1979: Piran 1980). Like fluid dynamics (Roache 1976), general relativity has begun to yield its secrets to numerical modelling. We cosmologists are now able to simulate physical processes in the early Universe numerically and thereby to test chaotic cosmology in a meaningful fashion (Centrella 1977, 1979: Nadejin et al. 1978: Novikov and Polnarev 1979). In these lectures we will show how to defeat the supposed intractability of relativity by demonstrating how to build dynamic inhomogeneous and anisotropic cosmological models on a computer.

2. ORIENTATION: THE FRW MODELS (Misner, Thorne and Wheeler 1973: Weinberg 1972)

We begin our exposition of numerical relativity with a discussion of the homogeneous and isotropic FRW models. These metrics are familiar, in some form, to most cosmologists, and we will use them repeatedly to illustrate the concepts in numerical relativity.

The metric of a spacetime gives us information about the distances between events in a four-dimensional manifold. Using the symbol ds for the invariant proper distance interval we write the metric of the FRW models as (Weinberg 1972)

$$ds^2 = -d\tau^2 + S^2(\tau)\{dr^2/(1-kr^2) + r^2d\theta^2 + r^2\sin^2\theta d\phi^2\} \tag{1}$$

$$= {}^{(4)}g_{\mu\nu}dx^\mu dx^\nu \ .$$

Here τ is the cosmic time coordinate and measures proper time for observers at constant spatial coordinates (r,θ,ϕ). We take the metric signature to be $(-+++)$ and set the velocity of light $c = 1$. The constant k takes the values

$$k = \begin{cases} +1 & \text{positive curvature} \\ 0 & \text{flat} \\ -1 & \text{negative curvature} \end{cases} \tag{2}$$

If $k = 0$ or $k = -1$, space is infinite: for $k = +1$, space is finite but unbounded and looks like a sphere with radius $S(\tau)$ in four-dimensional Euclidean space. In all the FRW models, $S(\tau)$ gives the scale of the geometry and is called the cosmic scale factor.

The metric (equation 1) contains all possible spatially homogeneous and isotropic spacetimes. Its form can be derived from geometrical arguments alone, with no mention of source terms. The spatial homogeneity and isotropy are most easily seen in the $k = 0$ model. Transforming from spherical to Cartesian coordinates we find

$$ds^2 = -d\tau^2 + S^2(\tau)(dx^2 + dy^2 + dz^2). \tag{3}$$

Clearly, the spatial distances given by $S^2(\tau)$ at any one time are independent of position and direction along the three axes. The spatial homogeneity and isotropy of the $k = \pm 1$ models are not evident at first glance but can be deduced using arguments from group theory and differential geometry. We refer the reader to the literature for details (Ryan and Shepley 1975: Weinberg 1972). We remark only that if 'typical' galaxies have constant comoving coordinates (r,θ,ϕ) in all the models (equation 1), then the distribution of galaxies as viewed from any of these galaxies is independent of position and direction at any constant cosmic time τ.

The dynamics of the FRW cosmologies is embodied in the scale factor $S(\tau)$. To get the functional form of $S(\tau)$ we turn to the field equations

$$G_{\mu\nu} = T_{\mu\nu} \ , \tag{4}$$

where we have set the Newtonian gravitational constant $8\pi G = 1$ and the cosmological constant $\Lambda = 0$. The Einstein tensor is defined by

$$G_{\mu\nu} = {}^{(4)}R_{\mu\nu} - \frac{1}{2}{}^{(4)}g_{\mu\nu}\,{}^{(4)}R. \qquad (5)$$

The superscript $^{(4)}$ indicates that these quantities refer to the full four-dimensional spacetime. $^{(4)}R_{\mu\nu}$ is the Ricci curvature tensor and $^{(4)}R$ the Ricci scalar of the spacetime: both these quantities are constructed from the metric $^{(4)}g_{\mu\nu}$ and its first two derivatives. We use Greek indices $\mu,\nu = 0,1,2,3$.

We assume the energy-momentum tensor $T_{\mu\nu}$ is that for a perfect fluid,

$$T_{\mu\nu} = (\rho + \varepsilon + P)u_\mu u_\nu + P\,{}^{(4)}g_{\mu\nu}. \qquad (6)$$

Here, P is the pressure, ρ is the rest mass energy density, and ε is the internal energy density. The vector u_μ is the four-velocity of (observers comoving with) the fluid. For the FRW models, ρ , ε, and P are functions of time alone, and $u^\mu = (1,0,0,0,)$. The equations of energy and momentum conservation are

$$T^\mu{}_{\nu;\mu} = 0, \qquad (7)$$

where ';' is the covariant derivative in the spacetime.

Upon solution, (Weinberg 1972) equations (4) and (7) reveal that all FRW models expand from an initial state of infinite density S = 0 at $\tau = 0$. The k = 0 and k = −1 models expand forever. The k = +1 model model reaches a point of maximum expansion $\dot{S} = 0$ (where $\dot{} = d/d\tau$) and recollapses back to a 'big crunch' (a term attributed to John A Wheeler) S = 0 in the future. We note that

$$\mu \propto S^{-3},\ P \ll \mu \quad \text{'matter dominated'}$$
and
$$\mu \propto S^{-4},\ P = 1/3\mu \quad \text{'radiation fluid'}, \qquad (8)$$

where $\mu = \rho + \varepsilon$ is the total energy density.

Before we look at these models in the formalism of numerical relativity, we will make a few remarks about topology (Misner, Thorne and Wheeler 1973). The requirement of spatial homogeneity and isotropy determines the local geometry of the spacetime completely by requiring that it take the form (1). The global topology of the spatial hypersurfaces at any time τ is still undetermined. Consider for example the k = 0 FRW model (eq. 3). The three-dimensional spaces at constant τ are usually taken to have the Euclidean topology $\mathbb{R}^3$, with the spatial coordinates taking the values

$$
\begin{aligned}
-\infty &< x < +\infty \\
-\infty &< y < +\infty \\
-\infty &< z < +\infty
\end{aligned}
\qquad (9)
$$

We are however, perfectly free to use cyclic spatial coordinates

$$
\begin{aligned}
0 &< x' < \ell \\
0 &< y' < \ell \\
0 &< z' < \ell
\end{aligned}
\qquad (10)
$$

where the points zero and ℓ are identified on each axis. This is equivalent to taking a cube of coordinate edge length ℓ and identifying opposite faces. The homogeneous and isotropic three-spaces are still described geometrically by equation 3 but are now three-tori having finite volume. This procedure is analogous to imposing periodic boundary conditions and will be used repeatedly in the numerical examples that follow. [We note that there are special models in which periodic boundary conditions are incompatible with the imposed symmetry conditions].

3. THE '3 + 1' FORMALISM AND THE KINEMATICS OF NUMERICAL RELATIVITY (Misner et al. 1973: Smarr and York 1978b: York 1979)

Einstein has taught us to treat space and time on an equal footing in a four-dimensional spacetime manifold. In order to build spacetimes numerically we will separate spacetime into a stack of three-dimensional spacelike slices and a set of (one-dimensional) curves giving a time direction at each point. In this '3 + 1' we look at the Cauchy problem of general relativity and set initial data on a spacelike slice. We then use the equations of motion to evolve this data into the future, thereby constructing a four-dimensional cosmological model slice by slice. We now set out the basics of this formalism and use it to look at the kinematics of relativity.

Our first step is to slice spacetime into a family of three-dimensional spacelike hypersurfaces $\mathcal{T}$. Each slice is labelled by a monotonically increasing function t, with t = constant on each slice. The timelike curves that thread these slices orthogonally have unit timelike tangent vectors $\hat{n}^\mu$, normalized so that

$$ {}^{(4)}g_{\mu\nu}\hat{n}^\mu\hat{n}^\nu = \hat{n}^\mu\hat{n}_\mu = -1. \qquad (11) $$

On any single slice $\mathcal{T}$ we measure proper distances using a three-metric γ_{ij}, where Latin indices i,j = 1,2,3. From γ_{ij} we

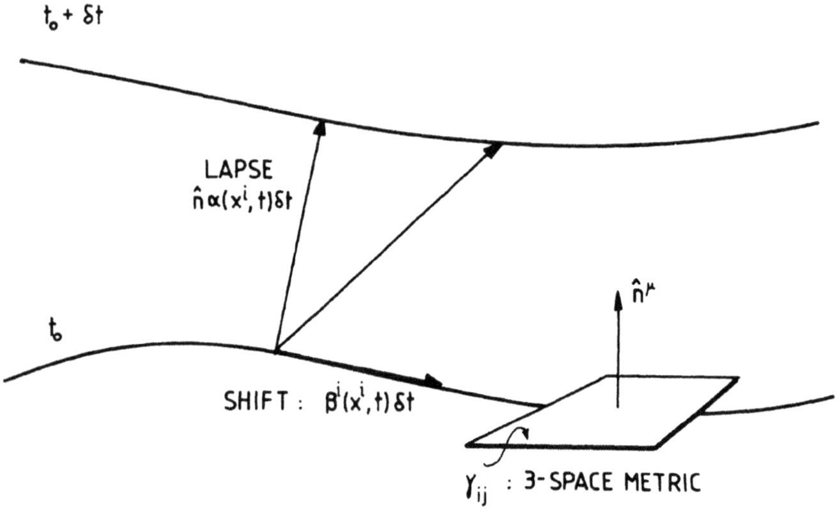

Figure 1. The lapse function α measures proper time $\alpha\delta t$ between slices $t = t_0$ and $t = t_0+\delta t$ along the normals to the slices. The shift function β^i moves coordinates in the slice surfaces.

can form a covariant derivative '$|$', a Ricci tensor R_{ij}, and a Ricci scalar $R = R^i{}_i = \gamma^{ij}R_{ij}$ to represent the curvature properties intrinsic to each slice.

In order to construct a four-dimensional spacetime we must be able to go from one slice to the next. Look now at slices at $t = t_0$ and $t = t_0+\delta t$ as shown in Figure 1. The simplest way to get from slice to slice is to advance a distance $\alpha\delta t$ along the normal n^μ at each point. The function $\alpha(t,x^i)$ measures the lapse of proper time along the normals to the slices and is called the lapse function.

We also require a method of moving our coordinates around as we evolve into the future. Therefore we introduce a shift vector β^i, which moves the coordinates a spacelike distance $\beta^i\delta t$ as we evolve to the new slice. This situation is shown in Figure 1. This part of the metric is given by

$$d\sigma^2\big|_{t_o} = \gamma_{ij}(dx^i + \beta^i dt)(dx^j + \beta^j dt). \tag{12}$$

Remembering that special relativity is valid locally we have, for the four-dimensional interval,

$$ds^2 = d\sigma^2\big|_{t_o} - \alpha^2 dt^2. \tag{13}$$

This gives (Misner 1973)

$$^{(4)}g_{\mu\nu} = \left\| \begin{matrix} -\alpha^2 + \beta_i\beta^i & \beta_k \\ & \\ \beta_i & \gamma_{ik} \end{matrix} \right\| \tag{14}$$

Note that γ_{ij} is used to raise and lower the index on β_k:

$$\beta_i = \gamma_{ik}\beta^k: \quad \beta^k = \gamma^{ik}\beta_i. \tag{15}$$

Also note that the inverse of the three-metric, γ^{ik}, is <u>not</u> the same as $^{(4)}g^{ik}$ (Misner <u>et al</u>. 1973):

$$^{(4)}g^{\mu\nu} = \left\| \begin{matrix} -1/\alpha^2 & \beta^k/\alpha^2 \\ & \\ \beta^i/\alpha^2 & \gamma^{ik} - (\beta^i\beta^k/\alpha^2) \end{matrix} \right\| \tag{16}$$

which gives the desired relation

$$^{(4)}g^{\mu\nu}\,{}^{(4)}g_{\mu\sigma} = {}^{(4)}\delta^{\nu}{}_{\sigma}$$

We stress that we are free to <u>choose</u> the quantities α and β^i. They tell how our coordinate system develops in time and embody the kinematical choices in relativity.

The remaining information we need to construct a spacetime is contained in the extrinsic curvature tensor K_{ij}, which tells how a slice curves relative to the enveloping four-dimensional spacetime. In our notation,

$$K_{ij} = 1/2\alpha\ [\beta_{i|j} + \beta_{j|i} - \partial\gamma_{ij}/\partial t] \tag{17}$$

The trace of K_{ij} will also be very useful:

$$\mathrm{tr}K \equiv \gamma^{ij}K_{ij} = K^i{}_i. \tag{18}$$

Let us now take another look at the FRW models in the form of equation (1). Comparing these metrics with (14) we see immediately that the lapse $\alpha = 1$ and the shift $\beta^i = 0$. This condition is known as geodesic slicing since the observers moving along the normals $\hat{n}^\mu$ are freely falling. The three-metric is

$$\gamma_{ij} = S^2(\tau) \begin{Vmatrix} \frac{1}{1-kr^2} & 0 & 0 \\ 0 & r^2 & 0 \\ 0 & 0 & r^2\sin^2\theta \end{Vmatrix} \tag{19}$$

and the components of the extrinsic curvature are

$$K^r{}_r = K^\theta{}_\theta = K^\phi{}_\phi = -\dot{S}/S, \tag{20}$$

and

$$\mathrm{tr}K = K^r{}_r + K^\theta{}_\theta + K^\phi{}_\phi = -3\dot{S}/S. \tag{21}$$

Therefore, in the FRW models $-K^i{}_j$ is the Hubble parameter, which is the same in all directions and at all points.

Another interesting cosmology is the Kasner metric, which can be written in the form

$$ds^2 = -dt^2 + (t/t_o)^{2p_1}dx^2 + (t/t_o)^{2p_2}dy^2 + (t/t_o)^{2p_3}dz^2 \tag{22}$$

This model is a spatially homogeneous but anisotropic generalization of the k = 0 FRW metric. It satisfies the <u>vacuum</u> field equations, which give (Landau and Lifshitz 1975)

$$p_1 + p_2 + p_3 = 1 \tag{23}$$

and

$$p_1^2 + p_2^2 + p_3^2 = 1. \tag{24}$$

Equations (23) and (24) show that one of the anisotropy parameters p_i, p_3 say, is free and obeys $-1/3 \lesssim p_3 \lesssim 1$. The extrinsic curvature components are

$$K^x{}_x = -p_1/t \quad K^y{}_y = -p_2/t \quad K^z{}_z = -p_3/t \tag{25}$$

and thus the t = constant slices have constant trace K_{ij}, where

$$\mathrm{tr}K = -1/t. \tag{26}$$

In this example the trace of K_{ij} can be used in place of t as a time parameter, a choice known as York time (York 1972).

We now write the Einstein equations in the '3 + 1' formalism. Four of these equations, $G^o{}_\mu = T^o{}_\mu$, constrain the data on each slice Υ. The equation

$$G^o{}_o = T^o{}_o$$

becomes the <u>Hamiltonian constraint</u>

$$R + (trK)^2 - K^i{}_j K^j{}_i = 2\rho_H, \tag{27}$$

where $\rho_H = T_{\mu\nu}\hat{n}^\mu\hat{n}^\nu$.

The three other equations, $G^o{}_i = T^o{}_i$, become the <u>momentum constraints</u>

$$K^j{}_{i|j} - \partial_i(trK) = j_i, \tag{28}$$

where $j^i = -\gamma^{i\mu}T_{\mu\nu}\hat{n}^\nu$. The equations (27) and (28) are elliptic and must be satisfied on each time slice. In particular, they must hold on the initial slice in a numerical evolution, and thus they serve as the initial value equations for relativity. Note that these equations are independent of α and β^i.

The remaining field equations $G^i{}_j = T^i{}_j$ are dynamical and give the evolution of the basic field variables $(\gamma_{ij}, K^i{}_j)$ away from the initial slice. In our notation the <u>evolution equations</u> are

$$\partial_t\gamma_{ij} = -2\alpha\gamma_{ik}K^k{}_j + \gamma_{jk}\beta^k{}_{|i} + \gamma_{ik}\beta^k{}_{|j} \tag{29}$$

and

$$\begin{aligned} \partial_t K^n{}_j = {} & \beta^c\partial_c K^n{}_j - K^c{}_j\partial_c\beta^n + K^n{}_c\partial_j\beta^c \\ & + \alpha(trK)K^n{}_j + \alpha R^n{}_j - \gamma^{in}\alpha_{|ij} \\ & - \alpha\gamma^{in}S_{ij} + 1/2\,\alpha T\delta^n{}_j, \end{aligned} \tag{30}$$

where $\partial_\mu = \partial/\partial x^\mu$. The source terms are

$$T = tr(T_{\mu\nu}) = T^\mu{}_\mu \tag{31}$$

and

$$S_{ij} = \gamma^\mu{}_i\gamma^\nu{}_j\, T_{\mu\nu}. \tag{32}$$

Note that the dynamical equations do not contain any time derivatives of the lapse and shift, which means that the kinematical considerations used to fix α and β on any slice depend only on the state of the field at that time. We now examine some of the gauge choices currently used.

The FRW models in the form of equation (1) exhibit three features common in cosmology. The gauge choice is geodesic slicing, $\alpha = 1$ and $\beta^i = 0$. This choice leads to

$$\mathrm{tr}K = K_0(t) = \text{constant}$$

on each time slice and to coordinates that comove with the fluid. In more general inhomogeneous models it is not possible to impose these three conditions simultaneously. We therefore examine each of these choices in turn to see what benefits and limitations they hold for cosmology.

Geodesic slicing is by far the easiest gauge condition to impose. We simply set $\alpha = 1$ and $\beta^i = 0$ at the beginning of the calculation: spatial derivatives of the lapse and shift drop out of the equations, thus simplifying the numerical code. However, these coordinates can cause problems in a strong field region, such as the vicinity of a black hole (Smarr and York 1978b). The attraction of gravity causes the geodesics to converge to a coordinate singularity on a freefall time scale. Once this occurs, the computer evolution grinds to a halt.

Some cosmological spacetimes (Centrella 1979, 1980a,b) have been constructed using geodesic slicing and will be exhibited in Section 5. There are no black holes in these models. In fact, the expansion of the Universe is sufficient in these cases to overcome the tendency for these coordinates to focus to a caustic. Finally, since most analytic work in cosmology has been carried out using geodesic slicing, these evolutions can readily be compared with analytic estimates (Centrella 1980b: Centrella and Matzner 1982).

Another popular gauge choice in cosmology is comoving coordinates. By choosing α and β^i so that coordinates follow the fluid, we hope to gain good resolution in the areas of the numerical grid where most of the hydrodynamical action occurs. One benefit derived from comoving coordinates is that the fluid equations are easily integrated. But, as might be expected, the condition that the coordinates move with the fluid leads to problems (Smarr, Taubes and Wilson 1980) when (1) shocks develop in the fluid and the coordinates become discontinuous, (2) black holes form and the time slicing becomes singular, and (3) vortices develop and the coordinates become wound up and tangled. In general, comoving coordinates are not recommended:

instead, grid velocity may be introduced to give good spatial resolution of the flow.

Almost a decade of work on numerical black hole spacetimes suggests the use of trK = 0, or maximal, slices. Physically, for any t, trace $K_{ij}(t)$ is just the convergence of the geodesics having tangent vector n^{μ} on the slice at time t:

$$\mathrm{tr}K = -n^{\mu}{}_{:\mu}. \tag{33}$$

Thus, requiring trK = 0 means that these geodesics will not converge. This gauge works beautifully in black hole spacetimes (Eppley 1979: Piran 1979: Smarr 1979b) because the maximal slicing slows down the evolution near the hole yet allows the equations to be integrated forward in time elsewhere on the slice. This has the effect of 'wrapping' the slice around the hole so that the gravitational wave pattern in the radiation zone can be computed (Eardley and Smarr 1979).

Maximal slicing is probably not the preferred gauge choice for cosmology, however. Families of maximal slices exist in asymptotically flat spacetimes (Eardley and Smarr 1979) but not in general closed cosmologies. The situation for general open cosmologies is unclear and is currently under investigation by York, Cantor and Tipler (1981). We see from equation (21) above that the FRW metric admits a maximal slice only when S = 0: this occurs only in the k = +1 model as it halts before recollapse.

All the FRW models are, however, foliated by trK = $K_0(t)$, or constant mean curvature, slices. Such slices also have 'singularity avoidance' (Eardley and Smarr 1979) properties similar to those of maximal slices, and so they are favoured for strong field calculations in the early universe (Centrella and Smarr 1978: Centrella 1978: Centrella and Wilson 1982a: Piran 1980).

Since trK = $K_0(t)$ is a condition on the slicing, we use the lapse α to impose this gauge. Taking the trace of (30) and using the Hamiltonian constraint (27) we get the <u>lapse equation</u>

$$\partial_t(\mathrm{tr}K) = \alpha(K^i{}_j K^j{}_i + \frac{1}{2}(\rho_H + S)) + \beta^i \partial_i(\mathrm{tr}K) - \gamma^{ij}\alpha_{|ij} \tag{34}$$

where

$$S = ({}^{(4)}g^{\mu\nu} + n^{\mu}n^{\nu})T_{\mu\nu} . \tag{35}$$

This is a second-order linear equation for α and is solved on each slice. The function $\partial_t(\mathrm{tr}K)$ may be specified in advance or it may be determined by the boundary conditions on α, as will be discussed in Section 5. Notice that if $\mathrm{tr}K = K_0(t)$, the term $\beta^i\partial_i(\mathrm{tr}K)$ vanishes and the specification of the shift is decoupled from the lapse equation.

Because a computer evolution stops once a slice runs into a singularity, the choice of slicing, and hence α, is considered the most fundamental gauge choice in numerical relativity (Smarr and York 1978b). We still have the freedom to choose a shift vector β^i, however, and we now briefly examine some of the choices currently in use.

The choice $\beta^i = 0$ is the simplest one. We remark that although this choice doesn't consistently cause problems, it does mean that we are discarding some of the gauge freedom that ought to be used to improve our numerical evolutions (Wilson, private communication). A minimal distortion shift vector condition has been suggested by Smarr and York (1978a) to reduce coordinate shear and to serve as part of a radiation gauge for general relativity. Although this choice does reduce the problems of separating coordinate and physical effects in numerical spacetimes, the equations that must be solved for this shift vector are very lengthy and complex. At the present time, this condition is not considered practical for multi-dimensional codes (Eppley 1979).

The use of β^i to simplify the metric or extrinsic curvature was first suggested and implemented by Wilson (Wilson 1979; Smarr 1979b), and is known as the Wilson Gauge. For example, it might be desirable to use the shift vector to set $\gamma_{xx} = \gamma_{yy}$. To preserve this condition in time we need $\partial_t\gamma_{xx} = \partial_t\gamma_{yy}$, which reduces to an elliptic equation for one of the shift vector components β^i on each slice. This means we are shifting our coordinates as we evolve in order to keep the metric simple. The particular choice of simplified three-metric is determined by experimenting with different sets of variables to see which set gives the simplest and most reasonable equations (Centrella and Wilson 1982a). This procedure reduces the number of metric functions that must be evolved. Experience has shown that codes using elliptic (constraint) equations for variables whenever possible, with a minimum of evolution equations, are stabler than codes that simply propagate all the basic variables using the evolution equations (Centrella and Wilson 1982b; Piran 1980). We will explore this point further in Section 5.

4. THE INITIAL VALUE PROBLEM (Centrella and Matzner 1979; O'Murchadha and York 1974: York 1979)

We saw above that four of the Einstein equations constrain the data on spacelike slices and therefore serve as initial value equations. Any initial data for a numerical cosmological model must satisfy these constraints, which are a set of four coupled elliptic equations. We now investigate several methods of solving these initial value equations.

There exists a constructive algorithm due to York (O'Murchadha and York 1974) for solving the initial value problem in asymptotically flat or closed three-spaces. The procedure begins with 'guess' data (γ'_{ij}, K'_{ij}) and (ρ'_H, j'^i) which are freely specifiable and motivated by the physical system being studied. The initial value equations are then written as a set of four quasilinear elliptic equations that determine four constrained potentials: a function ϕ and a vector W^i. The guess data are then dressed by appropriate factors of ϕ and W^i to produce good initial data (γ_{ij}, K_{ij}) and (ρ_H, j^i) which satisfy the initial value equations.

We point out that if $trK = K_o(t)$ the equations for ϕ and W^i decouple. These equations have been solved numerically for maximal $K_o = 0$ slices by Eppley (1977), Dykema (1980) and Piran (York and Piran 1981). More work needs to be done applying York's algorithm to the cosmologically interesting case of $trK = k_o(t) = 0$ for closed three-spaces. Very little is known about the solutions to these equations in general open models (York, Cantor and Tipler (1981). We need to direct more of our attention to the numerical solution and physical implications of these equations as we develop more realistic numerical cosmologies.

There is another way of solving the initial value equations that has produced interesting cosmological initial data (Centrella and Matzner 1979). Begin by considering two different Kasner models, labelled A and B, in the form of (22). In each model, choose t_o and select the slice $t = t_o$: note that $t_{oA} = t_{oB}$. Then match these two slices together at $z = 0$. We now have a composite initial slice for an anisotropic cosmological model inhomogeneous in the z direction.

Since we take our spatial coordinate system to be continuous across $z = 0$, we must make sure the three-metric is also continuous across the join. We see this is true since the choice $t = t_{oA}$ in model A and $t = t_{oB}$ in model B means $\gamma_{ij} = \delta_{ij}$ everywhere on the slice. Also note that the two Kasner pieces fit smoothly together across the $x - y$ plane at $z = 0$ since each of these slices is intrinsically flat.

What constraints do the initial value equations impose on the anisotropy parameters p_i in these two Kasner models? Note first that $\gamma_{ij} = \delta_{ij}$ implies $R = 0$: the Hamiltonian constraint (27) is therefore satisfied pointwise everywhere on the slice and imposes no further restriction on the data. The momentum constraints (28) reduce to one nontrivial equation containing derivatives which must be matched across the join. It imposes the condition

$$\left.\frac{p_3 - 1}{t_o}\right|_A = \left.\frac{p_3 - 1}{t_o}\right|_B \tag{36}$$

on the two models. To satisfy this equation choose the cosmology ages t_{oA} and t_{oB}, and one of the anisotropy parameters p_{3_A}. The momentum constraint (36) then determines p_{3_B}.

Notice that the slices from models A and B each have $\mathrm{tr}K = K_o(t)$ individually: the composite initial slice therefore does not have constant trace K_{ij}. In fact, the extrinsic curvature is jump discontinuous at $z = 0$. This discontinuity produces δ function discontinuities in the Riemann tensor of the spacetime which propagate at the speed of light out of the join surface (Centrella and Matzner 1979, 1980b). We will show numerical evolutions of these gravitational shocks in the next section.

We can generalize this matching procedure to produce an initial slice with a Kasner model at each point along the z axis. The ages of the models are chosen to be some function $t_o(z)$. The momentum constraint then becomes

$$\frac{p_3(z)-1}{t_o(z)} = C \tag{37}$$

where $C < 0$ is a constant and $0 \lesssim t_o(z) \lesssim -4/3\ C^{-1}$. This prescription for setting initial data can also be extended to models with perfect fluid matter sources (Centrella 1979; Centrella and Matzner 1979).

The final method of determining initial data that is currently in use (Centrella and Wilson 1982a,b: Piran 1979) was pioneered by Wilson. The Einstein equations are written in terms of freely specifiable, physically interesting variables on the one hand and constrained variables on the other. On the initial slice, we specify values for the physical variables and 'good guesses' for the constrained variables. The set of constraint equations are then iterated until the solution converge. This procedure has worked well in practice for one-dimensional (one spatial dimension plus time) codes, but it may be necessary to use the more sophisticated York algorithm for more complex situations and mutli-dimensional codes (Dykema 1980).

5. EVOLVING A NUMERICAL COSMOLOGICAL MODEL (Centrella 1979; Wilson 1979; Centrella and Wilson 1982a,b)

We now examine the problem of evolving cosmological initial data into a four-dimensional spacetime. This process involves numerically solving the evolution equations (29) and (30) coupled to the hydrodynamical equations (7). There are a number of detailed accounts of this task in the literature (Smarr 1979a; Piran 1980). We therefore summarize the major results and highlight a few of the subtleties of this process, and refer the reader to the literature for details.

Several general remarks about the Einstein evolution equations (29) and (30) are in order. In the early days of numerical relativity the extrinsic curvature tensor was evolved with both of its indices down, i.e. in the form K_{ij}. Subsequent work on maximal slicing showed that the mixed tensor $K^i{}_j$ is a better variable to use (Eppley 1979; Wilson, private communication). This is due to the fact that $\mathrm{tr}K = K^i{}_i$ is simply a sum of the variables $K^i{}_j$; written in terms of K_{ij}, trK involves metric factors as well. If we evolve $K^i{}_j$ and then solve the lapse equation for maximal slicing, trK remains zero identically because the sum of the evolution equations is equivalent to the lapse equation in differenced form:

$$\partial_t K^1{}_1 + \partial_t K^2{}_2 + \partial_t K^3{}_3 = \partial_t(\mathrm{tr}K).$$

The evolution equations for K_{ij} are not in general consistent with the lapse equation in differenced form, and errors are introduced which make

$$\gamma^{1j}K_{1j} + \gamma^{2j}K_{2j} + \gamma^{3j}K_{3j} \neq 0.$$

This same behaviour occurs for $\mathrm{tr}K = K_0(t)$. The use of the variables $K^i{}_j$ has thus become standard practice in numerical relativity.

The centering of variables in space and time is crucial to developing a stable numerical scheme (Wilson 1979). Note, for example, that the field equations (29) and (30) contain spatial derivatives of the shift vector β^i. To mimic the structure of terms like $\partial_t K^n{}_j = K^n{}_c \partial_j \beta^c$ we therefore store the variables $K^i{}_j$ at, say, zone edge and β^i at zone center. The spatial derivative of β^i is then located at the spot where $K^i{}_j$ is defined. This spatial derivative is differenced using a method similar to that applied to the transport terms in the hydro equations (46), (47), and (49), since β^i acts like a velocity for the coordinate system. The equations are also centered in time since terms like $\partial_t \gamma_{ij} = -2\alpha\gamma_{ik}K^k{}_j$ tell us to store the three metric on the full steps M and the extrinsic curvature at M + 1/2.

We now turn our attention to the evolution of the source terms $T_{\mu\nu}$. This formulation of general relativistic hydrodynamics was developed by Wilson (1979) and is highly successful. We take the source to be a perfect fluid with energy-momentum tensor $T_{\mu\nu}$ given by equation (6). The energy-momentum equations of motion (7) are supplemented by an equation for the conservation of rest mass

$$(\rho u^{\mu})_{:\mu} = 0, \tag{38}$$

which can be written in the form

$$\frac{1}{\sqrt{-g}}\,\partial_{\mu}\,(\sqrt{-g}\;\rho u^{\mu}) = 0, \tag{39}$$

where $g = \det({}^{(4)}g_{\mu\nu})$, and by the normalization equation for the four-velocity u^{μ},

$$u^{\mu}u_{\mu} = -1. \tag{40}$$

We also need an equation of state, which we take to be that of a perfect gas,

$$P = (\Gamma - 1)\varepsilon \tag{41}$$

We take our basic variables to be (Wilson 1979)

density	$D = \rho u^{t}$	(42)
energy	$E = \varepsilon u^{t}$	(43)
momentum	$S_{\mu} = (\rho + \varepsilon + P)u_{\mu}u^{t}$	(44)
three-velocity	$V^{\mu} = u^{\mu}/u^{t}.$	(45)

In these variables the hydro equations look similar to the equations of motion for Newtonian fluids. The particle conservation equation (38) becomes the density equation

$$\partial_{t}D = -\frac{D}{\sqrt{-g}}\,\partial_{t}(\sqrt{-g}) - \frac{1}{\sqrt{-g}}\,\partial_{i}(DV^{i}\sqrt{-g}). \tag{46}$$

The conservation equations (7) for $\nu = 1,2,3$ become the momentum equations

$$\partial_{t}S_{i} = -\frac{S_{i}}{\sqrt{-g}}\,\partial_{t}(\sqrt{-g}) - \frac{1}{\sqrt{-g}}\,\partial_{j}(S_{i}V^{j}\sqrt{-g}) - \partial_{i}P - \frac{1}{2}\,\frac{S_{\mu}S_{\nu}}{S^{t}}\,\partial_{i}({}^{(4)}g^{\mu\nu}). \tag{47}$$

Finally, the (thermal) energy equation is derived by projecting out the exponent of (7) along u^{μ}:

$$u^{\nu}T^{\mu}{}_{\nu\,;\mu} = 0. \tag{48}$$

We find

$$\partial_t E = -\frac{E}{\sqrt{-g}}\,\partial_t(\sqrt{-g}) - \frac{1}{\sqrt{-g}}\,\partial_i(EV^i\sqrt{-g})$$
$$-\frac{1}{\sqrt{-g}}\,P[\,\partial_t(u^t\sqrt{-g}) + \partial_i(u^tV^i\sqrt{-g})\,]. \tag{49}$$

These equations are differenced using techniques developed for Newtonian flows, and we refer the reader to the literature for details (Wilson 1979).

We now have all the basic equations necessary to build a code that will evolve inhomogeneous cosmological models. We are faced with the choice of gauge through α and β^i. We must also decide whether to evolve all the field variables using the evolution equations (a free evolution scheme), or to solve the constraint equations for some of the variables and to evolve only the remaining variables (a constrained scheme) (Piran 1979). We will discuss two cosmology codes that are currently in use, both using Cartesian coordinates (x,y,z). Both of these are one-dimensonal codes (one spatial dimension plus time: note that the spacelike slices are truly three-dimensional but are inhomogeneous in only one coordinate), and use periodic boundary conditions as described in Section 2. One uses a free evolution scheme (Centrella 1979, 1980a,b) while the other uses a fully constrained scheme (Centrella and Wilson 1982a,b). We now take a brief look at each of these codes in turn.

The first of these codes requires that both the three-metric and the extrinsic curvature be diagonal: $\gamma_{ij} = 0$ and $K^i{}_j = 0$ if $i \neq j$. This leaves six variables to be evolved using (29) and (30). The gauge choice is geodesic slicing, $\alpha = 1$ and $\beta^i = 0$. This code has been used to evolve gravitational wave perturbations (Centrella 1980a) and gravitational and fluid shocks using the initial data described in Section 4 (Centrella 1979, 1980b). The degree to which the constraint equations are satisfied serves as an independent check on the accuracy of the code. Typically, we find the relative error in the Hamiltonian constraint to be on the order of a few percent or less.

Geodesic slicing has worked well in the evolutions run on this code. The expansion of the Universe counteracts the tendency for the geodesics $\hat{n}^{\mu}$ to converge, and no coordinate singularities develop. For the case of gravitational shocks travelling in a vacuum cosmology, the choice $\alpha = 1$ and $\beta^{i} = 0$ makes it simple to compare our results with analytic estimates (Centrella 1980b: Centrella and Matzner 1982). This is due to the fact that the Kasner regions in front of the shocks naturally use geodesic slicing. A comparison of the computed amplitude of a gravitational shock with an analytic estimate of its size is shown in Figure 2.

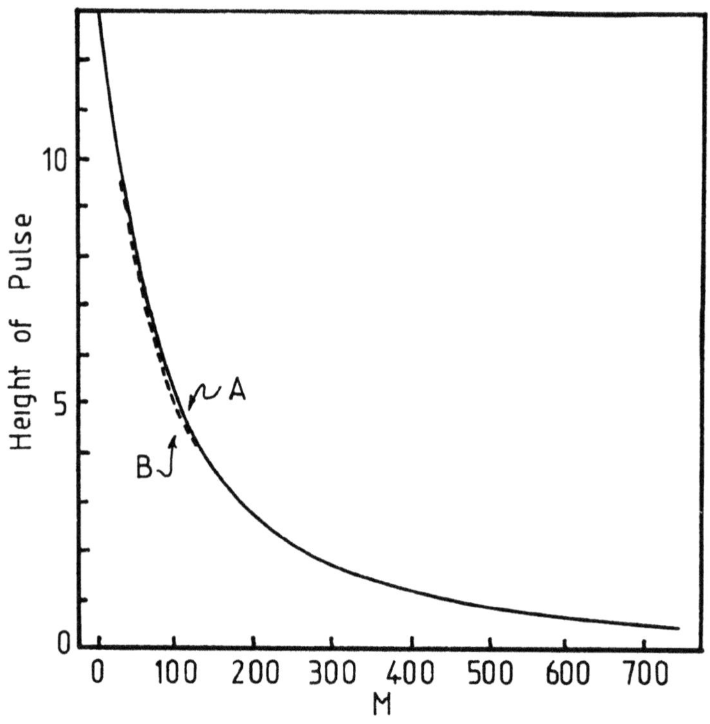

Figure 2. The gravitational shocks formed from the step-discontinuous initial data are sharp pulses in the curvature tensor. These pulses decay in amplitude as they propagate through an expanding cosmology. This graph shows the decay of the pulse versus timestep M. Curve A is calculated from the numerical evolution and curve B from the analytic estimate. (Figure reproduced from Centrella 1980b).

The other, and more recent cosmology code also requires a diagonal three-metric but allows an off-diagonal component of $K^i{}_j$. One metric component and one extrinsic curvature component are evolved using (29) and (30). These quantities were chosen as dynamical variables because they satisfy 'nice' coupled wave equations. One shift vector component is used to keep two of the metric components equal, as described in Section 3. Another shift vector component is used to keep the metric diagonal in the face of an off-diagonal $K^i{}_j$. The constraint equations are used to solve for the remaining variables on each slice.

This code also features $\mathrm{tr}K = K_0(t)$ slicing. The lapse equation (34) is solved on each time step subject to the periodic boundary condition $\alpha(0) = \alpha(\ell) = 1$. This boundary condition tells us that at $z = 0$ (which is equivalent to $z = \ell$) we are advancing a proper time $\alpha\delta t = \delta t$ in one step. Thus, this requirement selects the constant trK slice we will reach in one step. The lapse equation plus its boundary condition therefore give us α and $\partial_t(\mathrm{tr}K)$ on each slice. The constant mean curvature slicing will allow us to use this code to study strong field dynamics in the early universe and the structure of the Big Bang singularity (Centrella and Wilson, 1982a: Eardley 1979).

6. CONCLUSION

In these lectures we have presented the tools necessary to build numerical codes to study dynamical inhomogeneous cosmological models. These methods serve as a bridge between theory and observation by allowing us to simulate astrophysical processes in the early universe on a computer. Numerical cosmology thus takes its place as part of the burgeoning use of computer modelling in astrophysics.

Numerical relativity is a vigorous and expanding field. In addition to the '3 + 1' formalism discussed above, a '2 + 2' formalism has been developed which uses two null and two spacelike coordinates (Corkill 1980: Corkill and Stewart 1981: d'Inverno and Stachel 1978). Corkill and Stewart have used these techniques to evolve colliding plane waves, which focus to a singularity, with beautiful results.

We see that the two common misconceptions about general relativity are false. Einstein's theory is vital to our understanding of the early universe. Furthermore, today we have simple numerical codes which study gravitational and fluid shocks in the early universe. In the future we can expect to see cosmology codes with many interacting degrees of freedom, such as gravity waves, electromagnetism, and hydrodynamics. The results of multidimensional codes will be particularly useful in revealing the dynamics of inhomogeneities in the early universe.

To be sure, a numerical program requires stamina and determination. The rewards to be gained are a deeper understanding of this beautiful universe, our home.

Acknowledgements

I would like to thank my many colleagues and friends who have contributed so much to my understanding of numerical relativity and cosmology, particularly Pieter Dykema, Malcolm MacCallum, Richard Matzner, Larry Smarr, James York, and Jim Wilson. I would also like to thank Bernard Jones for sustaining me while I was a student groping for a way to study the early universe.

REFERENCES

Barrow, J.D. and Matzner, R.A. 1977. MNRAS, 181, 719.
Centrella, J. 1977. Inhomogeneous Cosmologies in the ADM Formalism, in Abstracts of Contributed Papers, GR8, Waterloo, Ontario.
Centrella, J. 1978. Talk at Battelle Seattle Workshop on Sources of Gravitational Radiation (1978) and at the Festschrift in Honor of Abraham Taub (1978).
Centrella, J. 1979. Ph.D. Thesis, Cambridge University.
Centrella, J. 1980a. Phys.Rev.D., 21, 2776.
Centrella, J. 1980b. Ap.J., 241, 875.
Centrella, J. and Matzner, R.A. 1979. Ap.J., 230, 311.
Centrella, J. and Matzner, R.A. 1982. Phys.Rev.D., 25, 930.
Centrella, J. and Smarr, L. 1978. Unpublished work, Institute of Astronomy, Cambridge.
Centrella, J. and Wilson, J.R. 1982a. Preprint.
Centrella, J. and Wilson, J.R. 1982b. In preparation.
Corkill, R. 1980. Smith Prize Essay.
Corkhill, R. and Stewart, J.M. 1981. In preparation.
d'Inverno, R.A. and Stachel, J. 1978. J.Math.Phys., 19, 2447.
Dykema, P. 1980. Ph.D. Thesis. University of Texas at Austin.
Eardley, D.M. 1979. Global Problems in Numerical Relativity. In Sources of Gravitational Radiation, ed. L. Smarr. Cambridge University Press.
Eardley, D.M. and Smarr, L. 1979. Phys.Rev.D., 19, 2239.
Ellis, G.F.R. 1971. Relativistic Cosmology in General Relativity and Cosmology, Proceedings of the International School of Physics 'Enrico Fermi', Course XLVII, ed. R.K. Sachs. Academic Press, New York.
Eppley, K. 1977. Phys.Rev.D., 16, 1609.
Eppley, K. 1979. Pure Gravitational Waves, in Sources of Gravitational Radiation. ed. L. Smarr. Cambridge University Press.
Jones, B.J.T. 1976. Rev.Mod.Phys., 48, 107.

Misner, C.W. 1968. Ap.J., 151, 431.
Misner, C.W., Thorne, K.S. and Wheeler, J.A. 1973. Gravitation. W.H. Freeman and Company, San Francisco.
Nadejin, D.K., Novikov, I.D. and Polnarev, A.G. 1978. Sov.Astr. 22(2), 129.
North, J.D. 1965. The Measure of the Universe. Clarendon Press, Oxford.
Novikov, I.D. and Polnarev, A.G. 1979. The Hydrodynamics of Primordial Black Hole Formation and the Dependence of the Process on the Equation of State. In Sources, ed. L. Smarr.
O'Murchadha, N. and York, J.W., Jr. 1974. Phys.Rev.D., 10, 428.
Peebles, P.J.E. 1971. Physical Cosmology. Princeton University Press.
Piran, T. 1979. Gravitational Waves and Gravitational Collapse in Cylindrical Systems in Sources of Gravitational Radiation. ed. L. Smarr. Cambridge University Press.
Piran, T. 1980. J.Comp.Phys., 35, 254.
Rindler, W. 1956. MNRAS, 116, 663.
Roache, P.J. 1976. Computational Fluid Dynamics. Hermosa Publishers, Albuquerque.
Ryan, M.P., Jr. and Shepley, L.C. 1975. Homogeneous Relativistic Cosmologies. Princeton University Press.
Smarr, L. 1979a. (Editor), Sources of Gravitational Radiation. Cambridge University Press.
Smarr, L. 1979b. Gauge Conditions, Radiation Formulae and the Two Black Hole Collision, in Sources of Gravitational Radiation. ed. L. Smarr. Cambridge University Press.
Smarr, L. 1979c. Basic Concepts in Finite Differencing of Partial Differential Equations, in Sources of Gravitational Radiation. ed. L. Smarr. Cambridge University Press.
Smarr, L., Taubes, C. and Wilson, J.R. 1980. General Relativistic Hydrodynamics: The Comoving, Eulerian and Velocity Potential Formalisms. In Essays in General Relativity: A Festschrift for Abraham Taub, ed. F. Tipler Academic Press, New York.
Smarr, L. and York, J.W., Jr. 1978a. Phys.Rev.D., 17, 1945.
Smarr, L. and York, J.W., Jr. 1978b. Phys.Rev.D., 17, 2529.
Weinberg, S. 1972. Gravitation and Cosmology: Principles and Applications of the General Theory of Relativity. John Wiley and Sons, Inc., New York.
Weinberg, S. 1979. Phys.Rev.Lett., 42, 850.
Wilson, J.R. 1979. A Numerical Method for Relativistic Hydrodynamics, in Sources of Gravitational Radiation. ed. L. Smarr. Cambridge University Press.
York, J.W., Jr. 1972. Phys.Rev.Lett., 28, 1082.

York, J.W., Jr. 1979. Kinematics and Dynamics of General Relativity, in Sources of Gravitational Radiation, ed. L. Smarr. Cambridge University Press.

York, J.W., Jr., Cantor, M. and Tipler, F. 1981. In preparation.

York, J.W.,Jr., and Piran, T. 1981. The Initial Value Problem and Beyond. In Spacetime Geometry: The Alfred Schild Lectures. eds. L.C. Shepley and R.A. Matzner. University of Texas Press, in press.

GRAND UNIFICATION AND COSMOLOGY

J.N. Fry

Astronomy and Astrophysics Center, Enrico Fermi Institute, University of Chicago, Chicago, Il 60637

1. INTRODUCTION

Most recent breakthroughs in understanding the physics of elementary particle interactions have been based on the framework of spontaneously broken gauge symmetries. The historically very successful quantum field theory of electromagnetism is a gauge theory based on the group U(1), complex phase rotation. More recently, the Weinberg-Salam SU(2) x U(1) model has become the accepted theory of 'electroweak' interactions, and SU(3) 'quantum chromodynamics' is the popular theory of strong interactions. From here, a Grand Unification (GU) of all these forces into a theory with a single large symmetry group is a natural outgrowth.

Although no specific Grand Unification model has achieved universal acceptance, all of the most popular have many features in common which are consequences only of the general considerations which <u>any</u> GU theory must satisfy. Given only these general features, there are still several kinds of qualitatively new behaviour which can be found early in a hot big bang. In the following we will develop briefly some relevant aspects of gauge symmetry and Grand Unification, and then we will discuss in this context some of these new phenomena: the dynamical production of a matter-antimatter asymmetry in an initially symmetric universe: the possibility (necessity) of new structures - domain walls, strings, or magnetic monopoles, which arise as topological 'knots' in the symmetry-breaking Higgs fields: the possibility of a de Sitter phase of exponential expansion of the scale factor of the expanding universe, which may be able to explain why the universe is nearly cosmologically flat (Ω~1) and homogeneous: the (unfortunate) problem of the

B. J. T. Jones and J. E. Jones (eds.), The Origin and Evolution of Galaxies, 65–79.

cosmological constant which appears in any Spontaneous Symmetry Breaking (SSB) theory (including Weinberg-Salam): and finally, some remarks on the origin and nature of inhomogeneities which can come out of the GU era of the early universe (Ellis et al. 1980).

2. GAUGE SYMMETRY

For the present purposes, gauge theories are useful in that the requirements of gauge symmetry and renormalisability severely limit the form of possible interactions. Nevertheless, the Lagrangian describing the universe, $\mathcal{L}_u$, still looks complicated and there is still a considerable amount of freedom (perhaps too much) in choosing some of the physical constants it contains. Without going into too much detail, $\mathcal{L}_u$ has the form

$$\mathcal{L}_u = \frac{1}{2} (\mathcal{D}_\mu \phi)^\dagger (\mathcal{D}^\mu \phi) - V(\phi) + \bar{\psi} i\gamma^\mu \mathcal{D}_\mu - m\psi\bar{\psi} - \frac{1}{4} \mathrm{Tr}[G_{\mu\nu} G^{\mu\nu}] + \Gamma \psi\bar{\psi}\phi$$

where

$$V(\phi) = \frac{1}{2} \mu^2 \phi^2 + \frac{1}{3} g\phi^3 + \frac{1}{4}\lambda\phi^4.$$

ϕ denotes generically scalar fields and ψ fermion fields, and these appear in representations of a Lie Group G (the gauge group). (The symbol $\mathcal{D}_\mu$ denotes a covariant derivative associated with G, and the $G_{\mu\nu}$ are derived from the gauge fields W_μ: see for example Abers and Lee (1973) and Taylor (1976) for details.) The remaining freedom is due to the constants λ, Γ, μ, g and m which are arbitrary and subject only to the constraints that the terms in which they appear reduce to scalars under the symmetry operations of G. We observe that the symmetry must somehow be broken, since at laboratory energies ($E < 10$ GeV today) the strong, weak, and electromagnetic interactions behave very differently. This is achieved via the Higgs mechanism: the coefficients in $V(\phi)$ are such that the classical minimum (and thus the quantum ground state) occurs for $|\phi| \neq 0$. For a specific example, consider ϕ in the N-dimensional (vector) representation of SO(N) (the rotation group in N-dimensions), and take

$$V(\phi) = -\frac{\mu^2}{2} |\phi|^2 + \frac{\lambda}{4}|\phi|^4.$$

The minimum occurs at $|\phi| = v = \mu/\lambda$, but is independent of (N-1) angular variables. The resulting physics does not depend on which values of these angles are chosen for the ground state, as all are equivalent under symmetry transformations under the residual symmetry group. We note also, at high energies $E \gg v$, interactions do not know about the breaking of symmetry (see Section 5 below).

3. GRAND UNIFICATION

In the spirit of gauge theories, with low energy physics ($E < 10^2$ GeV) described by the product

$$G_0 = SU(3)xSU(2)xU(1), \quad [(QCD)x(Weinberg\text{-}Salam)],$$

it is perhaps only natural to ask what happens if G_0 is only a residual of a larger, compact gauge group G_0 whose symmetry is restored at higher energies. This is Grand Unification (Pati and Salam 1973: Georgi and Glashow 1974: Georgi, Quinn and Weinberg 1974: Gursey and Sikivie 1976: Buras et al. 1978: and Proceedings of the Second Workshop on GU, (referred to as SWOGU 1981)). See Langacker (1981) for a general review.

The motivations for this are numerous. (1) There has been an historical impetus to describe physics in unified terms as much as possible: starting for gauge theories with Maxwell electromagnetism, through the Weinberg-Salam electroweak unification, this is a natural outgrowth. (2) Electric charge is quantized [$Q_\nu - Q_e = Q_u - Q_d$, where ν = neutrino, e = electron, u = up-quark, d = down-quark]. This implies that the U(1) factor of G_0 is a residual symmetry, for a true U(1) freedom would allow arbitrary real charges, not necessarily small integral multiples of a fundamental unit. Also, there is evidence here for unification of quarks and leptons, for this relationship is not obviously true in an arbitrary theory. (3) The total charge of a 'family' vanishes [$3Q_u + 3Q_d + Q_e + Q_\nu = 0$]. This is another relationship among both quark and lepton charges, and is necessary for so-called 'triangle anomalies' to vanish. (A similar relation also holds for weak isospin.) (4) Renormalized coupling strengths appear to be converging to a common value $\alpha \sim 1/45$ when extrapolated to high energies [$E > 10^{14}$ GeV]. If G_0 is really a free product of (strong) x (weak) x (electromagnetic), then there are three independent coupling strengths.

In Grand Unified theories, all of the above properties are necessarily true. In addition, there is the aesthetic impetus that in the simplest model, SU(5) of Georgi and Glashow (1974), the $\underline{5}$ and $\underline{10}$ multiplets contain only and exactly the fermions we observe in nature (the up- and down-quarks, in three colours and two helicity states, two helicities of electron, and only left-handed neutrinos). In SO(10) the fermions are in a single $\underline{16}$ (the extra particle is the right-handed neutrino, which can also exist in SU(5) as a singlet under all group operations).

Observationally, to this point, the evidence for Grand Unification is only suggestive and indirect. All GU theories of a large class (including those mentioned above) predict the same value of the Weinberg angle, which is in very close agreement with the experimental value (see SWOGU 1981). Fermion mass ratios (e.g. m_d/m_e) offer similar indirect tests. A definitive prediction, since leptons and quarks are placed in the same multiplets and thus are coupled by gauge interactions, is that the proton is unstable to decay into, say, $e^+\pi^0$. Dimensionally,

$$\tau_p \sim \alpha^{-2} M_x{}^4/m_p{}^5 .$$

For $M_x \sim 6 \times 10^{14}$ GeV (Goldman and Ross 1979: SWOGU 1981), and with more detailed calculations,

$$\tau_p = 10^{31\pm1} \text{ yr},$$

which is accessible with the sensitivity of currently planned experiments. In fact one group at the Kolar Gold Field claims to have detected proton decay at $\tau_p = 3 \times 10^{30}$ yr (SWOGU 1981). Another 'observation', the ratio of baryon number density to specific entropy kn_B/s, is discussed below (Section 4.)

4. THE BARYON ASYMMETRY

The observed universe is baryon-antibaryon asymmetric, with no evidence for any significant amounts of antimatter. With reasonable limits on Ω_0, H_0, and T_γ (assuming $T_\nu = (4/11)^{1/3}\, T_\gamma$ and 3 neutrino species), the ratio of baryon number to specific entropy is

$$(kn_B/s)_0 = 10^{-10\pm1}. \tag{2}$$

At high temperatures, thermally produced quarks and anti-quarks will be present in numbers essentially equal to photons and this number translates into a very small difference between the two. It is a characteristic of GU theories that this asymmetry can arise dynamically in an initially baryon-antibaryon symmetric universe: $n_B = n_b - n_{\bar{b}} = 0$. (For the 'standard' scenario see Weinberg 1979: Toussaint et al. 1979, and references therein).

There are three ingredients necessary to produce $kn_B/s \neq 0$. (1): Baryon number - violating interactions. These are naturally contained in GU theories, which necessarily place quarks and leptons in the same multiplets. For example, in SU(5), the superheavy X couples to two $\bar{u}$-quarks (B = -2/3) or to a d-quark and an electron (B = +1/3). [We will borrow the notation X to denote any generic baryon number violating species.] (2): C- and CP-violations. These provide a necessary 'arrow', a direction for the sign of the asymmetry. We know that CP violations exist from the $\bar{K}^0$-K^0 system and the Nobel award experiment of Fitch and Cronin. (3): A departure from thermal equilibrium. This appears in the early universe when the Hubble expansion rate exceeds particle reaction rates as the temperature falls through $M_X \sim 10^{14}$ GeV. The key point here is that at high temperatures the X's (and all particles) are super-relativistic, and thus exist in numbers comparable to photons, $n_X \sim T^3$, while for low temperatures the equilibrium number density is suppressed by a Boltzmann factor $\exp(-M_X/T)$. However, X's can only disappear by interactions with other particles, which happens at a finite rate. There are two cases to consider, depending on the relative magnitudes of the X decay rate Γ_X and the cosmic expansion rate. Γ_X is given by

$$\Gamma_X \sim \alpha M_X$$

where $\alpha \sim 1/45$ is the GU coupling constant. H is given by

$$H \sim (G\rho)^{1/2} \sim g_*^{1/2} M_X^2/m_{pl}$$

where m_{pl} is the Planck mass $m_{pl} \sim (hc/G)^{1/2} \sim 10^{19}$GeV and g_* is the effective spin degeneracy

$$g_* = \Sigma\, g_{boson} + \frac{7}{8} \Sigma\, g_{fermion} \sim 10^2$$

If $\Gamma_X < H$, then X's and $\bar{X}$'s do not decay until $T \ll M_X$, when inverse reactions are ignorable.

The CP violation appears as different branching ratios into different decay channels for the X, $\bar{X}$. By the CPT theorem, the total decay rates $\Gamma_{\bar{X}}$ and Γ_X are the same, but rates in specific channels need not agree. For example, in SU(5) we can write

$$\begin{aligned} X &\rightarrow \bar{u}\bar{u} & \Gamma_1 &= (1-r)\Gamma_X \\ X &\rightarrow de & \Gamma_2 &= r\Gamma_X \\ \bar{X} &\rightarrow uu & \bar{\Gamma}_1 &= (1-\bar{r})\Gamma_X \\ \bar{X} &\rightarrow \bar{d}e^+ & \bar{\Gamma}_2 &= \bar{r}\Gamma_X. \end{aligned} \qquad (3)$$

This leads to an average baryon number (weighted by rate) produced in the free decay of an X-$\bar{X}$ pair of $\Delta B = r-\bar{r} \equiv \varepsilon$. We shall discuss the value of ε below.

Since the entropy density is provided by the sum of all relativistic species, $s/k \sim g_*T^3$ and the baryon asymmetry is ε per X-$\bar{X}$ pair, or $n_B \sim \varepsilon T^3$, the prediction of GU is (for $\Gamma_X \ll H$ at $T = M_X$)

$$(kn_B/s) \sim \varepsilon/g_*. \tag{4}$$

When $\Gamma_x > H$ we must integrate the detailed Boltzmann equations, the essence of which can be described as follows. Write

$$X = n_x/T^3,\ B = n_B/T^3,\ z = M/T$$

and introduce a parameter $K = \Gamma_x/H$ evaluated at $T = M_x$ to describe the 'effectiveness' of the decay relative to the expansion rate at that time. Approximately

$$K \sim \Gamma_x/H \sim \alpha m_{pl}/g^{1/2}M_x \simeq \frac{3 \times 10^{17}\alpha}{M_x(\mathrm{GeV})}$$

The Boltzmann equations then take the form

$$\begin{aligned} \frac{dB}{dz} &= -\gamma_{ID}\ KzB - \varepsilon Kz\gamma_D(X-X_{eq}) \\ \frac{dX}{dz} &= -\gamma_D Kz(X-X_{eq}) \end{aligned} \tag{5}$$

(Kolb and Wolfram 1980a,b: Fry et al. 1980a,b). The quantities γ_D and γ_{ID} are the 'reduced' decay and inverse decay rates averaged over the thermal distributions:

$$\gamma_D = \frac{1}{2}\ z^2K_1(z),\ \gamma_{ID} = K_2(z)/K_1(z) \tag{6a}$$

(K_1 and K_2 are standard modified Bessel functions) and

$$X_{eq} = \gamma_{ID}/\gamma_D = \begin{cases} 1 &, z \ll 1 \\ z^{3/2}e^{-z}, & z \gg 1 \end{cases} \tag{6b}$$

We see in equations (5) the three 'ingredients' necessary to produce non-zero kn_B/s. With no baryon number violations, $\gamma_D = \gamma_{ID} = 0$. With no CP-violation $\varepsilon = 0$ and there is no source of B. Finally if $X = X_{eq}$ there is no source of B either.

The results of numerical integrations are plotted in Fig. 1. For $K \ll 1$, the results are as in (4), while for $K \gg 1$ it can be shown that asymptotically

$$(kn_B/s) \sim (\varepsilon/g_*)\quad (K\ \ln K)^{-1}. \tag{7}$$

Unfortunately, the CP-violation ε is very theory dependent. In 'minimal' SU(5), it is too small by several orders of magnitude, but with only slight increases in complexity, it can be increased to $O(\alpha_H) \sim 10^{-5}$ or perhaps O(1) (see Segre and Turner 1980). Fortunately, the CP violation which appears in the

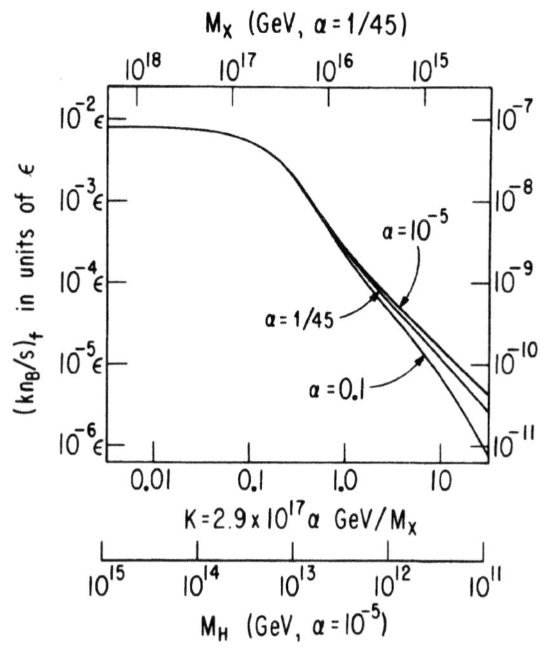

Figure 1. The final baryon number to entropy ratio kn_B/s exiting the GU epoch, as a function of $K = 3 \times 10^{17}$ GeV α/M. For $M \sim 6 \times 10^{14}$ GeV, $\alpha \sim 1/45$ (gauge bosons), $K \sim 10$ and we need $\varepsilon \sim 10^{-5}$. For $\alpha \sim 10^{-4}$ (Higgs bosons), $K < 1$, and we only need $\varepsilon \sim 10^{-7}$.

baryon asymmetry necessarily appears also in the neutron electric dipole moment (Ellis et al. 1981), and the calculations are subject to some observational test. We note that for gauge bosons, $\alpha \sim 1/45$, $M \sim 6 \times 10^{14}$ GeV, $K \sim 10$, and we need $\varepsilon \sim 10^{-5}$, which is typical of the value of Higgs couplings, while for Higgs bosons $K < 1$, and we need only $\varepsilon \sim 10^{-8}$.

The detailed numerical integrations have also shown that the final kn_B/s does not depend on initial conditions, as gauge boson interactions will quickly erase any initial asymmetry (to well below 10^{-10}, barring the possibility of 'accidentally' conserved modes). This is important, as initial conditions (or conditions leaving the quantum gravity era) are not known at all. Thermalizing reactions are slow until $T \sim 10^{15}$ GeV (Ellis and Steigman 1980), but this is still early compared to the time at which the final baryon asymmetry is established when

$T_f \sim M_X\ (K\ \ln K)^{-1}$ for $K > 1$,

$T_f \sim M_X K^{1/2}$ for $K < 1$.

(but see also Turner and Fry 1981).

5. TOPOLOGICAL STRUCTURES: DOMAINS, STRINGS, MONOPOLES

A characteristic of all spontaneous symmetry breaking models which start from a simple gauge group is the creation of topological knots in the Higgs order parameter $\langle\phi\rangle_0$ in the early universe. The SU(5) potential as a function of temperature when radiative corrections are included is (Kennedy, Lazarides and Shafi 1981)

$$V(\phi) = a/4\ [\mathrm{Tr}(\phi^2)]^2 + b/2\ \mathrm{Tr}(\phi^4) + c/3\ \mathrm{Tr}(\phi^3)$$

$$+ 1/2\ (\sigma T^2 - \mu^2)\ \mathrm{Tr}(\phi^2) \tag{8}$$

$$\sigma = 1/60\ (130a + 94b + 75g^2)$$

where ϕ (in the SU(5) <u>24</u> multiplet) is represented by a 5x5 matrix and a,b,c, and μ^2 are the only parameters allowed by the symmetries of the theory. We expect $a \sim b \sim g^2$, $c \sim g\mu$ (g is the gauge coupling constant, $\alpha = g^2/4\pi$).

The behaviour of this potential as a function of T is plotted in Figure 2. For large T, there is a single minimum at $|\phi| = 0$: the symmetry is unbroken. At $T = T_1$, a local minimum at $|\phi| \neq 0$ appears at $V > V(\phi = 0)$. At $T = T_c$ this minimum is at the same value of potential as $\phi = 0$. It is here that the symmetry breaking phase transition would occur, if not inhibited by a possible barrier. At $T = T_2$, $\phi = 0$ is no longer even a metastable equilibrium point, while at $T = 0$ the universe settles to the broken symmetry vacuum at $V = 0$. (For particular values of the parameters not all types of behaviour need be present.) We emphasize again that although $\langle|\phi|\rangle_0 = v$ is specified by the potential, there are still many <u>directions</u> among which to choose, which are not determined at all (cf. Section 2).

The key point now is that there is a finite length, the horizon, beyond which these direction variables <u>cannot</u> be correlated. (In a FRW universe $r_H = 2t$ at early times. See MacCallum, this volume.) Thus, $\langle\phi\rangle_0$ cannot avoid becoming tangled on larger scales. Although the 'direction' at any point can be changed by a gauge transformation, a continuous gauge transformation can not remove any topological singularities, which will arise with (calculable) probability per horizon volume.

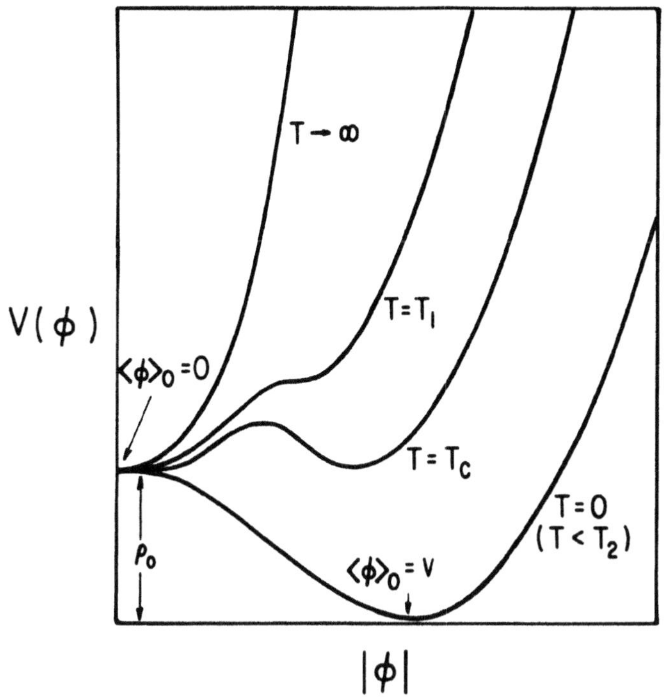

Figure 2. Behavior of the Higgs potential $V(\phi)$ as a function of temperature. At $T = T_1$ a local minimum appears away from $|\phi| = 0$. At $T = T_c$ this local minimum has lowered to the same value as $V(\phi = 0)$: it is here that the symmetry breaking transition should occur. At $T = T_2$, the point $\phi = 0$ is no longer even metastable (not plotted): at $T = 0$ there is a single minimum, $|\phi| = v$, at $V(\phi) = 0$.

The type of singularity depends on the dimension of the order parameter. Surface singularities, or domain walls, from 1-dimensional order parameter, appear to be ruled out by observations. Line singularities, from a 2-dimensional order parameter (in circulation, e.g.) or 'strings' are possible, but require the manifold $\mathcal{M}_0 = \{\phi | V(\phi) = 0\}$ be not simply connected. In GU theories, the most likely topological singularities are point-like. It has been shown (t'Hooft 1974: Polyakov 1974: Dokos and Tomores 1980) that an unexpected bonus of these point defects is that they also carry magnetic charge.

These objects are all of finite thickness, $r \sim M_x^{-1}$, the Compton wavelength of the superheavy gauge boson. The mass per length of a string is $\mu \sim \alpha^{-1}M_x^2$, while the mass of a monopole is $M_M \sim \alpha^{-1}M_x$. Strings may be useful for producing density fluctuations (see Hogan, this volume: Zel'dovich 1980: Vilenkin 1981), but monopoles have proved an embarrassment. Calculations of their production and subsequent evolution (Kibble 1976: Zel'dovich and Khlopov 1978: Preskill 1979: Einhorn, Stein and Toussaint 1980: Guth 1981 and references therein) indicate that about 10^{-9} monopoles per photon, roughly the number of baryons, by coincidence, ought to survive from the GU era to the present. Since $M_M \sim 10^{16}$ GeV, about 10^{16} times the nucleon mass, we would have $\Omega_M \sim 10^{16}\Omega_N \sim 10^{14}$. Although we do not know exactly the density parameter, this seems to be excluded by the observations. Many unsatisfactory 'solutions' have been proposed, but none are compelling.

6. THE 'INFLATIONARY UNIVERSE'

In the debate over whether the universe is open or closed, it is sometimes forgotten how close it really is to being cosmologically flat (Glashow has expressed this as, the universe is neither open nor closed, but 'ajar': see also Dicke and Peebles 1979). In the early universe, the only inherent time scale is the Planck time,

$$t_{pl} \sim G^{1/2} \sim 10^{-43}s.$$

This is the only natural time scale for an open universe to become wide open ($\Omega \ll 1$), or for a closed universe to recollapse. The present age of the universe, $t_o \sim 10^{17}s$, is a large number of Planck times (the large number 10^{40} raised to the power 3/2).

As a measure of the degree of implausibility that $\Omega \sim 1$ now (but $\Omega \neq 1$), we note that from the dynamical equation for the scale factor in a FRW universe,

$$(\dot{a}/a)^2 = 8/3\ \pi G\rho - k/a^2\ , \tag{9}$$

The density parameter $\Omega = \rho/\rho_c$ is given by

$$(\Omega^{-1}-1) = (\Omega_o{}^{-1}-1)\ (a_o/a)^2\ (\rho_o/\rho), \tag{10}$$

where the subscript 'o' refers to the present. Thus, at $T_i = 10^{17}$ GeV (chosen to be comfortably below the Planck mass 10^{19} GeV, but above the GU scale $T_{GU} \sim 10^{14}$ GeV), for $10^{-2} < \Omega_o < 10$, we need $\Omega_i = 1 \pm O(10^{-55})$ (or equivalently, for a given density we need the initial expansion rate adjusted to a part in 10^{55}). Guth (1981) has expressed the view, held by many, that this is too much of a 'fine-tuning' to be accidental, and he has proposed the following physical explanation.

His basic assumption is that the GU symmetry breaking transition is strongly first order, leaving the universe in a supercooled 'false vacuum' state which is metastable all the way to $T = 0$. Thus, there is a constant vacuum contribution to the energy density, ρ_0 (see Figure 2), which becomes the only significant contribution as matter terms redshift away. Thus, from (9),

$$a(t) \sim \exp(\chi t)$$
$$\chi = (8/3\ \pi G\rho_0)^{1/2} \sim (10^{-35}s)^{-1} \ . \qquad (11)$$

The energy density remains constant, ρ_0 = 'vacuum energy', but from (9) we see that the curvature rapidly becomes unimportant and $\Omega \to 1$. It is obvious that we do not live in a supercooled SU(5) universe, so the transition must have taken place at some point. Let $z = a(t_f)/a(t_i) \sim \exp(\chi\Delta t)$ be the factor by which the universe expands in the false vacuum state. Then to solve the 'flatness problem' we need $z > 10^{28}$, or $\chi\Delta t > 60$. This is easily achieved: Guth points out that it is just as reasonable to have

$$z \sim 10^{10^{10}},$$

and $\Omega = 1$ to high precision today.

As an added bonus, this model also bears on the 'horizon problem', that the universe today, in the standard model, was composed of some 10^{83} causally unconnected regions at $T_i \sim 10^{17}$ GeV, and yet is highly homogeneous and isotropic, at least on large scales (e.g. $\Delta T/T \sim 10^{-3}$ in the microwave background radiation). In Guth's inflationary model, although the comoving horizon rapidly approaches a constant limiting value $x_H \sim \chi^{-1}$, the physical horizon $r_H \sim a(t)x_H$ grows as $\exp(\chi t)$, and the final volume as z^3. The vacuum energy is converted into entropy (thermal matter) when the symmetry breaking transition occurs, and the universe is reheated to essentially the temperature it had when the exponential phase began (but this time on the right side of the energy barrier), so that the matter energy density is the same as before GU breakdown but now the horizon is a factor z^3 larger. (It is curious that the horizon problem also requires $z > 10^{28}$.)

The problem with this model, as also pointed out by Guth, is finding a 'graceful exit'. If the false vacuum loses metastability at some temperature T_2, it is natural for T_2 T_c. To have $z > 10^{28}$ is not likely. If the false vacuum remains metastable to $T = 0$, then the transition proceeds by nucleation of bubbles by quantum tunnelling. Here the problem is: to have $z > 10^{28}$ the nucleation rate must be small, in which case the bubbles never meet (in the jargon, 'percolation' does not occur): all the energy released in the transition will remain in the bubble walls, and the universe will be grossly inhomogeneous. (There is also the problem that the de Sitter universe does not single out a preferred comoving frame, nor is there a preferred time for the transition to happen.)

Press (1981) has considered an extension of this model. In the de Sitter universe an event horizon appears, which is the source of thermal 'Hawking radiation', with temperature $T_H \sim \chi$. Considering the interaction of this thermal radiation and the particle physics, he concludes that the universe can still grow exponentially, while the temperature falls as a power of time: thus the vacuum can lose stability at $T_2 \sim T_c$, and still have $z \gg 1$. The details of this model are not yet finalised: currently the author proposes that single-particle quantum fluctuations before the phase of exponential growth can correspond to galaxies afterwards.

7. COSMOLOGICAL CONSTANT

The shift in the vacuum energy density at the GU phase transition also raises the question of the cosmological constant today. Here, by cosmological constant, we mean a constant contribution to ρ which is independent of temperature, expansion, etc. At the Grand Unification phase transition, the vacuum energy density changes by $\sim T_c^4 \sim 10^{60}$ GeV^4. By contrast, we know today $\rho_0 \sim 10^{-29}$ g/cm^3 $\sim 10^{-46}$ GeV^4. Thus we need $\rho_0 \sim 10^{60}$ GeV^4 before the transition, so that the lowering of the vacuum will exactly cancel, and we need this constant adjusted correctly to one part in 10^{106} or so, an even more extreme 'fine-tuning' than that in Section 6 (Guth 1981).

This is a problem, but not one restricted to the lofty energies of GU. Even in Weinberg-Salam, where SU(2)xU(1) is broken at $T_c \sim 10^2$ GeV, this effect requires an adjustment to one part in 10^{54}. In fact, in the general universe, with an arbitrary number of symmetry breaking transitions, the initial cosmological constant must be (carefully) adjusted to cancel the sum of all vacuum shifts. This only points out the incompleteness of the theory: perhaps in a complete theory, which includes gravity, the freedom to choose the zero-level of energy will be fixed instead. It is also necessary to point out that there is no universal agreement that 'vacuum stress energy' gravitates in the same way as 'matter' (see section 8 below).

8. DENSITY FLUCTUATIONS IN GU

GU is probably not the answer to the <u>origin</u> of structure in the universe. Strings (see Hogan, this volume: Zel'dovich 1980: Vilenkin 1981) are not likely to appear in the more popular models, but are not ruled out. Phase transitions can produce anything from complete uniformity to gross inhomogeneities in which the universe never looks like ours (Guth 1981) or from 'single particle fluctuations' producing galaxies (Press 1981) to 'spontaneous production of the Zel'dovich spectrum' (Press 1980). Baryon production in GU theories produces (kn_B/s) = (function of fundamental constants) i.e. only adiabatic fluctuations. Even production of comparably-sized isothermal and adiabatic perturbations would not help, since adiabatic fluctuations (true curvature fluctuations) would grow until 'entering the horizon', when pressure response makes them oscillate, while isothermal fluctuations are 'frozen in'. Also, in the standard model, the number of baryons within a causally connected region is

$$N_B \sim 10^4 \text{ at } T = M_x,$$

far smaller than any object one would wish to make from fluctuations. Also any process which occurs when GR is valid (stress-energy conserved) produces a spectrum $\langle|\delta_k|^2\rangle \sim k^4$, and $n = 4$ seems to be ruled out by observations. Although schemes exist for producing isothermal fluctuations in GU (Guth 1981), none are compelling.

9. CONCLUSION

The general features inherent in Grand Unification are likely to remain with us for some time, no matter what final model emerges. Independent of any specific model is a framework for <u>calculating</u> (kn_B/s), rather than inserting this as an (arbitrary) initial condition. This is one of the most pleasing results so far. Less pleasing, but also an outcome of the framework and not of a specific model, is the implication that magnetic monopoles, relics of the early universe, also exist. More model dependent is the possibility of 'strings' while completely model dependent (in that no working detailed model has yet been completed) is the possibility of an 'inflationary phase' of exponential growth. It is worth emphasizing that the considerations presented here are consequences of the spirit of Grand Unification and are likely to survive any new developments in the 'desert' between 10^2 and 10^{14} GeV, barring a whole reformulation of elementary particle theory.

The uncertainty in the CP-violation parameter ε and in the contortions to get the cosmological constant right today point out that even GU theories are not complete. Too much freedom remains, which we hope will be removed in the final 'correct' theory. Meanwhile, the newly found interaction between elementary particle physics and cosmology in the early universe continued to provide interesting areas of research for all sides.

REFERENCES

Abers, E.S. and Lee, B.W. 1973. Phys.Rep. 9C, 1.

Buras, A.J., Ellis, J., Gaillard,M.K. and Nanopolous, D.V. 1978. Nucl.Phys., B135, 66.

Dicke, R.H. and Peebles, P.J.E. 1979. In General Relativity: An Einstein Centenary Survey, ed. S.W. Hawking and W. Israel (Cambridge University Press).

Dokos, C.P. and Tomoras,T.N. (1980) Phys.Rev., D21, 2940.

Einhorn,M.B., Stein, D.L. and Toussaint, D. 1980. Phys.Rev., D21, 3295.

Ellis, J., Gaillard, M.K. and Nanopolous, D.V. 1980. In Unification of the Fundamental Particle Interactions, ed. Fermana, Ellis and Van Niewenhuizen (Plenum, 1980).

Ellis, J., Gaillard, M.K., Nanopolous, D.V. and Rudaz, S. 1981. Phys.Lett., 99B, 101.

Ellis, J. and Steigman, G. 1980. Phys.Lett., 89B, 186.

Fry, J.N., Olive, K.A. and Turner, M.S. 1980a. Phys.Rev., D22, 2953.

Fry, J.N., Olive, K.A. and Turner, M.S. 1980b. Phys.Rev., D22, 2977.

Georgi, H. and Glashow, S.L. 1974. Phys.Rev.Lett., 32, 438.

Georgi, H., Quinn, H. and Weinberg, S. 1974. Phys.Rev.Lett. 333, 451.

Goldman, T.J. and Ross, D.A. 1979. Phys.Lett., 84B, 208.

Gursey, F. and Sikivie, P. 1976. Phys.Rev.Lett., 36, 775.

Guth, A. 1981. Phys.Rev., D23, 347.

t'Hooft, G. 1974. Nucl.Phys., B79, 276.

Kennedy, A., Lazarides, G. and Shafi, Q. 1981. CERN Report TH 2944.

Kibble, T.W.B. 1976. J.Phys.A.: Math.Gen., 9, 1387.

Kolb, E.W. and Wolfram, S. 1980a. Phys.Lett., 91B, 217.

Kolb, E.W. and Wolfram, S. 1980b. Nucl.Phys., B172, 224.

Langacker, P. 1981. Phys.Rep., 72, 185.

Pati, J.C. and Salam, A. 1973. Phys.Rev., D8, 1240.

Polyakov, A. 1974. JETP Lett., 20, 194.

Preskill, J. 1979. Phys.Rev.Lett., 43, 1365.

Press, W. 1980. Physica Scripta, 21, 702.

Press, W. 1981. Proceedings Moriond Astrophysics.

SWOGU 1981. Proceedings of the Second Workshop on Grand Unification (SWOGU), Ann Arbor, MI, April 1981. In press.

Segre, G. and Turner, M.S. 1980. EFI Preprint 80-43.

Taylor, J.C. 1976. Gauge Theories of Weak Interactions. (CUP).

Toussaint, D., Treiman, S.B., Wilczek, F. and Zee, A. 1979. Phys.Rev., D19, 1036.

Turner, M.S. and Fry, J.N. 1981. EFI Preprint, submitted to Phys.Rev.Lett.
Vilenkin, A. 1981. Phys.Rev.Lett., 46, 1169.
Weinberg, S. 1979. Phys.Rev.Lett., 42, 850.
Zel'dovich, Ya.B. 1980. MNRAS, 192, 663.
Zel'dovich, Ya.B. and Khlopov, M.Yu. 1978. Phys.Lett., 79B, 239.

THEORY AND EVIDENCE ABOUT THE ORIGIN OF COSMOLOGICAL STRUCTURE

Craig J. Hogan

Enrico Fermi Institute, University of Chicago

1. INTRODUCTION

I adopt the view that exact uniformity is a natural symmetry for the Universe near the singularity. There is, however a fundamental graininess of matter on quantum scales, and these lectures explore the possibility that this graininess can lead to the observed astronomical inhomogeneity up to galaxy cluster scales. During a phase transition, quantum discreteness can be amplified by many orders of magnitude. I discuss three specific examples - formation of 'strings' at the unification epoch (10^{-20}sec) in a hot Universe, bubble formation in vacuum phase transitions of spontaneously broken gauge theories, and shattering of a pion condensate at nuclear density (10^{-5} sec) in a cold Universe. The latter process has the advantage of using only known physics and naturally producing a generation of pregalactic stars without any arbitrary initial fluctuations, but the disadvanage of abandoning the elegant predictions of hot Big Bang nucleosynthesis. These theories and others may be distinguished by observations of the microwave background anisotropy, which may in principle show directly whether fluctuations existed in the initial state, and indirectly whether the entropy is truly primordial. Various mechanisms for producing anisotropy, and their signatures in $(\Delta T/T)_\theta$, are reviewed here.

Since the early work of Lifshitz (1946) the conventional approach to studying inhomogeneities has been linear perturbation theory (see e.g. Peebles 1980). The initial state is described as a nearly-uniform perfect fluid with small-amplitude density fluctuations on large scales. Physics tells us how these fluctuations develop in time up to the point where they become

B. J. T. Jones and J. E. Jones (eds.), The Origin and Evolution of Galaxies, 81–95.

non-linear, at which point approximate physics takes over. The character of the initial perturbations determines whether one has 'pancakes' or 'hierarchical clustering', the two competing paradigms for galaxy formation which have emerged over the last decade and which are described in detail elsewhere (Rees 1975: Peebles, Shandarin, this volume).

The motivation for these models lies in their mathematical tractability: we can do linear perturbation theory. On the other hand astrophysics on galactic scales is definitely non-linear and may have been so since before galaxies formed. There is no reason to suppose that the linear approach is the one Nature takes. In these lectures I investigate the possibility that structure developed in a series of non-linear processes ultimately originating at the quantum level. My attitude is that the development of structure may not be simple, but the initial conditions ought to be. We should be suspicious about introducing a priori structures similar in scale to those we wish to explain.

One key idea goes back to Zel'dovich (1963) and was enlarged by Layzer (1975). They suggested that a cold expanding Universe could shatter into fragments as it passed through a phase transition. This process offers a way, in principle, to go from perfectly smooth matter to a system which has large, astrophysically active objects. A more recent suggestion (Zel'dovich 1980) utilizes exotic objects called 'strings' to generate structure spontaneously in a hot Universe. I describe these mechanisms as examples of processes which cause spontaneous large-scale lumpiness.

Any non-linear theory is bound to be complicated - indeed I maintain that these complications may be very important. How is one to test such a theory? The simplest way is to observe features which are independent of details: an example I describe in some detail is the large-scale anisotropy in the microwave background. Simple arguments from causality and transparency tell us that we are looking at regions of space-time which were disconnected when they emitted the radiation, and thus the angular spectrum of anisotropies $(\Delta T/T)_\theta$ tells us about the origin of the radiation and the inherent smoothness of the initial state. Present observations already strongly constrain theory: we might hope that one or more of the ideas discussed below can be ruled out by the next generation of detectors.

2. GALAXIES FROM A PHASE TRANSITION?

I begin with an unconventional model which applies Zel'dovich's shattering idea to a phase transition which occurs in matter near nuclear density. The following series of events illustrates one possible way galaxies could be consequences of non-linear processes.

Suppose the Universe were initially cold. At nuclear density it would shatter into large pieces ($\sim 10^{-6}\ M_{\odot}$) due to a phase transition (see Peierls, Singwi and Wroe 1952). These fragments would probably melt back into a nearly-uniform soup of degenerate particles, but residual fluctuations left behind would cause stars to form at a much lower (but still high) redshift, $z \gg 100$. These stars might then trigger collapse of galactic-size systems, by an astrophysical instability (e.g. Ostriker and Cowie 1981: Doroshkevich, Zel'dovich and Novikov 1967). The production of the microwave background and helium is presumed to follow in principle once stars are formed: independent evidence for pregalactic stars is discussed elsewhere in this volume (see also Carr 1981: Rees 1978). In spite of many uncertainties, a clean test of this 'cold Big Bang' theory can be made by examining large-scale anisotropies in the microwave background – a characteristic dependence of $(\Delta T/T)$ on θ distinguishes a non-primordial origin for the radiation. I avoid discussing the complicated issues of pregalactic astrophysics and focus instead on two aspects of the problem which are fairly straightforward: the process by which shattering occurs, and the evidence in the microwave background.

We now describe the process of 'shattering' in expanding matter during a phase transition (details may be found in Hogan 1982a). Figure 1 shows a schematic equation of state for matter near a first-order phase transition (see Zel'dovich and Novikov 1978). The derivative of the internal energy density E with respect to inverse volume n is pressure: the second derivative is compressibility. At ρ_0 the pressure goes negative and there is a tension on the medium: it is metastable, and if cut it flies apart, propagating at about the sound speed as a crack or bubble. Matter surrounding the crack relaxes back to equilibrium density above ρ_0 where the pressure is positive again: this relaxed region also spreads at the sound speed. If the crack meets another crack, the region between them is relaxed, condensed matter – a fragment.

In real matter there is nothing to start the cracks going except nuclei themselves, and in fact it is possible for nuclei to quantum-tunnel out of the metastable phase at isolated points, and nucleate cavities which spread like the cracks just described once they reach a minimum size. There is a barrier-penetration involved because the matter is metastable: there is a positive surface tension, and therefore for a small cavity the work done in creating an interface exceeds the energy gained in releasing tension. Metastability vanishes at ρ_{crit} when the

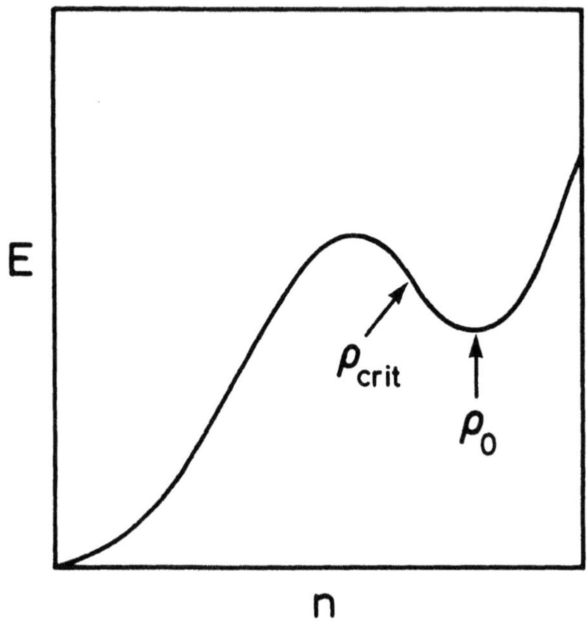

Figure 1. Schematic equation of state, showing a first-order phase transition. Between ρ_{crit} and ρ_0, expanding matter is metastable and large-scale structure can be produced.

compressibility vanishes. The size of lumps to come out is the mean separation between the first set of nucleation sites whose cracks meet.

The mean separation between cavities $L_N(t)$ is calculable from quantum mechanics. If the distance travelled by a sound wave or crack between t_1 and t_2 is $L_s(t_1,t_2)$, then the size of the first cracks to travel their own mean initial separation is given by

$$L = L_s(t_1,t_2) = L_N(t_1)$$

and the cracks meet at t_2, leaving fragments of size L.

It turns out (Hogan 1982a) that L_N can be calculated in a general way for cold matter, and that L always turns out to be a sizeable fraction ($\sim 10^{-3} - 10^{-2}$) of the sound horizon size (sound speed x Hubble time), which at nuclear density is $\sim 10^{19}$ nucleons in length and contains nearly a solar mass. The reason the pieces are so big is that the nucleation of cavities is very rare at first and increases in the right way - not too quickly (too many cracks) and not too slowly (not enough cracks - leading to widespead rupture on nucleon scales because cracks never meet). This process provides an example of how structure can arise spontaneously on astronomical scales - it amplifies quantum inhomogeneity by ~ 50 orders of magnitude in mass. Since the largest inhomogeneities are presently about $\sim 10^{70}$ m_p, the largest hurdle is already overcome! More significant is the fact that the objects eventually produced are astrophysically active (i.e. exploding stars instead of asteroids).

We now turn to the formation of stars at $z > 100$ which is a direct consequence of the shattering just described. The next 10 orders of magnitude of amplification come from the standard Lifshitz instability. Figure 2 shows a series of parallel lines which illustrate the redshift z at which $(\delta\rho/\rho)_M \sim 1$ (if no pressure forces intervene to stabilize growth) on mass-scale M for various values of the initial fragment size m:

$$M = (z/10^{14})^{-6/7}\ m$$

(see Peebles 1980 section 28: Zel'dovich 1965). The smallest mass for which pressure is always unimportant is the maximum Jeans mass,

$$M_{J,max} \sim (2 - 10)(S + 1)^2\ M_\odot$$

where S is the entropy. Thus the first things to form if $S \sim 1$ have a stellar mass and a high density: we appeal to these objects to trigger structure on much larger scales by explosions (Ostriker and Cowie 1981) or inhomogeneous ionizations (Rees 1981) because the natural clustering mass by pure Lifshitz instability is only

$$M_o \sim (10^{14})^{6/7}\ m \sim 10^{12}\ m \sim 10^{6}\ M_\odot.$$

Let us generalize this model to 'hot' matter with initial entropy $S_p > 1$. Suppose matter passes through a first-order phase transition at temperature T_{GeV} by randomly nucleating bubbles through quantum tunnelling. This may happen, for example, in gauge theories during symmetry-breaking. Let the length scale of lumpiness be L_m, the comoving mass-scale m, and the expansion rate t_1^{-1}. As above, the exponential in quantum tunnelling generally leads to bubbles with size of order $L_m \sim ct_1/\ln[L_m/\ell]$ (where ℓ is some microscopic unit) when they collide. The amount

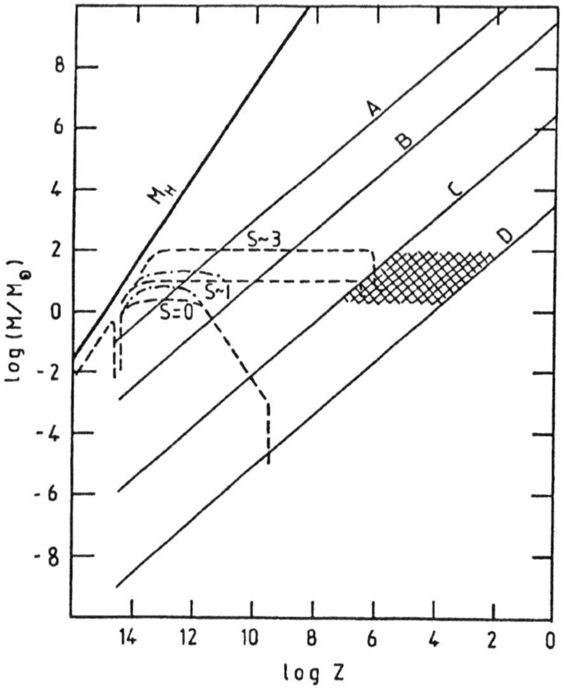

Figure 2. Growth of structure in cold universe (s < 1). M_H denotes the horizon mass. At $z \sim 10^{14.5}$, fragments form spontaneously, with > 10^{-9} $M_{\odot}$. This creates small amplitude fluctuations on larger scales (indicated by parallel solid lines A-D). Dashed lines indicate the maximum Jeans mass, below which growth of fluctuations is suppressed by pressure forces. Fragments in the range C-D cause stars to collapse in the hatched region.

of work done by the bubble walls divided by the total (matter + radiation) energy density is

$$\eta_m = (L_m/ct_1)^2$$

which is called by Bardeen (1980) the 'entropy perturbation'.

On a large scale with N bubbles the dispersion in work is of order $N^{1/2}\eta_m$ since neither the size or positions of bubbles are correlated on scales larger than L_m. The entropy perturbation for baryon mass M is then

$$\eta_M = (M/m)^{-1/2}\ \eta_m$$

Press and Vishniac (1980) and Bardeen (1980) show that pressure gradients and peculiar-velocity gradients lead to variations in density of order η_M when M comes within the horizon. It is then possible to define a peculiar Newtonian potential $\delta\phi_M = G\delta M/R$ which obeys roughly

$$\delta\phi_M \sim \eta_M = (M/m)^{-1/2}\,\eta_m$$

which gives the binding energy of systems of mass M when they collapse.

The bubbles themselves disappear after a time ~ t_1 leaving an entropy $S > S_p$ per baryon and nearly smooth matter. As in the model above, the first systems to collapse are of order the maximum Jeans mass $M_J = S^2 M_\odot$. Now the criterion for any model like this to generate active objects (rather than linear fluctuations) is that enough fluctuation $\delta\phi(M_J)$ is caused on the scale M_J to cause collapse by the present epoch. The reader may verify that this is so only for

$$S < 10^5\, T_{GeV}^{-9/17}\,(L_m/ct)^{21/17}$$

Evidently, the cold-universe picture just sketched is not unique: if $S < 10^3$, a first-order transition at electroweak symmetry breaking kT = 300 GeV (Witten 1981) may have similar effects, with objects of mass $< 10^6 M_\odot$ collapsing in place of stars. The main problem with $1 \ll S \ll 10^8$ is that excessive helium production seems unavoidable without a 'cold' neutrino component.

Is it possible for <u>any</u> mechanism to produce structure spontaneously in a hot universe with $S_p \sim 10^8$? Probably not using <u>known</u> physics. Probably yes, if we are allowed to speculate about physics at the Planck time (e.g. Press 1980: Peebles 1968). In between these is the 'possibly' known physics of grand unification, and there is a possible mechanism using exotic remnant structures of the GUT epoch: astronomically large, topologically stable, infinitely long, very thin structures called 'strings'

In certain gauge theories strings are necessarily formed during a vacuum phase transition (see Fry's contribution in this volume). They form a net which is always kinky on ~ the horizon scale (at least one string passes through any horizon volume) and annihilates on smaller scales by reconnection and/or formation of closed loops. Strings have a fixed mass per unit length

$$M/L \sim m_H^2$$

where m_H is the Higgs mass of the gauge theory. GUT strings with $m_H \sim 10^{-3}$ in Planck units introduce a stochastic perturbation in gravitational potential of order $(v/c)^2 \sim 10^{-6}$, which is appropriate for forming clusters of galaxies. Thus they may imitate a primordial $n = 1$ or constant-curvature spectrum of primordial fluctuations. The astrophysics of strings is still uncertain, as indicated by very different scenarios for their evolution described by Zel'dovich (1980) and Vilenkin (1981). For a description of strings, see Kibble (1980).

3. ANISOTROPY IN THE MICROWAVE BACKGROUND

3.1 Special angular scales

We assume that $\Omega = 1$, and that

$$z_1 < z_{rec} < z_{eq}.$$

Here z_1 is the redshift of the last scattering surface ($e^{-\tau}d\tau$ peaks in the region of optical depth $\tau = 1$, over roughly one e-fold in redshift), z_{rec} is recombination in the standard model, and z_{eq} is the epoch of equal matter and radiation densities. The assumption $z_1 < z_{rec}$ is not standard but is conservative in the pessimistic sense: unless the universe is remarkably smooth at $z \sim$ a few hundred, a residual ionization sufficient to smear out the last scattering surface occurs (Hogan 1980).

Each redshift z is associated with an angle $\theta_H(z)$, which is the angle subtended by the horizon at redshift z. For $\Omega = 1$,

$$\theta_H(z) \sim 30[(1+z)^{1/2}-1]^{-1} \text{ degrees}$$

which is roughly 1° at z = 1000, 3° at z = 100 and 15° at z = 10. Last scattering probably lies in the range $10 < z_1 < 200$, giving

$$5° < \theta_1 < 15°.$$

For example, the optical depth from Compton scattering back to redshift z is

$$\tau_{es} = 0.05h[(1+z)^{3/2} - 1]\, \Omega_{HII}$$

where Ω_{HII} is the fraction of matter in ionized hydrogen and $h = H_0/100$ km s^{-1}Mpc^{-1} is Hubble's constant. For $\theta_1 = 15°$ we need $\Omega_{HII} = 1$, an absolute upper limit. The quadrupole-to-octopole ratio (that is, ~ 90° and 45°) is thus certainly unaffected by ionization. In the centimeter region, this is the most important source of opacity.

Thus it is almost certain that at $\theta > 15°$ the scaling in Fig. 3 should be valid and insensitive to the ionization history of the

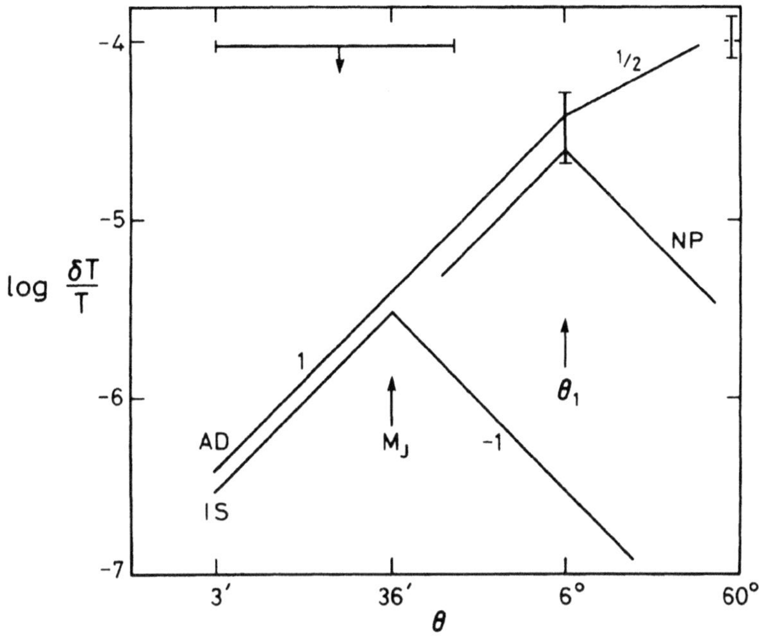

Figure 3. Angular spectrum of microwave anisotropy produced by pure isothermal perturbations (IS), adiabatic fluctuations with a flat (n = 0) power spectrum (AD), and fluctuations in non-primordial entropy production (NP), corresponding to hypotheses (b), (a), and (d) in text about origin of fluctuations. Observations are quadrupole ~ 10^{-4} (Boughn et al.), $(\Delta T/T)_{6°}$ ~ 3×10^{-5} (Melchiorri et al.), and upper limits reviewed by Partridge at arc-minute scales. Prospects are good for distinguishing NP and AD cases.

universe. That is what makes the large-scale anisotropy such a valuable tool for probing the symmetry of the initial state and the history of entropy production.

3.2 The temperature fluctuation

Recent observations (Boughn, Cheng and Wilkinson 1981) show a quadrupole moment in the radiation temperature of order $(\Delta T/T) = 10^{-4}$. If real, this is the first detection of very large-scale intrinsic structure. Tentative observations by Melchiorri's group (1981) show $(\Delta T/T)_{6°}$ ~ 3×10^{-5}, and well-established upper limits at $\theta < 1°$ (see Partridge, these proceedings) give $(\Delta T/T) < 10^{-4}$. What does all this tell us?

Consider $(\Delta T/T)$ produced by four distinct mechanisms. In Figure 3, $\Delta T/T$ is shown as a function of angular scale for each mechanism.

a) Fluctuations in primordial space-time curvature ('adiabatic')

Suppose that at the Planck time there were already large-scale fluctuations in energy density, with power spectrum $|\delta_k|^2 \propto k^n$, or perturbations on comoving baryon mass-scale M given by

$$(\delta\rho/\rho)_M \propto M^{-1/2-n/6}$$

These fluctuations must be laid down by some acausal and non-conservative mechanism involving quantum gravity (for example, Peebles 1968: Press 1980). At large θ we see the relic fluctuation in gravitational potential $\delta\phi \propto \theta^{1/2-n/2}$, or Sachs-Wolfe (1967) effect. Below θ_1 we see Doppler redshift of radiation scattering off moving electrons falling into lumps of matter within the last scattering surface (Sunyaev and Zel'dovich 1970): the spectrum steepens by 1/2 power. Finally at some small $\theta < \theta_{rec}$, the spectrum steepens further (Peebles 1980, section 92) because shorter wavelengths are stabilized by pressure as they come within the horizon during the radiation era.

The large θ observations are consistent with $\Omega = 1$ and with primordial $n = 0$ or $n = 1$, that is, $\delta\phi$ constant or increasing with scale. It is not yet clear whether a spectrum with a single n (i.e. a featureless power-law) can explain the anisotropy and the distribution of galaxies. If $\Omega < 1$, the fluctuations may be a power-law in one quantity (say, mass) and not another (say, angle), so there may be a feature in $(\Delta T/T)$ at a large angle $\sim \theta_H(z \sim \Omega^{-1})$. However, it is unlikely to be as strongly peaked as in effect (d) below, because (i) large associated peculiar velocities or inhomogeneities on scales > 100 Mpc would already have been noticed, and (ii) a sharp peak would require small-scale $n < -1$, which is probably not appropriate for making galaxies (Peebles 1981).

b) 'Pure' isothermal fluctuations

Let the density be initially uniform, and artificially introduce non-uniformities in the amount of net baryon charge without altering the total particle density. To do this we need to violate causality and conservation of baryon-number, but not conservation of energy: thus we need not involve quantum gravity.

To make galaxies, assume a fractional perturbation of order $\delta \sim 10^{-2}$ on galaxy scales, and let fluctuations be random 'white noise' on larger scales - a customary assumption leading to hierarchical clustering. Peebles shows that a random distribution of galaxies gives the correct quadrupole moment. The model just described does not produce a random distribution of galaxies at present on scales larger than the maximum Jeans mass $M_J \sim 10^{17}\ M_\odot$ (the horizon scale when the universe becomes matter-dominated). The distribution of matter at $M > M_J$ is smoother than random because radiation pressure gradients caused by entropy variations have very little effect on the motion of matter (Grishchuk and Zeldovich 1978). As a result the fluctuation in potential δ M (rather than density) is proportional to $(M/M_J)^{-1/2}$ for $M < M_J$, and proportional to $(M/M_J)^{1/6}$ for $M < M_J$: we thus expect to see a peak in $(\Delta T/T)$ at perhaps $\sim 30'$ (where upper limits exist which are less than the quadrupole amplitude). Pure isothermals do not produce a significant quadrupole unless we artifically add features to the power-spectrum.

c) 'Impure' isothermal fluctuations

Suppose we generate isothermal fluctuation by adding baryons to an otherwise uniform neutral soup of radiation at temperature kT. (This involves violating energy conservation, so we require $kT \gg 1$ GeV). Let the 'pure' isothermal fluctuation have amplitude δ_I on (say) a galaxy cluster scale. But there are also adiabatic density fluctuations with amplitude $\delta_A = S^{-1}\delta_I \sim 10^{-8}\delta_I$ and these grow like T^{-2} in the radiation era. Thus by recombination ($kT \sim 1$ eV) the adiabatic component has an amplitude at least 10^{11} greater than the isothermal one!

We are back to case (a) of just primordial adiabatic density fluctuations. Of course it is possible in principle to contrive a situation between (b) and (c) where the growing and non-growing components have comparable amplitudes at recombination, but it requires extraordinary 'fine-tuning' to do so. Thus the model of Gott and Rees (1975) (which makes constant-curvature fluctuations by adding baryons), although natural if the fluctuations originate near t_{eq}, seems very unnatural if they originate near the Planck time. It would seem an extraordinary coincidence if the isothermal amplitude were large enough to make galaxies, and the adiabatic amplitude were large enough to make the quadrupole distribution of the microwave background.

d) Finally, we note a new effect which is noticeably different from (a) and (b), namely the fluctuation expected from non-primordial entropy production. Assume that the microwave background is generated in pregalactic objects and thermalized by interaction with dust and gas at $z_1 \sim 20$-100. Assume that the luminosity per unit mass mass is a stochastic function which varies independently of the large-scale linear density fluctuation, and which is in fact an uncorrelated or 'white noise' process above some 'lump' scale D. Temperature fluctuations at $\theta > \theta_1$ will then have

$$(\Delta T/T)_\theta \propto N_\theta^{-1/2}$$

where $N_\theta = 4\theta^2 \Lambda^{-3} (1+z_2)^{-1/2}$

is the number of radiating lumps of comoving size $\Lambda = D(1+z_2)/ct_o$ in a redshift interval $(z_2, 2z_2)$ in a beam of size θ. These temperature fluctuations dominate even those produced by the gravitational effects of the lumps, which are of order (Hogan 1982b)

$$(\delta T/T) \sim \delta\phi \sim \Lambda^{7/2} \theta^{-3/2} (1+z_2)$$

Now consider smearing of fluctuations by scattering and absorption. At z_2, the optical depth is high, and a photon executes a random walk away from its source. At first this may be described as a diffusion process but as the medium becomes more transparent the flights become longer until at z_1 they exceed the Hubble length. The penultimate scattering is typically $\sim ct_1$ away from the scattering that finally delivers a photon into our line of sight. We view photons from a typical source distributed more or less smoothly over the angle θ_1. For simplicity approximate the sky as a plane and adopt a Gaussian with variance θ_1^2 for the 2-dimensional projection of the smoothing function. If sources are randomly distributed on the sky, the two-dimensional power spectrum of ΔT is

$$|\delta_k|^2 = \text{const.} \times e^{-k^2\theta_1^2}$$

The problem can be treated more precisely with specific assumptions about the nature and evolution of scattering or absorbing material, but the Gaussian approximation adequately models the general characteristics of the anisotropy.

Experiments at $\theta \gg \theta_1$ should observe the white-noise spectrum. This is probably the cleanest prediction of the non-primordial model. An experiment like that of Melchiorri et al. at $\theta < \theta_1$ may be modelled by a profile like

$$f_\sigma = \frac{\theta}{\sigma} \exp\{-\theta^2/2\sigma^2\}$$

which measures the difference between adjacent patches on angular scale σ. When this is convolved and averaged over the sky,

$$\langle \Delta T^2 \rangle = \int d^2k \; |\delta_k|^2 \, \{FT(f)\}^2$$

where FT denotes a Fourier transform. Thus the rms fluctuations are

$$(\Delta T/T)_\sigma = \text{const.} \times \frac{\sigma}{\theta_1{}^2+\sigma^2}$$

At large σ we again have shot noise behaviour, but the anisotropy peaks at angular scale $\sigma = \theta_1$, and decreases for smaller angles like $\Delta T \propto \sigma/\theta_1$ (see Figure 1). In the standard model this behaviour would seem contrived for reasons outlined above.

An important point is that because $\Delta T/T$ arises from a stochastic production of entropy, we are no longer constrained to have large mass perturbations within the last scattering surface to have a large quadrupole moment. If fluctuations of types (a) or (b) occur, such lumps are always necessary: this is a direct consequence of the primordial origin for the entropy because the radiation dominates the early stages of the expansion and there is no way to make radiation very lumpy without making matter lumpy as well. Therefore the spectrum of $(\Delta T/T)_\theta$ strongly peaked at large θ is a characteristic feature of non-primordial entropy production.

4. CONCLUSIONS ABOUT THE ORIGIN OF ENTROPY AND OF GALAXIES

The last remarks (d) should be regarded as predictions to be tested by the next generation of airborne detectors. However even with existing published data shown in Fig. 3 several interesting conclusions arise.

(1) It is impossible for a single-n spectrum of 'pure' isothermal fluctations to produce galaxies (or galaxy clusters) <u>and</u> the observed quadrupole moment without also badly contradicting existing upper limits on $\Delta T/T$ at ~ a few arc minutes - degrees. If one chooses high enough n to come within the small θ prediction (with the normalization constrained to give $(\delta\rho/\rho)_{rec} \sim 10^{-3}$ on cluster scales), then the quadrupole is too small. This conclusion follows from simple scaling arguments. It contradicts the detailed calculations of Silk and Wilson (1981): possibly their computational isothermal modes were 'impure', and contained an admixture of adiabatic modes.* This might be considered simply a matter of definition, but the above thought-experiment (c) suggests that if 'impure' isothermals form very early (kT > 1 GeV), the adiabatic (growing) part would dominate at late times. Therefore it would take fine tuning to produce the modes which Silk and Wilson call isothermal.

*This explanation is consistent with the relative amplitudes found by Silk and Wilson for quadrupole, octopole, and hexadecapole moments from isothermals, which obey the scaling predicted for <u>adiabatic</u> fluctuations - as shown by Figure 3, the scaling ought to differ by θ^2.

(2) As Peebles (in his lectures) has shown, a primordial $n = 0$ spectrum of adiabatic fluctuations is consistent with the microwave data at $\sim 6^\circ$, the quadrupole moment, and with the present day scale of galaxy clustering. (However, $n = 0$ may not be appropriate for generating that clustering because of the effects of the fireball (Peebles 1981).) This hypothesis predicts a monotonic increase of $(\Delta T/T)_\theta$.

(3) If galaxy formation is accomplished without large-scale primordial fluctuations then the only way to get a quadrupole as large as that observed is to have significant fluctuations in entropy production, and hence a significant fraction of the entropy must be non-primordial. Thus we probably already have evidence at hand for either primordial fluctuations in large-scale spacetime curvature from very near the singularity, or a non-primordial origin of the entropy. We do not yet know which, but these hypotheses have sharply divergent predictions for the intermediate scales $6^\circ \lesssim \theta \lesssim 90^\circ$, where the non-primordial hypothesis predicts a strongly peaked angular spectrum $(\Delta T/T)_\theta$ and the primordial hypothesis predicts a monotonic or approximately constant $(\Delta T/T)_\theta$.

(4) If the last scattering surface is not due to free-electron scattering but to dust or molecules (such as those which may thermalize the radiation) then the peak and small θ amplitude of $\Delta T/T$ should depend on wavelength λ, even if the optical depth at high redshifts is great enough to make the spectrum nearly thermal. The reason is that the width θ_1 of the smoothing function has a frequency-dependence if smoothing is accomplished with dust.

Measurements of large-scale statistical anisotropies in the microwave background, at several frequencies and angular scales, may decisively settle the two most fundamental questions about the universe near the singularity: (i) what was it made of i.e. pure entropy or cold matter: and (ii) how smooth was it i.e. were curvature fluctuations identically zero initially, or did they diverge at large scales?. These conclusions are independent of a detailed theory of galaxy formation, a phenomenon which could well remain obscure.

REFERENCES

Bardeen, J.M. 1980. Phys.Rev.D., 22, 1882.
Boughn, S.P., Cheng, E.S. and Wilkinson, D.T. 1981. Ap.J.Lett., 243, L113.
Carr, B.J. 1981. MNRAS, 195, 669.
Doroshkevich, A.G., Zel'dovich, Ya.B. and Novikov, I.D. 1967. Sov.Astr., 11, 233.
Grishchuk, L.R. and Zeldovich, Ya.B. 1978. Sov.astron., 122, 125.
Hogan, C.J. 1980. MNRAS, 192, 891.

Hogan, C.J. 1982a. Ap.J., in press.
Hogan, C.J. 1982b. Ap.J.Lett., in press.
Kibble, T.W.B. 1980. Phys.Rep., 67, 183.
Layzer, D. 1975. In Galaxies and the Universe, (eds. A. Sandage, M. Sandage and J. Kristian). University of Chicago Press.
Lifschitz, E.M. 1946. Sov.J.Phys., 10, 116.
Melchiorri, F., Melchiorri, B.O., Ceccarelli, C. and Pietranera, L. 1981. Ap.J.Lett., 250, L1.
Ostriker, J.P. and Cowie, L.L. 1981. Ap.J.Lett., 243, L127.
Peebles, P.J.E. 1980. The Large Scale Structure of the Universe. Princeton University Press.
Peebles, P.J.E. 1981a. Ap.J., 248, 885.
Peierls, R.E., Singwi, K.S. and Wroe, D. 1952. Phys.Rev., 87, 46.
Press, W.H. 1980. Physica Scripta, 21, 702.
Press, W.H. and Vishniac, E. 1980. Ap.J., 239, 1.
Rees, M.J. 1975. In The Structure and Evolution of Galaxies, (ed. G. Setti) Reidel: Dordrecht.
Rees, M.J. 1981. Private communication.
Rees, M.J. 1978. Nature, 275, 35.
Rowan-Robinson, M., Negroponte, J. and Silk, J. 1979. Nature, 281, 635.
Sachs, R.K. and Wolfe, A.M. 1967. Ap.J., 147, 73.
Silk, J. and Wilson, M.L. 1981. Ap.J.Lett., 244, L37.
Sunyaev, R.A. and Zel'dovich, Ya.B. 1970. Astrophys.Space Sci., 7, 1.
Vilenkin, A. 1981. Phys.Rev.Lett., 46, 1169.
Witten, E. 1981. Nucl.Phys.B., 177, 477.
Zel'dovich, Ya.B. 1963. Sov.Phys. JETP 16, 1395. (Original JETP 43, 1982 (1962)).
Zel'dovich, Ya.B. 1965. Adv.Astr.Astrophys., 3, 241.
Zel'dovich, Ya.B. 1980. MNRAS, 192, 663.
Zel'dovich, Ya.B. and Novikov, I.D. 1978. Relativistic Astrophysics, vol. I, (ed. K.S. Thorne and W.D. Arnett.) University of Chicago Press.

ADIABATIC VS. ISOTHERMAL: TWO PICTURES OF GALAXY ORIGIN

S.A. Bonometto

Istituto di Fisica Galileo Galilei,
Padova 35100, Italy.

1. INTRODUCTION: THE COSMIC MIDDLE AGES

It has become apparent in the last few years that many fundamental problems of cosmology will find their ultimate solution in the framework of recently formulated elementary particle interaction theories. Thus the question of why the initial conditions for galaxy formation were what they were will be answered by considering the earliest instants of cosmic evolution when the cosmic temperature exceeded the temperature for grand unification, or even earlier when we enter the domain of quantum gravity at the Planck time. Nevertheless, the period during which the cosmic temperature falls from ~ 10^{11} °K to ~ 10 °K is of fundamental importance. It is the domain of well-established physics and from it we have learned the important relationship between cosmic nucleosynthesis and the microwave background radiation temperature. We have also learned a good deal about the evolution of inhomogeneities during this period, and hence this cosmic 'middle age' is an essential link between the Universe observed today and the era of ultra high energy physics.

Already, the study of these 'middle ages' has highlighted the questions to be answered when we come to consider the hadronic and earlier eras. We recognise that we can form galaxies from either adiabatic or isothermal initial fluctuations (both ideas have their adherents), but we must turn to the earliest times to decide which kind of perturbation is to be preferred. A related question is why the entropy per baryon has the value it does. However, we must not get the impression that all the problems of the cosmic middle ages are solved: there are some outstanding

B. J. T. Jones and J. E. Jones (eds.), The Origin and Evolution of Galaxies, 97–119.

problems foremost among which is the fate of adiabatic perturbations at the time when the cosmic plasma recombines. I shall spend a lot of time on this since the adiabatic density perturbation theory depends very much on the precise details of what happens at recombination.

The 'decoupling epoch' occurs when the cosmic radiation temperature drops below ~ 4000°K, and electrons and protons combine and form H atoms. This is far from being an instantaneous process. The ionization is still ~ 1 at z_d ~ 1600: the final value ranging between 10^{-4} and 10^{-5} (the actual value depends on the density parameter Ω) is not attained before z_{de} ~ 900. This means that, for an observer sitting at the end of the decoupling, nearly two thirds of his past would have been taken up by the decoupling process itself. The absence of small scale temperature fluctuations in the microwave background radiation sets stringent limits on how lumpy the Universe can have been at that time (see lectures of Hogan and Partridge).

Let us start by considering the main components of the Universe at such times: there were protons, electrons, deuterons, α particles, very small quantities of other nuclides, and many, many photons. In principle one should consider the distribution function for each kind of particle in their phase space. However, the only distinction which needs to be made is between matter and radiation: the mean free paths for collisions among protons, electrons and atoms are so small that the matter can be considered as one fluid. (I am deliberately ignoring neutrinos for the moment, their relevance will be discussed separately later on. Moreover, we shall ignore the presence of He^4 and other nuclides: this leads to only minor quantitative errors since most of the inertia and energy density of the matter lies in the nuclei.)

The discussion of the physics of the radiation is, on the other hand, a far more delicate matter. The electrons are the agent which couples the matter to the radiation. At the simplest level it is possible to consider matter and radiation together as a single fluid whose viscosity and thermal conductivity is mostly due to the presence of the radiation (Silk 1968: Weinberg 1971). This hides a number of peculiar features, but these can be understood by considering matter and radiation as coupled fluids. However, the radiation is certainly not a fluid after decoupling. Hence, during the decoupling era, the two-fluids description gradually becomes less and less adequate. A satisfactory description can only be obtained by considering the distribution of photons in their phase space, and working out its dynamical evolution (Chibisov 1972: Silk 1974).

2. KINETIC TREATMENT FOR RADIATION

The evolution of density perturbations through the decoupling era was first studied by Peebles and Yu (1970) who considered matter as a fluid and the radiation by means of the photon distribution function. In the absence of fluctuations this function has the familiar form

$$f_o\ (\underline{x},t,\underline{n},p_o) = (\exp\frac{p_o}{kT} - 1)^{-1} \tag{1}$$

where $\underline{n} = \underline{p}/p_o$ is the direction of propagation of a photon of momentum $\underline{p}$ and energy p_o. Because of scattering, f changes with time and it is supposed that the actual distribution function can be written in terms of f_o as

$$f = f_o\ (1 + \varepsilon) \tag{2}$$

where $\varepsilon << 1$. The evolution of f is described by the Boltzmann equation which is linearised by the introduction of ε. It is convenient to introduce the energy density of the 'background' component of the radiation field corresponding to f_o:

$$\rho_o^r = 4\pi \int p_o f_o p^2 dp \tag{3}$$

and the fluctuating component

$$\delta^r = \frac{1}{\rho_o^r} \int \varepsilon p_o f_o p^2 dp \tag{4}$$

δ^r is a function of $\underline{x}$ and t (since the energy-momentum dependence has been integrated out), and we can simplify the description of δ^r, if desired, by defining the moments

$$\alpha^r = \frac{1}{4\pi} \int \delta^r\ d\Omega, \qquad \beta_i{}^r = \frac{1}{4\pi} \int n_i \delta^r\ d\Omega,$$

$$\gamma_{ik}{}^r = \frac{1}{4\pi} \int n_k n_i \delta^r d\Omega, \text{ etc.} \tag{5}$$

The fluctuations in the gravitational field can be calculated in the usual weak field approximation for general relativity by writing the background metric as, setting c = 1,

$$ds^2 = dt^2 - a^2(t)\eta_{ij}dx^i dx^j \tag{6}$$

and writing the fluctuating metric as

$$g_{oo} = 1, \quad g_{oi} = 0, \quad g_{ij} = -a^2(t)(\eta_{ij} - h_{ij}) \tag{7}$$

The Einstein field equations then relate h_{ij} to the distribution of matter and radiation. In what follows we shall only need the equation for

$$h = \sum_{i=1}^{3} h_{ii} \tag{8}$$

By taking moments of the perturbed Boltzmann equation (see for example Peebles (1980) section 92) we arrive at the basic equation for the evolution of δ^r:

$$\frac{\partial \delta^r}{\partial t} + \frac{n_i}{a}\frac{\partial \delta^r}{\partial x_i} - 2n_i n_j \frac{\partial h_{ij}}{\partial t} = n_{e1}\sigma_T(\alpha^r + 4n_i v_i - \delta^r) \tag{9}$$

where n_{e1} is the free electron number density and σ_T is the Thomson cross section. v_i is the matter velocity relative to the background time-orthogonal coordinates. It should be noted that this equation neglects the small anisotropies of Thomson scattering, and is only valid for zero mass particles (it would therefore need modification to deal with massive collisionless neutrinos). Integrating (9) over $d\Omega$, $n_i d\Omega$, $n_i n_k d\Omega$ etc. produces a hierarchy of equations fo. the moments α^r, $\beta_i{}^r$, $\gamma_{ik}{}^r$,... of the radiation field fluctuation δ^r:

$$\frac{\partial \alpha^r}{\partial t} + \frac{1}{a}\frac{\partial \beta_i{}^r}{\partial x_i} = \frac{2}{3}\frac{\partial h}{\partial t} \tag{10}$$

$$\frac{\partial \beta_k{}^r}{\partial t} + \frac{1}{a}\frac{\partial \gamma_{ik}{}^r}{\partial x_i} = n_{e1}\sigma_T(\tfrac{4}{3}v_k - \beta_k) \tag{11}$$

etc.

This set of equations offers no more insight into the problem than (9), unless we can make a justifiable physical assumption eliminating higher moment equations. Thus if we make the assumption that

$$\delta^r = \alpha^r + 3n_i\beta_i{}^r \tag{12}$$

and that no higher components need be considered, then

$$\gamma_{ik}{}^r = \frac{\alpha_r}{4\pi}\int n_i n_k d\Omega = \frac{1}{3}\alpha^r\delta_{ik} \tag{13}$$

and (11) simplifies to

$$\frac{\partial \beta_i{}^r}{\partial t} + \frac{1}{3}\frac{1}{a}\frac{\partial \alpha^r}{\partial x_i} = n_{e1}\sigma_T(\tfrac{4}{3}v_i - \beta_i) \tag{11'}$$

Combining (10) and (11') leads to a variant of the standard diffusion approximation. Note that there are important cases in which the assumption (12) is not justified: we shall return to that shortly.

3. COUPLED FLUIDS TREATMENT

We have disregarded the equations for the matter and the gravitational field until now. We shall give the full set of equations for radiation, matter, and the gravitational field under the hypotheses that δ^r depends only on one coordinate (let us call it x, θ will be angle between p, a vector and the x direction) and that (12) is true. The former could easily be dropped and is made here for simplicity. The latter condition will then read

$$\delta^r = \alpha^r + 3\beta^r \cos\theta. \tag{14}$$

Altogether the dynamical equations become

$$\frac{\partial\alpha^r}{\partial t} + \frac{1}{a}\frac{\partial\beta^r}{\partial x} = \frac{2}{3}\frac{\partial h}{\partial t}$$
$$\frac{\partial\beta^r}{\partial t} + \frac{c_s^2}{a}\frac{\partial\alpha^r}{\partial x} = \frac{1}{\tau}\left(\frac{4}{3}\beta^m - \beta^r\right) \tag{15}$$

$$\frac{\partial\alpha^m}{\partial t} + \frac{1}{a}\frac{\partial\beta^m}{\partial x} = \frac{1}{2}\frac{\partial t}{\partial t}$$
$$\frac{\partial\beta^m}{\partial t} + \frac{v_s^2}{a}\frac{\partial\alpha^m}{\partial x} = \frac{1}{\eta}\left(\frac{3}{4}\beta^r - \beta^m\right) \tag{16}$$

$$\frac{\partial^2 h}{\partial t^2} + \frac{2}{a}\frac{\partial a}{\partial t}\frac{\partial h}{\partial t} = 8\pi G[\rho_o{}^m\alpha^m + 2\rho_o{}^r\alpha^r]. \tag{17}$$

The following definitions are used:

$\tau = 1/\sigma_T n_{e1} c$: time for momentum transfer from photons

$\eta = \tau\ 3\rho_o{}^m/4\rho_o{}^r$: time for momentum transfer from electrons.

$\rho_o{}^m$: background matter density

$\alpha^m = \delta\rho^m/\rho_o{}^m$: matter density fluctuation

$\alpha^r = \delta\rho^r/\rho_o{}^r$: the radiation fluctuation

$\beta^m = v$

$c_s = c/\sqrt{3}$: sound velocity in 'pure radiation'

$v_s = (\gamma\ kT/m_p)^{1/2}$: sound velocity in a fluid whose adiabatic index is γ and whose particles have a mass m_p (protons).

The adiabatic sound speed in the cosmic plasma is

$$c_{sd} = c_s(1 + \eta/\tau)^{-1/2} \tag{18}$$

The nice symmetry of equations (15) and (16) is related to assumption (14), which is valid while the matter is able to inhibit the formation of higher harmonics in the radiation distribution. Nevertheless, equations (15), (16), (17) certainly give a better insight into the evolution of fluctuations before and during decoupling than a pure viscous fluid model (they also include the results obtainable in such a model).

4. ADIABATIC AND ISOTHERMAL MODES

An integration of the system (15)-(17) is difficult as most of the coefficients depend on time. This is true both because the scale factor a is time dependent and because τ (and hence η) depend on t through the free electron density, n_{el}, which varies quite strongly when protons and electrons combine into atomic hydrogen. This latter dependence is certainly the most important.

We are generally interested in fluctuations whose characteristic timescale (growing time or period) is short compared with the cosmic expansion timescale.

Before decoupling begins the ionisation is constant and it is a safe approximation to consider the coefficients to be constant and, in particular, to set a = 1. Then the system (15)-(17) can be simplified to

$$\begin{aligned}\left[\frac{\partial^2}{\partial t^2}+\frac{1}{\tau}\frac{\partial}{\partial t} - c_s^2\frac{\partial^2}{\partial x^2}\right]\delta\rho^r - \frac{1}{\eta}\frac{\partial}{\partial t}\delta\rho^m &= 4\pi G\,\frac{4}{3}\,\rho_o^r\,[\delta\rho^m + 2\delta\rho^r]\\ \left[\frac{\partial^2}{\partial t^2}+\frac{1}{\eta}\frac{\partial}{\partial t} - v_s^2\frac{\partial^2}{\partial x^2}\right]\delta\rho^m - \frac{1}{\tau}\frac{\partial}{\partial t}\delta\rho^r &= 4\pi G\,\rho_o^m[\delta\rho^m + 2\delta\rho^r]\end{aligned} \tag{19}$$

When the times τ and η are very short (19) yield the approximation

$$\frac{\delta\rho^r}{\delta\rho^m} = \frac{\tau}{\eta} = \frac{4}{3}\,\frac{\rho_o^r}{\rho_o^m}$$

Then all time dependent fluctuations have $\alpha^r = 4/3\ \alpha^m$. This is an 'adiabatic' fluctuation as should be expected in a viscous fluid treatment.

In general (19) is a good starting point for an analysis of fluctuation modes. If we look at specific perturbations of the form

$$\delta^r = A^r \exp i(\omega t - kx), \quad \delta^m = A^m \exp i(\omega t - kx) \tag{20}$$

we obtain a pair of algebraic equations for the amplitudes A^r and A^m. These only have solutions when the quantity y

$$y = i\omega / kc_s \tag{21}$$

satisfies the somewhat complicated dispersion relation (DR)

$$\begin{aligned} y^4 &+ (1+\eta/\tau)(kc_s\eta)^{-1} y^3 + .. \\ &..+ [1 - 4\pi G\rho_o{}^m (1+2\tau/\eta)/k^2 c_s{}^2] y^2 + .. \\ &..+ [1 - 4\pi G\rho_o{}^m (3+2\tau/\eta - \eta/\tau)/k^2 c_s{}^2](kc_s\eta)^{-1} y - .. \\ &.. - 4\pi G\rho_o{}^m / k^2 c_s{}^2 = 0 \end{aligned} \tag{22}$$

(see equation (6) of Bonometto and Lucchin, 1979 for more details). As should be expected, the system made of two coupled fluids admits four fluctuation modes. If the fluids were decoupled there would be a pair of modes for each fluid (two propagations towards positive and negative x, or an unstable and a damped fluctuation). In the presence of gravitation and Thomson coupling there remain four modes, though their characteristics are different.

Solutions of (22) can be obtained both numerically as well as analytically on the basis of suitable approximations. There are essentially two kinds of solutions: those which are characterized by $|y| \sim 1$ and those by $|y| \ll 1$, and these turn out to be just the adiabatic and isothermal modes of perturbation (see below).

Before decoupling three solutions with $|y| \sim 1$ exist, after decoupling there are two. Amongst the first three there may be a pair of complex conjugate solutions: these describe two waves propagating along the x axis with velocity $v^+ = c_s|\mathrm{Im}\ y|$ and damping rate $\mu = c_s\ \mathrm{Re}\ y$. For such complex conjugate solutions to exist it is necessary that at least one of the conditions

$$kc_{sd}\eta \lesssim 2 \tag{23}$$

$$kc_s\tau \gtrsim 1/2. \tag{24}$$

should be satisfied. (These conditions are only good approximations: see equations (21) and (24) of Bonometto and Lucchin, 1976. In this latter paper the region where neither inequality is satisfied is referred to as the Zero Sound Band (OSB) since acoustic modes do not exist under those circumstances.) The two fluid model allows an acoustic mode in the 'pure' radiation field that exists after decoupling. This is clearly unphysical since the photon gas is essentially collisionless. However, sound propagation is possible in the pure matter after recombination, though on much smaller length scales than prior to recombination.

In figure 1 the redshift (z) dependence of the solutions $|y| \sim 1$ is plotted, for a number of cases. Recall that k fixes a wavelength scale $\lambda = 2\pi/k$ with which a characteristic mass

$$M = \frac{4}{3}\pi\ (2\pi/k)^3\rho^m \tag{25}$$

is associated. This is the mass scale referred to in the figures. Solutions also depend on the value of Ω (density parameter) and $H = 100\ h\ kms^{-1}\ Mpc^{-1}$.

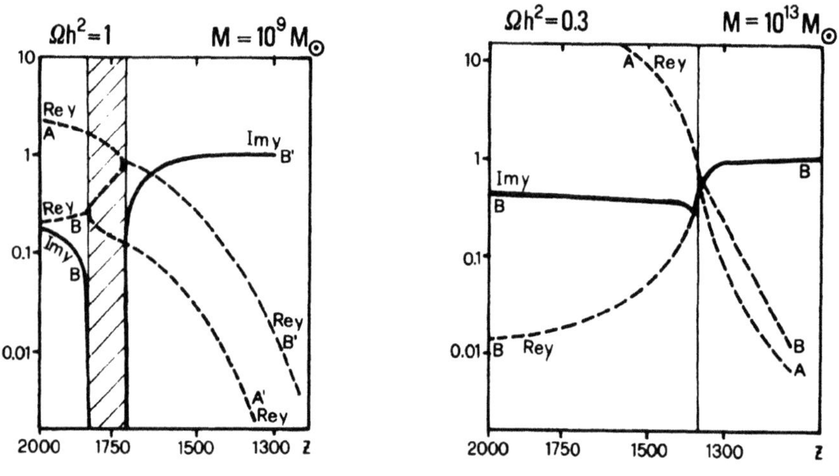

Figure 1. z dependence of the roots of the dispersion relation (22) for two cosmological models. The shaded area in the $\Omega h^2 = 1$ diagram is where no compressive modes propagate.

The cases $|y| \sim 1$ and $|y| \ll 1$ have important physical interpretations as ADIABATIC and ISOTHERMAL modes. To see this we note that when the dispersion relation (22) is satisfied, the ratio of the amplitudes of the matter and radiation fluids (19) can be calculated. When this is done, the result is a relationship of the form

$$\alpha^r = \frac{4}{3} A(y)\alpha^m \tag{26}$$

where $A(y)$ is a complicated function of y and the quantities k, τ, η, c_s, v_s and ρ_o^m.

In the case $|y| \sim 1$, if in addition the wavenumber of the perturbation, k, satisfies

$$k^2 c_s^2 \gg \begin{cases} 4\pi G \rho_o^m & \text{if } \rho_o^m > \rho_o^r \\ \frac{2}{3}\pi G \rho_o^m & \text{if } \rho_o^m < \rho_o^r \end{cases} \tag{27}$$

$A(y)$ simplifies to

$$A(y) = 1 + k\eta c_s y = 1 + i\omega\eta \tag{28}$$

In the limit $|\omega\eta| \to 0$ the modes are the familiar adiabatic modes $\alpha^r = 4\alpha^m/3$, and the quantity $i\omega\eta$, being complex, measures the departure from adiabaticity. That $i\omega\eta$ is complex means that there is a phase difference between the temporal oscillations of the matter and radiation components: this phase difference is associated with damping. The condition (27) on the wavelength merely states that the wavelengths should be much smaller than the horizon, and in any case our approximations regarding a/a become invalid on larger scales.

In the case $|y| \ll 1$, when in addition

$$k^2 c_s^2 \gg \begin{cases} 4\pi G \rho_o^m \,(1+2\tau/\eta) & \rho_o^m > \rho_o^r \\ 4\pi G \rho_o^m \,(3+2\tau/\eta-\eta/\tau) & \rho_o^m < \rho_o^r, \end{cases} \tag{29}$$

we find that

$$|A(y)| \ll 1 \text{ and } A(y) < 0. \tag{30}$$

Hence the perturbation involves mostly matter, the small perturbation in the radiation being present simply to balance the nearly negligible matter pressure fluctuation. These are just the familiar isothermal modes. The condition (29) in the radiation dominated case only states that the perturbation should be smaller than the horizon, and, like (27) in the adiabatic case, imposes no further restrictions on the modes. However in the matter dominated regime (29) is equivalent to a wavelength restriction

$$\lambda < \lambda_I = c_{sd}\left(\frac{\pi}{G\rho_o{}^m}\right)^{1/2}\left(1 + \frac{8\rho_o{}^r}{3\rho_o{}^m}\right)^{-1/2}, \qquad \rho_o{}^m > \rho_o{}^r \tag{31}$$

The scale λ_I is just smaller than the Jeans lengthscale, and above such scales strictly isothermal modes cannot exist. An isothermal mode imposed on a scale $\lambda > \lambda_I$ during the matter dominated era would not remain isothermal: because of gravity the fluctuation would start to collapse, dragging the radiation along with it.

This behaviour is an example of what happens to any fluctuation which does not belong to one of the above 4 modes: in practice it must be resolved (projected) into a sum of contributions from each of the four modes. Then each contribution (projection) evolves independently of the others.

5. PASSAGE OF FLUCTUATIONS THROUGH DECOUPLING

Let us begin by considering isothermal perturbations since these are the simpler: their value of y starts off small and gradually increases only when the decoupling is nearly over, when these fluctuations begin a phase of collapse. If $\Omega h^2 > 0.03$ the epoch when $\rho^m = \rho^r$ comes before the decoupling and all pure mass fluctuations entering the horizon after $z_e \sim 4\text{x}10^4\Omega h^2$, have a collapse phase if their scale length exceeds the one set by (31). When they enter the decoupling stages they involve both matter and radiation and have length scales exceeding

$$\lambda_{1000} \sim \left[\frac{c}{\sqrt{3}}\left(1 + \frac{3\rho^m}{4\rho^r}\right)^{-1/2}\left(\frac{\pi}{G\rho^m}\right)^{1/2}\right]_{z=1000} \sim 5\text{x}10^{22}(\Omega h^2)^{-1}\text{cm}. \tag{32}$$

This amounts to a fraction $f \sim c_{ds}/c \sim \frac{1}{6}(\Omega h^2)^{-1/2}$ to the horizon at $z \sim 10^3$, and corresponds therefore (see Weinberg 1972) to an angular scale

$$\Theta \sim \Theta_H/f \sim 0.1\ (\Omega h)\ \text{degrees.} \tag{33}$$

(Θ_H is the angular scale corresponding to the horizon at $z \sim 10^3$.) The mass involved in such fluctuation is

$$M_{1000} \sim 5\text{x}10^{15}\ (\Omega h^2)^{-2}\ M_\odot \tag{34}$$

We should bear this value in mind in order to compare it with the adiabatic fluctuation case.

The main disadvantage with isothermal fluctuations is that, after the plasma has combined into atomic H, we start with a fully developed spectrum of density fluctuation possibly extending from 10^6 $M_\odot$ up to nearly the horizon. No scale is characterized by damping processes.

Let us now turn to the case of 'adiabatic' fluctuations. It has been known for many years that, even before decoupling begins, they suffer a severe damping. However, the most important damping effect occurs during the decoupling itself: it is during the decoupling that the adiabatic mode dies out. The first published computation of the 'transmission factor' for adiabatic fluctuations through the decoupling was due to Peebles and Yu 970). (Michie (1967) had performed a similar computation earlier, without being able to publish it.) Altogether published results on the transmission factor are due to Peebles and Yu (1970), Bonometto and Lucchin (1979), Press and Vishniac (1980), Silk and Wilson (1980). We shall refer to these papers as PY, BL, PV, SW.

PY and SW base their computation on the equations (9), (16) and (17). This amounts to considering arbitrary deformations of the radiation field. They avoid making any assumption like (12) or (14) and, therefore, do not find the spurious post-decoupling pure radiation waves. They compute a transmission factor for matter and radiation amplitudes. However they both make the assumption that adiabatic fluctuations enter the horizon exaltly at rest. This amounts to fixing the initial phase for the adiabatic mode and to assuming a super-position of the two propagating modes generating a standing wave: the fluctuations thus enter the decoupling era with a scale-dependent phase. As a consequence the amplitude of the residual fluctuation shows a peculiar scale dependence. Of course there is no a priori reason to demand that the fluctuations have zero peculiar velocity on entering the horizon, so this feature is merely an artefact of an unnecessary hypothesis.

On the other hand the results of BL and PV are based on the system (15), (16) and should be therefore taken with caution as far as the amplitude of radiation density fluctuations after recombination is concerned. In fact they allow for the presence of purely radiative waves which are not real. Moreover the ionization curve taken by BL is not completely accurate, as the paper is more concerned with the details of the transfer of stimulus amongst the different modes, rather than with giving quantitative results.

In Figure 2, a number of transmission functions worked out in the above papers are plotted (I took one or two cases from each paper) and they show an essential agreement on the following points: i) There is a minimum mass scale under which fluctuations are damped. ii) This mass scale is increasing as Ωh^2 decreases.

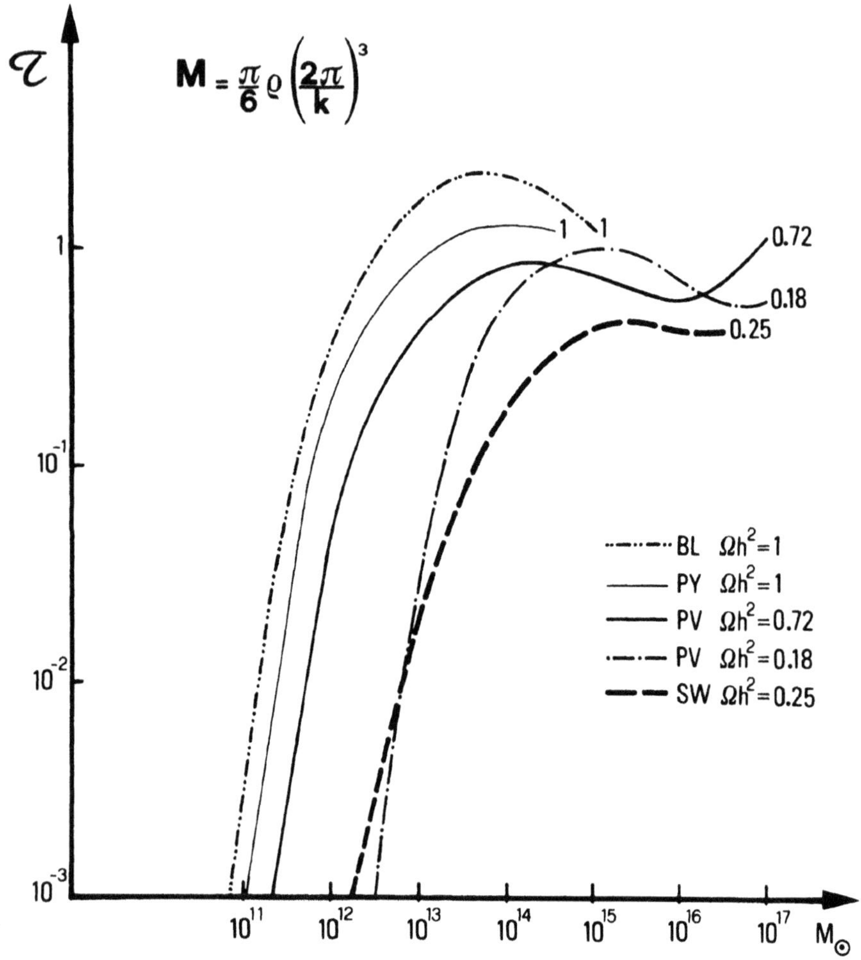

Figure 2. Transmission factors τ calculated by different authors for various cosmological models.

The masses M_D at which the transmission factors reach a value $\sim 1/\sqrt{10}$ can be obtained from the relation (see Figure 3).

$$M_D = 10^{12.25}\ (\Omega h^2)^{-1.4}\ M_\odot \tag{35}$$

which fits quite well the PY, BL and PV results, while the SW results for M_D are greater by ~ 1 order of magnitude.

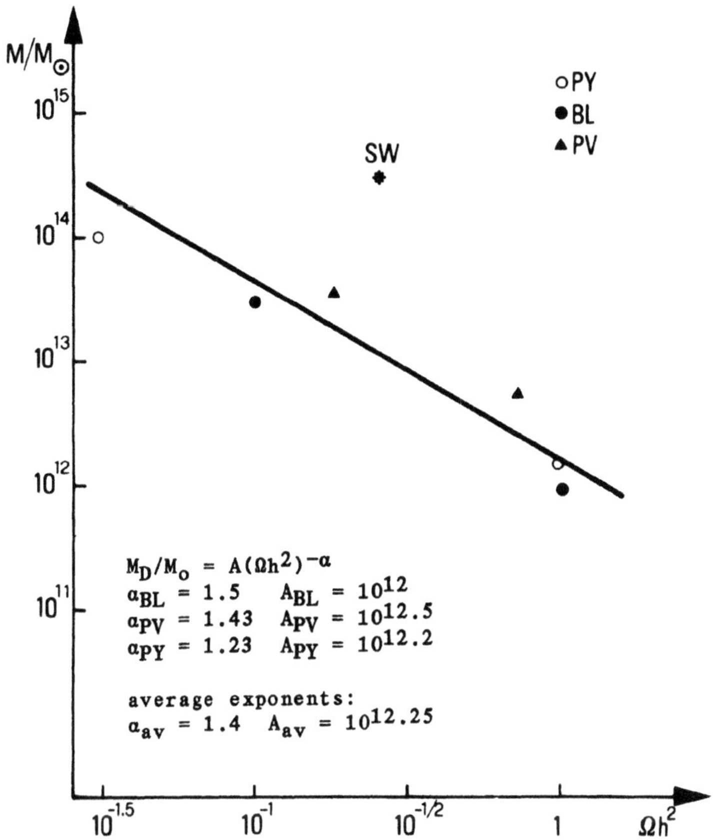

Figure 3. Damping mass as a function of cosmological model calculated by different authors.

It is more difficult to compare the amplitude of the residual fluctuations in radiation. Most results given by PY and SW depend on the choice of the initial phases, though SW also give results obtained by integrating with a set of initial spectra of amplitudes. The results do not depend much on the assumed spectral index. In figure 4 they are plotted for $\Omega = 1$, $h = 0.5$. The normalization is the one needed to have the non-damped fluctuations form galaxies by the present epoch. For sake of comparison the transmission factor for radiation fluctuations worked out by BL is also plotted. The normalization here is that the amplitude of matter density fluctuation was 10^{-2} prior to

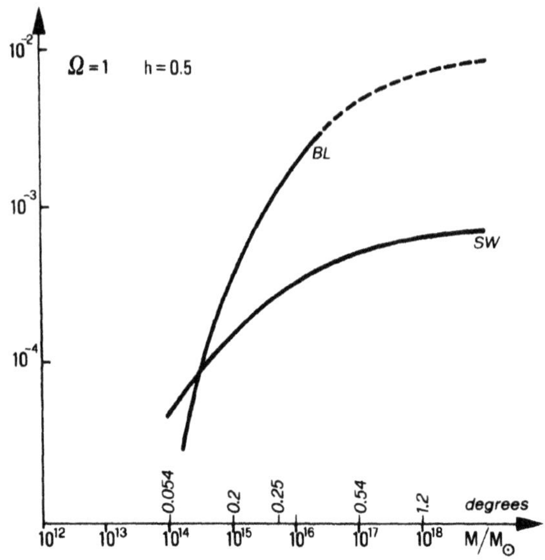

Figure 4. Temperature fluctuations as a function of mass or angular scale from work of BL and SW. Note that the normalisations of the curves are slightly different (see text).

decoupling. Over the relevant mass scales the two fluid approximation should be quite good. (In fact no spurious oscillation can occur for such large wavelengths before the end of the recombination.) Nevertheless, there is only qualitative agreement between the two curves. This does not mean that the results are really in contradiction as the quantities which are plotted are similar but not identical.

DIFFICULTIES WITH ADIABATIC THEORY

Difficulties for adiabatic theories come from two directions: 1) observed behaviour of n-point correlation functions (n=2....5): 2) limits on Microwave background (MWB) small scale fluctuations. Let us examine the former problems first.

From an analysis of Zwicky, Lick, and Jagallonian samples the n-point functions have been deduced, mainly by Peebles and colleagues (see Peebles 1980). The two point function $\xi(r)$ has the simple form

$$\xi(r) = \left(\frac{r_o}{r}\right)^{\gamma}, \quad r_o = 4h^{-1}\mathrm{Mpc}, \quad \gamma = 1.8$$

and this extends over a range $3h^{-1}$kpc $< r <$ $2r_o$ Mpc (Gott and Turner 1979). The higher order functions appear to be constructed simply from suitably symmetrised products of ξ's (see for example Peebles 1980).

The question must now be asked: where is the scale corresponding to M_D? It obviously cannot be above r_o, as then $\xi(r) < 1$ and this means that fluctuations are still in a linear regime. For galaxies to form in an adiabatic theory fluctuations over scale M_D must have fully recollapsed and fragmented. Hence M_D must be on some scale $< r_o$.

We can further constrain the location of M_D if we follow the collapse of the fluctuation after recombination. Gott and Rees (1975) considered spherical collapse and showed that at the time t_{max} of its maximum expansion a perturbation is a factor

$$\chi = (\pi/2)^2 \Omega^{-1} (H_o t_o)^{-2}$$

times denser than the background Universe where suffix o denotes present values. By the time $3t_{max}$ the cluster will be virialised and have a density 8 times that at maximum expansion but the Universe will have decreased in density by a factor 9 since then. The density contrast at $t_{virial} = 3\ t_{max}$ is thus

$$[\rho_{internal}/\rho_{external}]_{t_{vir}} = 72\chi \quad . \tag{36}$$

As an order of magnitude estimate, at 2 t_{max} the density contrast might have been intermediate between χ and 72, say 30χ .

In the adiabatic-pancake theory, however, one assumes that essentially only one dimension is involved in the collapse. Expansion might even continue in both of the two other directions. This leads to assuming a density contrast, for a mass which has undergone the collapse phase, never smaller than $(30\chi)^{1/3}$. In my opinion this is however quite an 'optimistic' estimate, corresponding to a density contrast of at least ~ 10.

From the 2-point function one has an immediate estimate of the typical density contrast over each scale for $\xi(r) >> 1$:

$$\rho_{int}(r)/\rho_{ext} \sim \xi(r) \tag{37}$$

(see, e.g. Gott and Rees 1975). Clearly one has $\xi(r) \sim 10$ for $r \sim 0.3\ r_o$. Hence I would conclude that M_D corresponds to a scale below $0.3\ r_o$, where $\xi(r)$ exceeds ~ 10.

The adiabatic theory then requires either that geometrical effects, evaporation, etc. produce the same r dependence of

$\xi(r)$ above r_0 as below $0.3\ r_0$, or that the actual shape of the n-point correlation functions depends very little on the initial data, the same dependence of r being generated on scales $> r_0$ and $< 0.3\ r_0$. In my opinion this is a serious difficulty for the adiabatic theory.

Let us now turn to the difficulties faced by the adiabatic theory in view of the observed temperature fluctuations of the cosmic microwave background radiation. Provided the Universe is not re-ionised at a later time it is in principle straightforward to calculate the temperature fluctuations due to a given spectrum of inhomogeneity at the decoupling epoch. Peebles in his lectures showed, for example, how the large scale quadrupole anisotropy could be interpreted as being due to the combined effects of inhomogeneities on relatively small scales. It is important to realise however that the smaller angular scales on which temperature fluctuations are observed nevertheless correspond to mass scales greater than M_D by one or two orders of magnitude. Thus a suitably sharp cut-off of the density fluctuation spectrum on scales above M_D could almost certainly be reconciled with the limits on the small scale anisotropy of the radiation. However, as mentioned above, it is difficult to reconcile such a cut off with the lack of features in the observed two point correlation function above $\sim 0.3\ r_0$.

Before concluding, let me stress what should now be an obvious statement: neither adiabatic nor isothermal theory leaves an imprint on the actual value of the galactic mass-scale. In either theory this scale is selected by phenomena occurring subsequent to decoupling.

7. MASSIVE NEUTRINOS

Some recent experiments (Lyubimov <u>et al</u>. 1980) suggest that neutrinos are massive particles with masses in the range

$$14\text{eV} < m_\nu < 46\text{eV} \tag{38}$$

(This range is claimed as a 99% confidence interval.)
If the number densities of the cosmic neutrino and cosmic radiation backgrounds are similar, then the neutrinos would be the dominant form of energy. Indeed, a mass $m_\nu = 30\text{eV}$ would make $\Omega \sim 1$. Furthermore since the mass density of neutrinos is $\rho^\nu = n^\nu m_\nu$ and the number density is $n^\nu \sim \rho^r/kT^r$, we have $\rho^\nu = \rho^r$ when the radiation temperature drops to $m_\nu = kT^r$ and this is when the neutrinos move from the relativistic to the nonrelativistic regime. This happens at redshift

$$z_e \sim 4 \times 10^4\ (m_\nu/30\ \text{eV}) \tag{39}$$

Unless the mass of the neutrino is very small, z_e falls well before recombination. Moreover, the physics of the Universe at redshifts $z > z_e$ differs little from the standard picture with zero mass neutrinos. The presence of massive neutrinos does however affect the evolution of cosmological perturbations at redshifts $z < z_e$ (Doroshkevich et al. 1980a) and has relevance to the dynamics of galaxies and clusters of galaxies (a point realised some time ago by Cowsik and McClelland (1973) and Szalay and Marx (1976)).

When ν-fluctuations enter the horizon their evolution is governed by the kinetic equation of the kind already worked out for radiation but with the difference that the zeroth order distribution f_0 is a Fermi distribution and no collisons are to be considered. For massless or relativistic ν's this equation then becomes

$$\frac{\partial \delta^{\nu}}{\partial t} + \frac{\cos\theta}{a}\frac{\partial \delta^{\nu}}{\partial x} - 2\frac{\partial h}{\partial t}\cos^2\theta = 0 \tag{40}$$

(initial dependence of δ^{ν} on only one spatial coordinate, x, is assumed). However since the Jeans scale is almost coincident with the horizon, the gravitational term can be dropped soon after the fluctuation comes inside the cosmic horizon. The resulting equation is easily solved (see for example equation (91.40) of Peebles (1980), or Stewart (1972)):

$$\delta^{\nu}(t,\cos\theta,x) = \delta^{\nu}[t_{in},\cos\theta,\ x-\cos\theta\int_{t_{in}}^{t} dt/a(t)] \tag{41}$$

Thus the initial x dependence generates a further dependence on $\cos\theta$ and after a time all spherical harmonics will be excited, though their amplitudes will be very small. In particular α^{ν} (defined analogously to the α^{r}: cf. equation (5)) rapidly tends to zero. In the relativistic limit the ν-fluctuations are erased because of the free ν-diffusion.

If a ν-fluctuation enters the horizon after z_e, the equation governing its evolution turn out to be

$$\frac{\partial^2 \alpha^{\nu}}{\partial t^2} + \frac{2}{a}\frac{\partial a}{\partial t}\frac{\partial \alpha^{\nu}}{\partial t} = \frac{64}{3}\pi G\rho^{\nu}\alpha^{\nu} \tag{42}$$

This equation shows that ν-fluctuations are unstable, for putting $\alpha^{\nu} \propto t^n$ one finds the usual growing solution $\alpha^{\nu} \propto t^{2/3} \propto z^{-1}$.

Consider now a Universe with $\Omega = 1$ having $\Omega_m = 0.1$ for the baryonic matter. Fluctuations entering the horizon earlier than z_e soon lose their neutrino component because of the neutrino diffusion described above. They then contain only matter and radiation and are damped unless they are more massive than $M_D \sim 10^{13.65}$ $M_\odot$. Fluctuations entering the horizon at z_{in} subsequent to z_e involve more than 10^{15} $M_\odot$ and consist of matter plus radiation and the neutrinos. The neutrino component grows in amplitude by a factor z_{in}/z_{dec} because of the effect of gravity (cf. equation (42)) while the matter plus radiation component oscillates. Thus the neutrino fluctuation has an amplitude z_{in}/z_{dec} times that of the matter, and after recombination forms a potential well into which the matter can fall. Thus δ^m rapidly approaches δ^ν without requiring a comparable δ^r, and thus the residual fluctuations in the microwave background are reduced by a factor z_{in}/z_{dec} on the relevant angular scales. The situation is depicted in Figure 5 and is described in more detail in the works of Doroshkvich et al. (1980a,b).

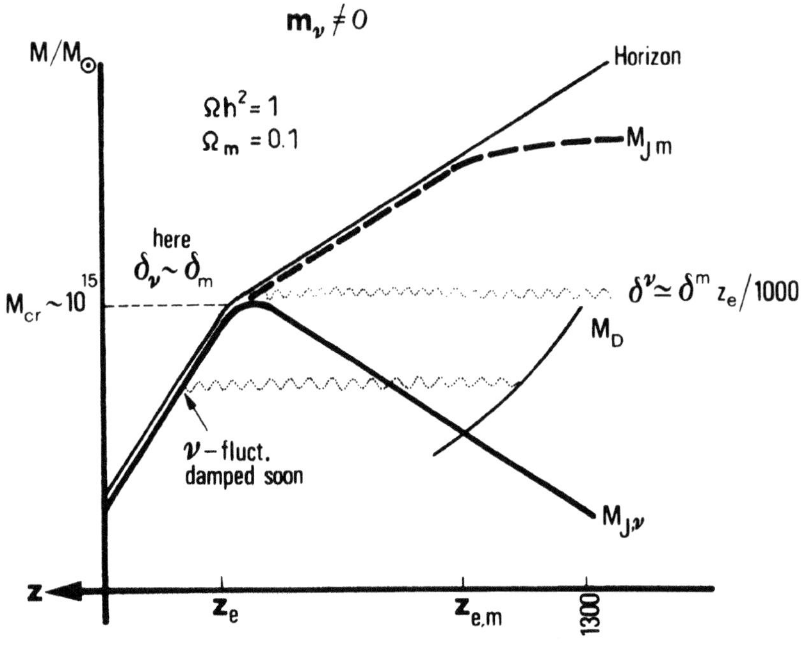

Figure 5. Critical mass scales in the evolution of perturbations as a function of z. The neutrino mass is taken such as to make $\Omega = 1$. The baryonic matter is supposed to have $\Omega_m = 0.1$. M_J is the Jeans mass in the neutrino component and $M_{J,m}$ is the standard Jeans mass for the baryon + photon plasma.

8. FLUCTUATIONS IN THE EARLY UNIVERSE

With recent advances in the theory of elementary particle interactions we may begin to speculate about the origin of fluctuations in the very early Universe. The possibility that at very early times baryon number was not a conserved quantity provides a mechanism with which we may understand the origin of adiabatic fluctuations and the reason for the value of the cosmic entropy per baryon. If this program were to prove successful we would have a situation where high energy physics had a direct impact on theories of the Universe. The Universe would then become the 'laboratory' for very high energy physics.

The present status of the theory of elementary particle interactions is the consequence of a search for unification, the same idea which once pushed Einstein to try to geometrize all forces. This unification, which did not succeed on a geometrical basis, seems now to have led somewhere on the basis of the following main ideas: i) all forces are carried by vector fields, ii) theories of interactions can be studied on a Lagrangian basis and are local and unitary. The reason for (i) is simply that, because it is so for the electromagnetic field, other fields must be the same. The point (ii) seems fairly obvious, its only real constraint being the requirement of locality. The main problem with vector fields is that they describe ∞^4 degrees of freedom. We can say roughly that, amongst these, ∞^3 refer to spin-one particles and the rest to a spin-zero component. This spin-zero component must be either set to zero (as is obtained in the electromagnetic case by means of the Lorentz condition) or be shown to be physically irrelevant (electromagnetic gauge invariance essentially means that the physical quantities are independent of the 'amount' of spin-zero component). Otherwise it causes a number of serious difficulties amongst which the most important is the non-unitarity of the theory.

A theory based on vector fields must also be short range. This can be achieved by adding a mass term. However, while this might allow a normal elimination of the spin-zero part in the case of a non-self-interacting vector field (e.g. a heavy photon) it cannot be reconciled with 'charged' vector particles such as are needed when an interaction has an internal symmetry group. Then, different solutions must be sought. However, this is not the place to debate the details of the way in which it has been possible to put together these three critical ingredients: vector interaction, short range and internal symmetry group. An excellent introductory review to gauge theories is that of Iliopoulos (1976). However, it is the role played by elementary particle interactions in the early universe that concerns us here. The exciting prospect is that in the near future, high energy physics may provide us with explanations for several outstanding cosmological mysteries.

The application to cosmology has been reviewed by Fry in this volume, and by Weinberg (1980). The essential aspect of these ideas, from the point of view of cosmology, is that above some critical temperature T_c elementary particles and their interactions look the same. The temperature for electro-weak unification is ~ 200 GeV, and below that temperature the electromagnetic and weak interactions take on their separate characters. It is as though the coupling constants were a function of energy, tending to a common value at high enough energies. The language used is to say that the gauge symmetry of weak and electromagnetic interactions is 'spontaneously broken' as the Universe cools below T_c.

An analogous idea has been proposed to unify weak, electromagnetic, and strong interactions. There are a number of reasons to believe that this is going to work, but no real physical effect has yet been detected which can be accounted for only on the basis of grand-unified theories ('GUTS'). The temperature at which grand-unification would occur is ~ 10^{15}GeV = $10^{-4}m_{pl}$ where m_{pl} is the Planck mass. Those temperatures are certainly out of the range of any present accelerator.

The main feature of GUTS is that quarks and leptons will belong to the same multiplet. Because of that, although the fermion number is still conserved, high energy, transitions from lepton into baryon states are possible and frequent. Even in the regime of symmetry violation these transitions are still possible, though rare. This leads to the prediction of proton decay with a lifetime of ~ 3 x 10^{-32} (m_x/GeV) yrs. Here m_x is the mass of a boson (the X-boson) whose exchange leads to the violation of baryon number conservation. This lifetime is above present limits on proton life by ~ 2 orders of magnitude. Experiments are in progress to try to improve these limits.

It has been suggested that grand-unification may provide an explanation of the cosmic number of photons per baryon. It may be worth remarking, however, that the presence of interactions which violate the baryon number conservation is far from being sufficient to cause an asymmetry (however low) between baryons and antibaryons. Indeed violations ocurring in single reactions would rapidly be compensated by other reactions, the average number of each species of particles being dictated by pure statistics unless we are out of statistical equilibrium.

It is possible that statistical equilibrium is violated if the X-boson is able to survive for a sufficiently long time after kT became smaller than m_x. Nevertheless, the generation of baryon number is still limited by the fact that particles and antiparticles have the same mass and lifetime.

The only possibility of having different decay rates relies on high order effects, and because of that, the difference between the branching ratios amongst different decay channels for particle and antiparticle will be very small. This smallness should correspond to a high value of entropy per baryon. The critical temperature for the decay of the X particle, and the net production of baryons ΔB resulting from the decay of the X particle, are both determined (in principle) by microscopic physics and it is in that way that the value of the cosmological entropy per baryon depends on the results of high energy physics.

Let us now consider within this framework, the possible history of fluctuations. It is possible, as Hogan and Fry have debated, that such fluctuations are simply generated in a phase transition, together with strings, monopoles, etc. according to the topology of the symmetry group. However, serious experimental limits already exist on the number density of monopoles and perhaps in future limits might also be deduced for strings. Until the situation has been clarified in this respect one is allowed to believe that such phase transitions may be the event in which fluctuations are generated, though we should always bear in mind the possibility that they may be later excluded on experimental grounds.

If phase transitions of this kind are responsible, the only kind of fluctuations which can be generated are adiabatic ones: there is no physical reason for separate regions having smaller or greater baryon content to exist. This is a strong point in favour of the adiabatic theory.

Suppose on the other hand that baryon density fluctations exist ab initio, from an era when interactions were grand-unified, then because of baryon number violation statistical equilibrium could be fully reconstituted and such fluctuations would have been erased. They might even stimulate density fluctuations, and thereby create adiabatic fluctuations, given that baryon deficits did not correspond to lepton excesses. This is another strong point in favour of the adiabatic theory.

What if we accept the possibility that fluctuations are not generated in phase transitions? In this case we shall probably look still further back in time to the quantum gravity realm to find the origin of fluctuations. However, small density fluctuations generated in the quantum gravity era would clearly turn into adiabatic fluctuations. So we must ask a final question: is there any possibility that isothermal fluctuations arose?

I want to consider here one possibility. Let us suppose that there is not only one X particle whose decay violates the baryon number conservation, but that another particle - say Y - does the same thing creating net baryon numbers ΔB_X and ΔB_Y respectively. The crucial point is then that X and Y differ because of a good quantum number which also discriminates between X and Y, and in the same direction. If this turns out to be possible and ΔB_y / ΔB_x, a possibility for generating isothermal fluctuations might still exist. Instead of assuming that density fluctuations existed at the end of the quantum gravity era, we may suppose that the Universe was homogeneous but fluctuations of the chemical potential associated with the above good quantum number were present. These fluctuations would result in local excesses of X or Y particles with respect to the average value. When X and Y decay, baryon density fluctuations would be generated. A check of this possibility with actual GUT is clearly necessary. X and Y could be either bosons or fermions: the possible violation of baryon number conservation in both cases has already been considered. A particularly relevant case would arise if X and Y are Majorana fermions.

We can think of other possibilities for isothermal fluctuation generation: i) anisotropic and inhomogeneous early expansion, leading to different decay rates in different directions and at different points: ii) turbulence, decaying necessarily well before decoupling, may leave residual isothermal perturbations, iii) use of variable cosmological constants.

9. CONCLUDING REMARK

I do not think that the final word in favour of either the adiabatic or the isothermal case can be said. Perhaps I can 'summarize' the situation by simply saying that whereas later events (MWB fluctuation, n-point correlations etc) seem to favour the isothermal picture the earlier events seem to favour the adiabatic picture from the point of view of the origin of the fluctuations.

REFERENCES

Anile, A.M., Danese, L., de Zotti, G. and Motta, S. 1976. Ap.J. Lett., 205, L59.
Bonometto, S.A. and Lucchin, F. 1976. Ap.J., 206, 391.
Bonometto, S.A. and Lucchin, F. 1979. MNRAS, 187, 611.
Cowsik, R. and McClelland, J. 1973. Ap.J., 180, 7.
Chibisov, G.V. 1972. Sov.A.J., 16, 56.
Dallaporta, N. and Lucchin, F. 1972. Astron.Astrophys., 19, 123.
Dallaporta, N. and Lucchin, F. 1973. Astron.Astrophys., 26, 325.

Doroshkevich, A.G., Zel'dovich, Ya.B., Sunyaev, R.A. and Khlopov, M.Yu. 1980a. Sov.Astron.Lett., 6, 252.
Doroshkevich, A.G., Zel'dovich, Ya.B., Sunyaev, R.A. and Khlopov, M.Yu. 1980b. Sov.Astron.Lett., 6, 257.
Gott, J.R. and Rees, M.J. 1975. Astron.Astrophys., 45, 365.
Gott, J.R. and Turner, E. 1979. Ap.J., 232, L79.
Jones, B.J.T. 1976. Rev.Mod.Phys., 48, 107.
Lyubinov, V., Novikov, E., Nozik, V., Tretyakov, E. and Kosik, V. 1980. Phys.Lett.B., 94, 266.
Michie, R.W. 1967. Kitt Peak National Obs.Contr.No. 440.
Ozernoi, L.M. 1978. IAU Symp. No. 79, p. 427.
Ozernoi, L.M. and Chernin, A.D. 1967. Sov.A.J., 11, 907.
Ozernoi, L.M. and Chernin, A.D. 1968. Sov.A.J., 12, 901.
Peebles, P.J.E. 1980. The Large Scale Structure of the Universe, (Princeton).
Peebles, P.J.E. and Yu, J.T. 1970. Ap.J., 162, 815.
Press, W.H. and Vishniac, E.T. 1980. Ap.J., 236, 323.
Silk, J. 1974. In Confrontation of Cosmological Theories with Observational Data, ed. M.S. Longair. (Reidel, Holland).
Silk, J. 1968. Ap.J., 151, 459.
Silk, J. and Wilson, M.L. 1980. Physica Scripta, 21, 708.
Stewart, J.M. 1972. Ap.J., 176, 323.
Szalay, A.S. and Marx, G. 1976. Astron.Astrophys., 49, 437.
Weinberg, S. 1971. Ap.J., 168, 175.
Weinberg, S. 1972. Gravitation and Cosmology, (Wiley).

STRUCTURE IN THE UNIVERSE AND FLUCTUATIONS IN THE COSMIC MICROWAVE BACKGROUND

R. B. Partridge

Haverford College, Haverford, PA. U.S.A.

1. DENSITY PERTURBATIONS AT THE EPOCH OF DECOUPLING t_d.

The very important work of Peebles (1981a and references therein) and his collaborators over the past decade tells us how luminous matter is distributed in the Universe now. On scales below ~ 10^{15} $M_\odot$, it is clearly clumped. The question we are faced with is how it acquired the structure it now has. Quite a number of theoretical suggestions have been made, in terms of both the physics and the philosophy invoked, these differ widely. All authors agree, however, that the structure we now see evolved. It follows immediately that any observation of the state and distribution of matter at an earlier epoch would be of great value in answering questions about the origin and evolution of galaxies - and, of course, of larger systems.

It is one such set of observations that I will treat in these lectures. Before turning to the observations, however, I need to set up a theoretical framework. It is of course similar in many ways to the scenarios of Peebles, Hogan and Bonometto but I will approach it in a slightly different manner for the sake of variety and perhaps even controversy.

As previous authors have, I will base my remarks on the hot Big Bang model, assumed to be exactly isotropic. The crucial feature of this model is a hot dense phase early in the history of the Universe, when conditions ensured that thermal equilibrium obtained. Evidence for such a phase is provided by the (largely) thermal spectrum of the cosmic microwave background.

B. J. T. Jones and J. E. Jones (eds.), The Origin and Evolution of Galaxies, 121–142.

A second property of this model is expansion. As the Universe expands, it cools. At an epoch some 10^6 years after the Big Bang, the temperature of the matter and radiation drops to ~ 3×10^3 K, and the primeval plasma recombines. The sudden disappearance of free electrons sharply reduces the (Thomson scattering) interaction between radiation and matter and these two components of the Universe decouple. Matter and radiation continue to cool, but at different rates. A second consequence of sudden decoupling at t ~ 10^6 yr is that the Universe becomes transparent, at least in the absence of any subsequent and universal re-ionization of the matter content of the Universe. The latter point in particular has important observational consequences, as we shall see.

Now let us introduce density perturbations into this model. They can be characterized by three parameters. The first, loosely speaking, is their type: the second, the amplitude, $\Delta\rho/\bar{\rho}$, and the third, a scale generally given as the mass of the perturbation, M, or its lengthscale ℓ at some fiducial time. (A convenient time is the epoch of decoupling which occurs at redshift $z_d \sim 1100$).

Four major types of perturbations have been discussed in the literature (for recent reviews, see Peebles' article in this volume, Sunyaev 1978, Ozernoi 1978 and references therein).

In adiabatic perturbations, both radiation and matter are perturbed, thus keeping the specific entropy n_λ/n_{p+} constant (Sunyaev and Zel'dovich 1970: Peebles and Yu 1970). In isothermal perturbations, as the name suggests, the temperature and hence radiation density are constant: only the matter is clumped (Sunyaev 1978: Gott and Rees 1975). Vortex or 'whirl' perturbation involve large-scale motions of the matter and radiation: this form of perturbation has been discussed by Ozernoi (1978), Ozernoi and Chibisov (1971, 1972), Chibisov (1972) and Chibisov and Ozernoi (1969). Finally, there is the possibility of source-free gravitational waves (see Dautcourt 1980).

The two other important parameters, mass and fractional amplitude, may be expressed by giving the index, α and normalising coefficients k and M_0 of a mass spectrum:

$$\Delta\rho/\bar{\rho} = k(M/M_0)^{+\alpha} \tag{1}$$

The index α can be related to the spectral power index n defined by $(\Delta\rho/\rho)^2 \propto \ell^{+n}$ as follows:

$$\alpha = -\frac{1}{2} + \frac{n}{6} . \tag{2}$$

If $\Delta\rho/\bar{\rho}$ is time dependent, the mass spectrum will be also.

A variety of theoretical arguments have been made to assign values to k and α. One suggestion with considerable appeal is due to Zel'dovich (1972): $\alpha = -2/3$ ensures that the perturbations are of constant curvature as they enter the horizon.* As we shall see, this requires $k > 10^{-4}$ for $M_0 \sim 10^{15}\ M_\odot$ at z_d. Other arguments suggest $\alpha = -1/3$ (Gott and Rees 1975), $\alpha = -1/2$ (Peebles, see these proceedings), or $\alpha = -5/6$ (Press, cited by Gott 1979) with values of k several orders of magnitude below or above 10^{-4}, respectively. A point of special importance is the link between particle physics and the index α. Predictions, rather than assumptions, can be made about the index α based on the particle physics of the early Universe (as discussed here by Fry and Hogan, and elsewhere by Gott 1979). Particle physics arguments may also bear on the type of perturbations generated early in the Universe: for instance, explanations of the baryon number of the Universe based on GUT's require adiabatic fluctuations (see Bonometto's lectures herein).

As I shall try to show later we may be able to determine the spectrum and even the type of primordial density perturbations from astronomical observations. The possibility of contributing to fundamental physics this way is an exciting prospect for an astronomer.

Unfortunately, while the initial mass spectrum holds the most interest for fundamental physics, we will probably not be able to observe it directly, except possibly at the large mass end. The reason is that a variety of physical processes affect the perturbations before the earliest epoch at which we can hope to observe them. Some of these processes are mass dependent, and hence will alter the mass spectrum. Fortunately, many of them are well understood.

To sketch the effect of these processes, I would like to focus attention on perturbations of three masses, $10^{11}\ M_\odot$, $10^{15}\ M_\odot$ and $10^{19}\ M_\odot$, which are, respectively, the representative mass of a galaxy, the largest aggregates of matter with $\Delta\rho/\bar{\rho} > 1$ now, and a rough estimate of the mass of the Universe at the important epoch of decoupling. In general, I will focus on adiabatic perturbations.

---

*At an epoch, t, the radius of the Universe is $\sim$ ct. As the Universe grows older, its radius and mass increase. At $t \sim 1$ yr, the mass within radius ct was about 10^9 to $10^{11}\ M_\odot$: larger masses were not causally connected then or earlier. (Harrison (1973) gives precise values for such masses and times in general Freidmann Universes.)

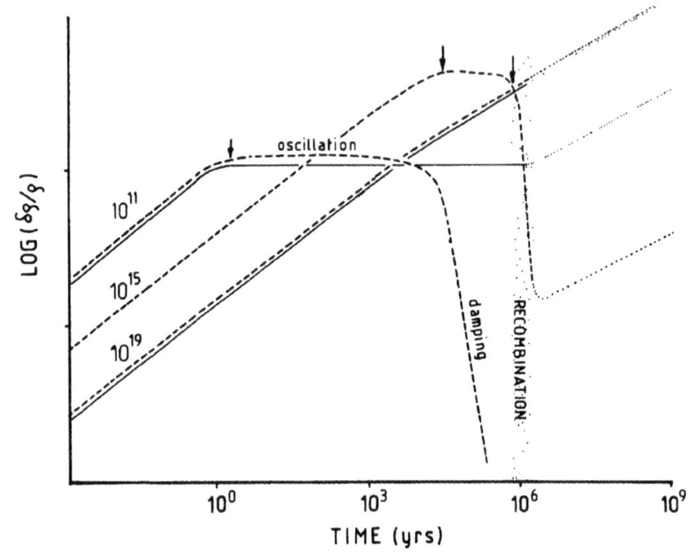

Figure 1. The evolution of the density fluctuation amplitude for perturbations of mass 10^{11}, 10^{15} and 10^{19} $M_\odot$. Solid lines represent isothermal modes and dashed lines adiabatic modes. The Jeans mass increases with time before t_d, and the epoch at which each perturbation mass equals the Jeans mass is marked with a tick.

$\underline{10^{11}\ M_\odot}$ Perturbations of this mass cross the horizon or become causally connected about 1 year after the origin of the Universe. Prior to that time, the perturbation amplitude can increase. Almost immediately after entering the horizon, a 10^{11} $M_\odot$ perturbation will stop growing and enter a phase of oscillation (since its mass has fallen below the Jeans mass, which in turn approximately equals the mass within the horizon - see Rees 1971). The oscillatory phase lasts until the epoch of decoupling when the Jeans mass suddenly drops to $\sim 10^5$ $M_\odot$ (Peebles and Dicke 1968: Field 1975). During this oscillatory phase, adiabatic perturbations of mass $< 10^{13}$ $M_\odot$ are strongly damped by photon drag, a process described first by Silk (1968, 1974). On the other hand, this process does not affect isothermal perturbations: for these $\Delta\rho/\bar{\rho}$ remains constant. Note the sharply divergent predictions about the amplitude of fluctuations on a scale of $\sim 10^{11}$ $M_\odot$ depending on the type of perturbation (Figure 1).

The fate of vortex perturbations of this mass has been

considered by Chibisov (1972), Ozernoi and Chibisov (1969) and reviewed by Jones (1976). Readers are referred to these papers for details.

<u>10^{15} $M_\odot$</u> In many ways, perturbations in this mass range have the simplest life histories. They enter the horizon well before the epoch of decoupling, then oscillate without appreciable damping until the epoch of decoupling. Observations of perturbations on this scale and larger, then, can tell us directly about the initial mass spectrum.

<u>10^{19} $M_\odot$</u> This value was chosen to be the mass just crossing the horizon at the decoupling epoch. Slightly less massive perturbations may or may not oscillate depending on the relation between the Jeans mass and the horizon mass. This in turn is model dependent: the Jeans mass will approximately equal the horizon mass if the Universe is radiation dominated until decoupling. For this, it is necessary to have the energy density in radiation larger than $\rho_m c^2$ until t_d, where ρ_m is the density of matter alone. In the open, low density, cosmological models now favoured, this condition does hold until decoupling.

Physical processes during the epoch of decoupling, when the fractional ionization is changing rapidly, are complicated and still not fully understood. In general, they work to reduce $\Delta\rho/\bar{\rho}$. Detailed discussions and calculations are given by Peebles and Yu (1970) and Doroshkevich <u>et al</u>. (1977), and the subject is reviewed here by Bonometto.

After decoupling, any perturbation with a mass greater than the Jeans mass of ~ 10^5 $M_\odot$ will grow (Peebles and Dicke 1968: Field 1975). However, as Lifschitz (1946) showed many years ago, the growth is slow

$$\Delta\rho/\bar{\rho} \propto (z + 1)^{-1} \tag{3}$$

as well as mass-independent. This phase lasts until $\Delta\rho/\bar{\rho}$ begins to exceed unity, and non-linear effects appear, including possibly dissipation and fragmentation*.

Note that pure Lifschitz growth would produce a spectrum of masses in the Universe today having the same mass-dependence as

---

*Or, in open models, until the epoch corresponding to $z = (\Omega^{-1}) - 1$ (Sunyaev 1971), Ω being the ratio of the present cosmic density ρ_0 to the critical density ρ_c.

the perturbations at decoupling. This fact presents a problem for those favouring adiabatic perturbations, since the Universe today is clumpier on galactic mass scales than on scales > 10^{13} $M_{\odot}$. One answer, supplied by Zel'dovich (1978) among others, is that pure Lifschitz growth of adiabatic perturbations with masses in excess of ~ 10^{13} $M_{\odot}$ is followed by collapse and fragmentation. 'Pancakes' of ~ 10^{15} $M_{\odot}$ collapse along a single axis and fragment to produce galaxies. In this scenario, then, the present mass spectrum reflects the mass spectrum at decoupling only for mass > 10^{15} $M_{\odot}$.

An alternative model is provided by Press and Schechter (1974): gravitational instability. In this picture, larger and larger masses form by gravitational 'agglomeration' of smaller masses. At present this process has produced density perturbations $\Delta\rho/\bar{\rho}$ ~ 1 on a scale of ~ 10^{15} $M_{\odot}$. This scheme requires the presence of smaller mass 'particles' which clump to form larger units, and thus appears to be consistent only with isothermal perturbations (see Peebles' discussion here).

Once again, note the clear distinction between the two theories. If the perturbations at t_d are primarily adiabatic, only masses > 10^{13} $M_{\odot}$ will be represented, the pancake theory or some variant of it is favoured: and galaxy formation is a secondary process. If the perturbations are primarily isothermal, the predominant mass scales will be ~ 10^5 $M_{\odot}$ and larger, and galaxies and clusters will form, in that order, by gravitational instability. We must now consider whether there are astronomical observations which can help distinguish between the two cases.

2. TEMPERATURE FLUCTUATIONS AT t_d

The epoch of decoupling marks an important change in the evolution of density perturbations: perturbations on all interesting astrophysical scales are free to grow in amplitude at $t > t_d$ since all exceed the Jean's mass. This epoch assumes even greater importance when we recognize that it may be the earliest epoch we can study directly, and that such a study can in principle provide answers to just the kind of questions being addressed in this School. The study is based on observations of the cosmic microwave background radiation.

When we make observations of the microwave background we observe in effect a surface of last scattering at some earlier epoch. After the time of last scattering, the microwave photons reach us with no further loss of energy or change in direction. The effect is similar to looking at a thick cumulus cloud: we receive no information about light sources behind the cloud because of scattering, but we can see the surface details of the cloud clearly.

For more than a decade, it has been recognized that the microwave background will be affected by non-uniform velocity fields or inhomogeneous distribution of matter on the surface of last scattering (Sunyaev and Zel'dovich 1970: Silk 1968). The net effect is to introduce intensity or temperature fluctuations $\Delta T/T$ in the microwave background. Thus a study of the fine scale anisotropies in the microwave background provides a 'snapshot' of the distribution of matter at a much earlier epoch, the epoch of last scattering of the microwave photons. In principle, such a 'snapshot' would permit us to determine both the slope and amplitude of the perturbation spectrum (equation 1).

To proceed further, we need to know what the redshift of the epoch of last scattering, z_s, is. Certainly $z_s < z_d$: before the primeval plasma recombined, Thomson scattering kept the mean free path of photons short. If there were no subsequent reionisation of the matter content of the Universe, we may take $z_s = z_d \sim 1100$. We will make this assumption for the remainder of this Section, then relax it later. Given this (quite reasonable) assumption, we find that observations of the microwave background can provide important information about the spectrum of density perturbations just as they become free to start their growth.

In keeping with the spirit of my enterprise, let me now turn to the observational questions - what are the amplitudes $\Delta T/T$ and angular scales of the fluctuations we may expect to observe?

Assuming that the surface of last scattering is at a large redshift, $z_d \sim 1100$ permits us, following Weinberg (1972), to write down a relation between the mass of a perturbation, assuming approximately spherical shape, and its observed angular diameter

$$\theta = \frac{\rho_o H_o}{2c\rho_c} \left(\frac{6M}{\pi\rho_o}\right)^{1/3} = 10(\Omega h)^{2/3} \left(\frac{Mh^{-1}}{10^{15} M_\odot}\right)^{1/3} \text{arcmin} \qquad (4)$$

where, as before, ρ_c is the critical density $3H_o^2/8\pi G$ and ρ_o is the actual mass density of the Universe.

Equation (4) above establishes a straightforward link between an observational parameter and the mass scale of a density perturbation. Unfortunately, the link between the amplitude of density perturbations and the fluctuation level $\Delta T/T$ is much more complicated. In the first place, fluctuations in the cosmic microwave background are produced by a number of physical processes, the more important of which will be described briefly below. In addition, because the recombination of the primeval plasma is not instantaneous, it follows that in many cases the observable fluctuations in the microwave background are averaged out (Sunyaev 1978: Sunyaev and Zel'dovich 1970).

The first effect producing $\Delta T/T$ fluctuations is the Sachs-Wolfe (1967) effect described here by Peebles (see also Peebles 1980b). For his particular model, the temperature fluctuations produced by curvature perturbations have $\Delta T/T \propto \theta^{1/2}$. As Kaiser (1982) points out , this angular dependence obtains only on scales less than the curvature scale, and hence does not apply on angular scales > 10° in open cosmological models.

The most straightforward connection between $\Delta\rho/\bar{\rho}$ and $\Delta T/T$ holds for purely adiabatic fluctuations: here $\Delta T/T = (1/3)(\Delta\rho/\bar{\rho})$. Unfortunately, this simple relationship between temperature fluctuations and density perturbations will apply only for adiabatic perturbations of mass > 10^{15} $M_\odot$, since smaller mass perturbations will be optically thin through the epoch of decoupling so that their temperature fluctuations must be averaged over a line of sight, and thus the observed level of $\Delta T/T$ will be reduced (Sunyaev 1978). The effect of gradual decoupling on $\Delta T/T$ is discussed more fully by Bonometto. For the present purposes, I wish only to note that for sufficiently large masses, M > 10^{15} $M_\odot$, these details may be ignored and the amplitude of adiabatic density perturbations at t_d is given exactly by $3\Delta T/T$.

All is not lost for lower mass perturbations and isothermal perturbations. For these scales, most of the observable temperature fluctuations are generated by the motion of matter. Density perturbations exert gravitational forces which cause the matter to move: the Doppler shift produces small temperature fluctuations $\Delta T/T = v/c$.

Here I may mention as an aside the third type of perturbation referred to above: vortex or 'whirl' perturbations involving large scale motions of matter and radiation. Models of galaxy formation based on 'whirls' require large velocities at t_d, and have large values of ΔT. These exceed (see Anile et al. 1976) the measured values, and this discrepancy and problems with primordial nucleosynthesis have killed off these models. Despite their appeal, I shall not consider them further. Instead, let us return to the two cases of adiabatic and isothermal perturbations.

For lower mass perturbations of all types, most of the observable temperature fluctuations are generated by the Doppler effect arising either from the motion of matter caused by the gravity of the perturbations at t_d (Sunyaev 1978: Silk and Wilson 1980: Wilson and Silk 1981), or directly from vortex perturbations (Chibisov 1972: Ozernoi and Chibisov 1971, 1972). The latter will be dominant if present.

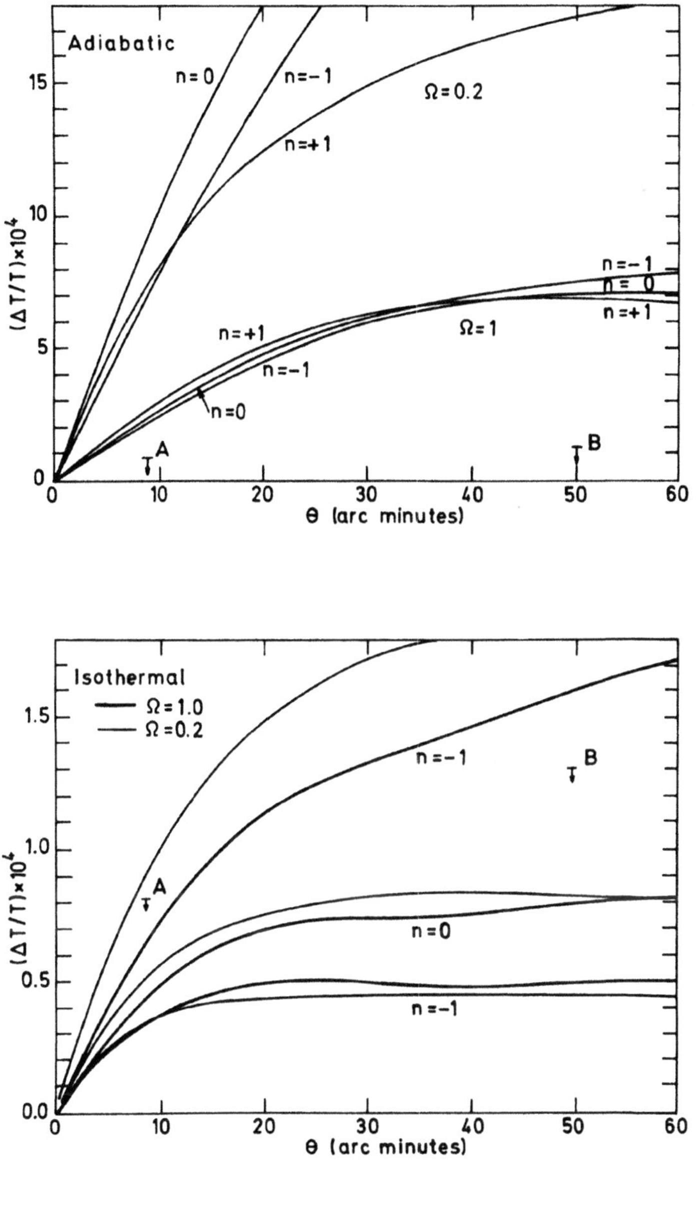

Figure 2. Model predictions of Silk and Wilson (1980) for observed anisotropy in two model universes for various values of n. (a): adiabatic modes, (b): isothermal models.

Calculations of the amplitude of observable fluctuations as a function of mass (or angular scale) have been made by several groups both analytically (Sunyaev 1978: Sunyaev and Zel'dovich 1970) and numerically (Peebles and Yu 1970: Doroshkevich et al. 1977: Silk and Wilson 1980: Wilson and Silk 1981), for many values of α. While there is some disagreement on the predicted amplitudes, especially for isothermal fluctuations (Doroshkevich et al. 1977: Silk and Wilson 1980: Wilson and Silk 1981), for all reasonable values of α, $\Delta T/T \sim 10^{-5} - 10^{-3}$. In the case of adiabatic perturbations, larger values of $\Delta T/T$ result for the lower values of ρ_0/ρ_c now favoured. As a representative example of calculated values of $\Delta T/T$, we reproduce the results of Silk and Wilson (1980) as figures 2(a), (b).

It is not unreasonable for an observer to ask his theoretical colleagues why they can not be more precise in predicting the scale and amplitude of fluctuations. One answer, of course, is the difficulty of 'following' the perturbations through the decoupling era. The other is the matter of normalizing the spectrum of fluctuations. Again, let us use the calculations of Silk and Wilson (1980) as an example. They normalize by forcing the two-point correlation function $\xi(r)$ to be unity at $\sim 5\ h^{-1}$ Mpc as required by the results of Peebles (1981a). This normalization seems reasonable, since non-linear effects may become important at smaller scales where ξ is greater. Unfortunately, the correlation function is less well pinned down at these larger mass scales. Peebles, in his lectures, uses a slightly different method.

3. OBSERVATIONS ON ARCMINUTE SCALES

An angular scale of 10' happens to fall nicely near the angular resolution of ordinary radio telescopes - those with the familiar parabolic antennas. A number of groups have used such antennas to search for temperature fluctuations. Two general remarks may be made about the results to date:- (i) no one has reliably reported detection of fluctuations which are indisputably primordial: only upper limits of $\Delta T/T$ on various angular scales are available. (ii) These upper limits, however, are tight enough to constrain the spectrum of density perturbations and even to eliminate certain types of perturbations entirely.

Observational upper limits available in mid-1979 are summarised and reviewed critically in Partridge (1980a). Table 1 is reproduced from that paper: the corrections referred to in the last column were made by the author, not by the observers, and include corrections for telescope efficiency and conversion to the 95% confidence level.

Table 1

A list of all published upper limits on the small-scale anisotropy of the cosmic microwave background. The results are generally expressed as limits at the 95% confidence level. The corrections are my own estimates, and may be ignored.

Observ.	Wavelength cm	Angular scale	Reported or published upper limit ΔT/T	Corrected upper limit ΔT/T
1.	2.80	10'	1.8×10^{-3}	3.5×10^{-3}*
2.	0.35	2'	9.0×10^{-3}	2.0×10^{-2}
3.	0.35	~1!5	3.7×10^{-3}	1.8×10^{-3}
4.	3.60	2'-1°	7.0×10^{-4}	
5.	2.8	3'-1°	3.0×10^{-5}	4.0×10^{-4}
6.	4.0	~12'x40'	5.0×10^{-5}	1.6×10^{-4}*
7.	11.0	8'-20'	1.5×10^{-4}	4.0×10^{-4}*
8.	0.13	30'	1.2×10^{-4}	
9.	4.0	5' to 150'	8.0×10^{-5} to 1.3×10^{-5}	4.0×10^{-4} to 7.5×10^{-5}
10.	0.9	~ 7'	8.0×10^{-5}	
11.	6.3	7'	$< 6\times10^{-4}$	
12.	2.8	11'	$< 2.5\times10^{-4}$	
13.	2.8	4!5	$< 2.5\times10^{-4}$	

* Converted to 2σ values and corrected for telescope efficiency. Not corrected for possible errors in statistical analysis.

1. Conklin and Bracewell (1967)
2. Penzias et al. (1969)
3. Boynton and Partidge (1973)
4. Carpenter et al. (1973)
5. Parijskij (1973a)
6. Parijskij (1973b)
8. Caderni et al. (1977)
9. Parijskij et al. (1977)
10. Partridge (1980b)
11. Ledden et al. (1980)
12. Seielstad et al. (1981)
13. Birkinshaw et al. (1981)

How were such measurements made? The facetious answer is - with much hard work. The more serious answer is - using either drift scans or beam switching or both.

Let me first justify the claim that hard work was involved. The aim is to measure a small temperature difference ΔT using a system which is itself noisy and emissive. The noise contributed by the receiver may be parametrized as T_R, the receiver noise temperature. Let us suppose that the receiver has a bandwidth of $\Delta\nu$ Hz and an overall efficiency ε, and that we integrate the signal from it for t seconds. One of the basic equations of radio astronomy tells us that the minimum detectable (brightness) temperature fluctuation is

$$\Delta T_{rms} = \frac{KT_R}{\varepsilon\sqrt{\Delta\nu\, t}},$$

where K is a constant of order unity, generally = 2.2 for the observations described below. We seek $\Delta T \sim 10^{-3} - 10^{-4}$ K: T_R is typically > 100 K*: and $\varepsilon \sim 0.5$ at best. Conventional radio telescopes employing superheterodyne receivers rarely boast of $\Delta\nu$ > 10^9 Hz. Putting these numbers together, we see that an hour or so is required to reach $\Delta T/T \sim 10^{-4}$ for each point observed. And we have so far neglected noise introduced by temporal and spatial variations in the emission of the earth's atmosphere. A fat cumulus cloud can contribute a temperature five orders of magnitude larger than the ΔT values we seek so can radio, radar interference, emission from the ground into side lobes, or the solar heating of the receiver or antenna.

For all these reasons, we generally seek to make differential measurements to permit us to subtract out some of the unwanted noise.

The simplest technique is to make drift scans: fix the telescope and allow a small strip of the sky to move through the beam at the rate 15° cos δ per hour. Generally, one then subtracts off a constant and a linear term from the measurements, stacks many such scans together, then looks at the fluctuation level. This technique has the advantage that side- and back-lobe contributions stay fixed, since the telescope does not move. It has the disadvantage that the statistics are less clear than for beam switching. For instance adjacent measurements are not truly independent because the beam of the telescope spans a non-zero angle. Care must be taken (Partridge 1980a) in deriving values of or upper limits on $\Delta T/T$ from drift scans.

In beam switching, the beam of the telescope is (rapidly) switched back and forth between two regions of the sky separated by some angle θ_s, 6° in the case of Fabbri et al. (1979), or

*Maser receivers, such as that now being employed by Wilkinson and his colleagues at Princeton, have lower values of T_R (and can have comparable values of $\Delta\nu$). Bolometric receivers (Melchiorri et al. 1981) reduce ΔT_{rms} by increasing $\Delta\nu$.

much smaller (Partridge 1980b) in the case where conventional radio telescopes are employed. As Boynton (Boynton 1980: Davis and Boynton 1980) has shown, one then measures fluctuations on angular scales θ_s, not the size of the telescope beam. The advantage of beam switching is that rapid differential measurements can be made, hence reducing possible sources of error due to time-varying atmospheric emission. Finally, one can of course combine beam switching with drift scans. A variant of this procedure was used by the Princeton group (Fabbri et al. 1980a,b: Boughn et al. 1981) to make the measurements on large angular scale to which I shall turn next.

4. OBSERVATIONS ON LARGE ANGULAR SCALES

First, I shall define 'large angular scale' to mean angles $> 1°$. This is a convenient dividing line for three reasons. First, $\theta \sim 1°$ corresponds very roughly to the horizon scale at decoupling. Second, as we shall see later, subsequent re-ionization may wipe out all fluctuations on smaller angular scales. Finally, ordinary radio telescopes rarely have beams $> 1°$ in angular size: thus specially designed equipment is needed for measurements at larger angular scales.

Let us consider the largest angular scale first - the dipole anisotropy (Smoot et al. 1977: Cheng et al. 1979: Gorenstein and Smoot 1981 and references therein) which has been known with good precision for several years. A recent review is provided by Smoot 1980). It is measured to be ~ 3 mK ($= 3 \times 10^{-3}$ K), but this quantity must be corrected for the 'known' motion of the Sun about the Galactic centre. With that known Doppler contribution to the dipole moment taken into account, we find $\Delta T/T \sim 2 \times 10^{-3}$ on an angular scale of 180°. There are, of course, many ways in which this number may be interpreted. It helps constrain homogeneous but anisotropic models of certain types: but it may equally well be regarded as a measure of the motion of the Galaxy relative to matter at large distances. What produces such motion? Here opinions differ. If it is a local region of overdensity, such as the Virgo supercluster, that local region is likely to have entered a nonlinear ($\Delta\rho/\bar{\rho} > 1$) phase of growth (Silk and Wilson 1980). Thus the connection between $\Delta T/T$ for the dipole moment and higher order moments is not clear (but see Wilson and Silk 1981). The matter is further complicated when we consider another measurement of the velocity of the Galaxy (Rubin et al. 1976) based on observations of a shell of galaxies at $z \sim 0.015$: the deduced velocity is of comparable magnitude to but in - roughly - the opposite direction from the microwave result.

As we move to the quadrupole moment, we encounter a clearer theoretical picture, but a slightly less crisp experimental one. Once again, anisotropic but homogeneous models can produce a

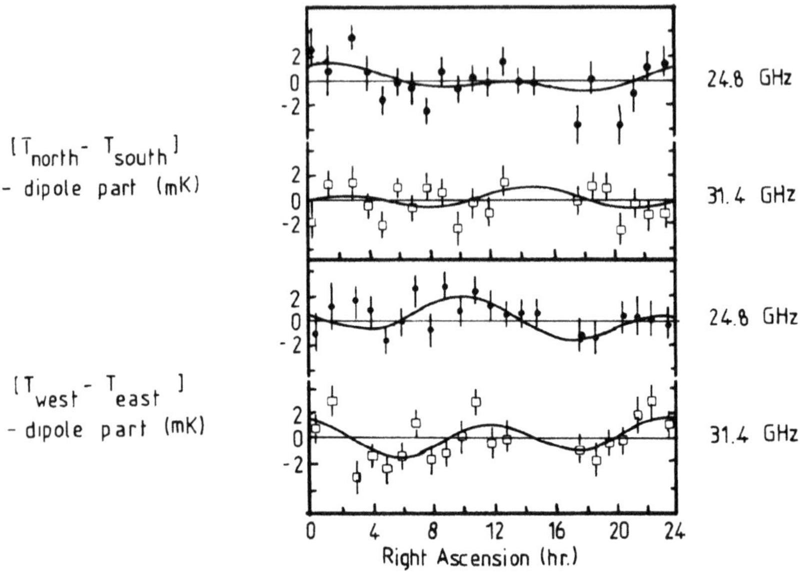

Figure 3. The anisotropy detected by Boughn _et al_. (1981). The curves represent their best fit models to the components of the quadrupole distortion.

quadrupole signature. For some anisotropic models, there is a fixed geometrical connection between the dipole axis and the quadrupole components. Since both have now been measured, one ought to be able to set tighter constraints on the kinds of models discussed here by MacCallum.

A different approach is taken here by Peebles (1981b and these lectures), who regards the quadrupole $\Delta T/T$ as a large scale fluctuation.

Now for the data. Until last year, only upper limits of roughly 3×10^{-4} were available (Cheng _et al._ 1979: Gorenstein and Smoot 1981: Smoot 1980). Then in 1980, Fabbri _et al._ 1980a,b) reported detection of a 'second harmonic' term of amplitude about 0.9 mK. This year, Boughn _et al_. (1981) reported a measurement of 4 of the 5 quadrupole components. These data, and their fits to it, are displayed as figure 3. The derived quadrupole $\Delta T/T$ is somewhat smaller than reported by Fabbri _et al._, but not seriously inconsistent with it. The reasonable agreement between independent measurements at two frequencies is encouraging. On the other hand, a pure quadrupole it is not the only polynomial fit which would reduce χ^2 or bring the observed points in line.

Perhaps the most cheerful news is that this experiment is being refined while we meet. The lower noise of the new maser receiver in use by Wilkinson and his colleagues should permit a more precise determination of the quadrupole components soon. In addition, I am told that the Italian data are being re-analyzed, with the expectation of providing measures of 4 of the 5 quadrupole components.

On a somewhat smaller angular scale, there are some poorish upper limits which emerge as a by-product of measurements of the dipole and quadrupole moments (Wilkinson and Partridge 1967: Smoot et al. 1977). These limit $\Delta T/T$ to $< 3 \times 10^{-3}$ on angular scales 7°–18°. A glance at figure 3 also suggests $\Delta T/T < 10^{-3}$ for $\theta \sim 15^{\circ}$.

Far better limits are now available at 6°, thanks to the work of Melchiorri and his colleagues (Melchiorri et al. 1981), who worked with a balloon-borne bolometric receiver operating in a wavelength band maximized at 600μ. On an angular scale of 6°, Fabbri et al. (1979) report $\Delta T/T < (3.5 \pm 0.7) \times 10^{-5}$. The authors recognize that at least some of the observed fluctuation at $\lambda \sim 600\mu$ must be produced by emission from warm dust clouds in our Galaxy. Hence, in a recent paper (Melchiorri et al. 1981), they have modelled and removed dust emission, and find

$$\Delta T = (8 \pm 2) \times 10^{-5}$$

or $\Delta T/T = (2.8 \pm 0.7) \times 10^{-5}$

This result stands both as the most sensitive measurement of ΔT to date and as the only search on small or intermediate angular scales to detect anisotropies. From a theoretical point of view the importance of this result lies in the fact that the corresponding mass scale is $> 10^{19}\ M_{\odot}$, the horizon mass scale at t_d (see section 1). Thus this observation samples the mass spectrum at t_d directly, uninfluenced by any causal physical process such as damping. From an observer's point of view, the results show the power of bolometric receivers with their very wide bandwidth (Boynton 1978). A group of us (Solheim et al. 1980: Sironi et al. 1981) is attempting to repeat such measurements on a large angular scale (in our case $\sim 3^{\circ}$), using conventional ground-based superheterodyne receivers at $\lambda \sim 3$cm. At present our results are more than an order of magnitude less sensitive than the Italian results (Melchiorri et al. 1981). It seems clear to me that the best future observations on 0.5°–20° scales will be made with bolometers, and will generally not be ground-based.

5. OBSERVATIONS ON ARCSECOND SCALES

Suppose we now consider measurements of the smallest angular scales, say below ~ 1'. For such observations, conventional filled-aperture antennas cannot be used - their resolution is > 1' because of diffraction. Hence observers (Goldstein et al. 1976, 1979: Martin et al. 1980: Partridge and Rattner in preparation: Martin 1981) have turned to radio interferometry or aperture synthesis. In addition to permitting observations on angular scales as small as 1 arcsec or less, aperture synthesis provides in principle a two-dimensional map of the temperature fluctuations. Unfortunately, aperture synthesis also has a strong disadvantage built in (Martin 1981) - much lower sensitivity for a given integrating time than a filled-aperture telescope. Essentially because an interferometer may be regarded as a very large antenna (of diameter ~ the longest baseline) of very low efficiency. For instance, 24 hours of observations with the Very Large Array has produced a tentative upper limit $\Delta T/T < 1\times10^{-2}$ on a scale of ~ 6 arcsec (see Table 2 below): this same integrating time resulted in an upper limit of 8×10^{-5} at ~ 7' (see Partridge 1980b). The use of shorter baselines at the VLA will improve the limits on $\Delta T/T$, but not by a large factor.

Table 2

Aperture synthesis (interferometric) seaches for fluctuations in the cosmic microwave background. See Goldstein et al. (1976, 1979): Martin et al. (1980) for references to the original papers.

Observers	Wavelength cm	Angular scale of synthesized beam	$\Delta T/T$
Goldstein et al.	21	~ 17 arcsec	$< 3\times10^{-2}$
Martin et al.	11	13 arcsec	$(7\pm4)\times10^{-2}$ *
Martin et al.	3.7	4 arcsec	$< 4\times10^{-2}$
Partridge et al.	6.0	6 arcsec	$< 1\times10^{-2}$ **

* Detected fluctuations probably due to discrete sources (Goldstein et al. 1976, 1979: Martin et al. 1980).
** Preliminary result.

The results of the aperture synthesis observations to date appear in Table 2. Given both the poor sensitivity and the fact that predicted values of $\Delta T/T$ discussed above drop rapidly as θ falls below ~ 10', it is reasonable to ask why people are pursuing these observations. One answer is that the theoretical predictions are model dependent and not always in perfect agreement: it seems prudent to check them. A second, and more interesting answer is that aperture synthesis provides a two-dimensional image of an area of the sky. Since pancakes, strings and other topological oddities may be involved in galaxy formation, a two-dimensional picture rather than a one-dimensional scan may be of value. Finally, the most compelling reason for pursuing these observations is laid out in the following section.

6. THE EFFECT OF RE-IONIZATION AFTER t_d

Up to this point in this article, I have assumed that the epoch of last scattering (the epoch we 'see' when we observe the microwave background) coincided with the epoch of decoupling at $z_d \sim 1100$. A number of models for the formation of galaxies and clusters suggest that this may not be so (see Rees 1978: Hogan 1980: and Hogan, J.E. Jones this volume). If the matter in the Universe is re-ionized at an epoch corresponding to $1100 > z > 10$ (perhaps by energy released in the process of galaxy formation) Thomson scattering from the free electrons will shift the surface of last scattering to a much lower redshift $z_s \sim 10$–20 (depending on the density ρ_0). On the other hand, re-ionization after an epoch of $z \sim 10$–20 will have no effect: the Universe will remain transparent so that we may again take $z_d \sim 1100$ as the surface of last scattering.

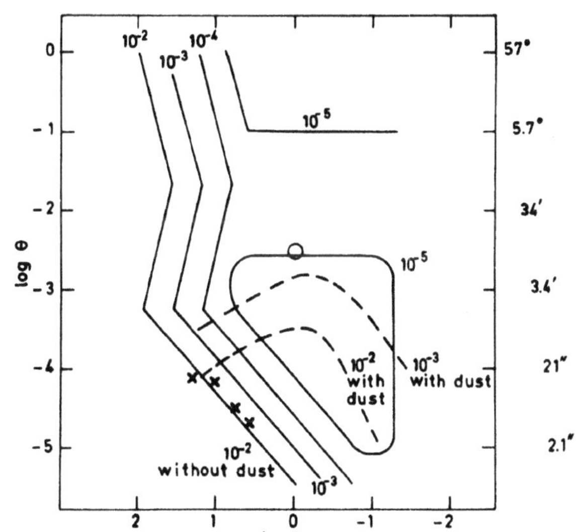

Figure 4. Calculated contours of $\Delta T/T$ with and without dust, according to Hogan (1980). The open circle is the filled aperture data (Partridge 1980), the crosses are the aperture synthesis upper limits of table 2.

If matter is re-ionized at z > 10-20, three important observational consequences follow. (1) obviously, observations of temperature fluctuations will no longer tell us anything directly about the initial mass spectrum at t_d. (2) While density perturbations on the surface of last scattering will still produce temperature fluctuations by the Doppler mechanism discussed in Section 2 above, the angular scale at which $\Delta T/T$ rises to its maximum value will be shifted from ~ 10′ to ~ 3° (see Davis 1980). (3) Perhaps the most intriguing new element introduced is the possibility of wavelength-dependent fluctuation amplitudes (Hogan 1980). Pure Thomson scattering (at these energies) is entirely independent of wavelength. However, there are other scattering and absorbing processes which may come into play during the epoch of galaxy formation, and many of these are wavelength-dependent. Extinction by dust, discussed by Puget and Heyaervts (1980) and Negroponte et al. (1981) is one example. Hogan (1980) has looked at these processes in some detail, and has predicted values of $\Delta T/T$ for a specific set of models. His results appear as figure 4 here. Note, first, the wavelength dependence of the contours of $\Delta T/T$, and, second, the large values of $\Delta T/T$ that these scattering processes produce on small angular scales. Indeed, comparison of Table 2 with figure 5 suggests that the available interferometric observations, insensitive as they are, already put the squeeze on Hogan's models. It seems likely to me that tighter constraints on the small scale anisotropy of the cosmic microwave background, combined with more precise measurements of its spectrum, can provide a critical test for those theories of galaxy formation which call for large releases of energy at epochs much later than t_d. Note that the available measurements discriminate most strongly against models containing dust.

7. WHAT HAVE WE LEARNED?

Now let me step back from the observations for a moment and try to summarize the impact they have had on theories of galaxy formation. Needless to say, this section will treat some of the same ground covered by other contributions to this volume.

First, there is a series of strong conclusions we may draw. Anisotropy in the cosmic microwave background has been detected on large angular scales. The size and orientation dipole moment and the existence of a quadrupole moment seem quite secure, and there is newly reported evidence for fluctuations on an angular scale of ~ 6°. On smaller angular scales, only upper limits are now available. These upper limits, however, have improved to the point where they are beginning to constrain theories of galaxy formation (if not theorists!).

Just how restrictive the constraints are on theories of galaxy formation depends very much on the parameters introduced into the theories. Hence we can draw only weak conclusions in this area. The work of Anile et al. (1976) suggests that upper limits on the fluctuations in the microwave background rule out all but a narrow range of possible vortex or 'whirl' models for galaxy formation. Even the rough measurements now available on angular scales $< 1'$ are in conflict with some of the models of Hogan (1980) described just above. Indeed, the measurements may prove to be important tests of any theory of early star formation in which large amounts of energy are injected into inhomogeneous matter at redshifts $\ll 1000$ (see J. Jones' contribution to this volume). As I have suggested, following the work of Silk and Wilson (1980, 1981) and Wilson and Silk (1981), the available upper limits are not consistent with adiabatic fluctuations unless the density of the Universe is high and the index n is large (see Figure 2). The agreement with their isothermal models is better, especially if $n > 0$. We must be careful, however: the predicted values of $\Delta T/T$ which we compare to the measurements are model dependent: not all authors predict values of $\Delta T/T$ as large as Silk and Wilson's. In addition, if the matter content of the Universe is reionized after decoupling, or if massive neutrinos play a dominant role in the dynamics of the Universe, then the predicted values of $\Delta T/T$ fall safely below present upper limits. At least one order of magnitude improvement in the upper limits on $\Delta T/T$ will be needed to put these more complicated scenarios to the test.

If these conclusions seem less striking than you might hope, recall that we observers are trying to measure fluctuations at a level of one part in 10,000 or less - roughly the ratio of the thickness of this print to the thickness of this book - in the cosmic microwave background, which is itself two orders of magnitude fainter than the terrestrial background.

8. ACKNOWLEDGEMENTS

These lectures, especially in their written form, owe much to congenial and helpful conversations I had with colleagues in Erice - a setting that encourages just this sort of interchange. The preparation of this report, and my observations described in it, were supported in part by grants to Haverford College from the U.S. National Science Foundation.

REFERENCES

Anile, A.M., Danese, L., De Zotti, G. and Motta, S. 1976. Ap.J. Lett., 205, L59.

Birkinshaw, M., Gull, S.F. and Northover, K.J.E. 1981. Submitted to MNRAS.

Boughn, S.P., Cheng, E.S. and Wilkinson, D.T. 1981. Ap.J.Lett., 243, L113.
Boynton, P.E. 1978. IAU Symposium 79. Ed. M.S. Longair and J. Einasto. p. 317. Reidel Publ.Co. Dordrecht, Holland.
Boynton, P.E. 1980. IAU Symposium 92. Ed. G.O. Abell and P.J. E. Peebles. p. 293. Reidel Publ.Co. Dordrecht, Holland.
Boynton, P.E. and Partridge, R.B. 1973. Ap.J., 181, 243.
Caderni, N., Fabbri, R., De Cosimo, V., Melchiorri, B., Melchiorri, F. and Natale, V. 1977. Phys.Rev., D16, 2424.
Carpenter, R.L., Gulkis, S. and Sato, T. 1973. Ap.J.Lett., 182, L61.
Cheng, E.S., Saulson, P.R., Wilkinson, D.T. and Corey, B.E. 1979. Ap.J.Lett., 232, L139.
Chibisov, G.V. 1972. Sov.A.J. 16, 235
Chibisov, G.V. and Ozernoi, L.M. 1969. Astrophys.Lett., 3, 189.
Conklin, E.K. and Bracewell, R.N. 1967. Nature, 216, 777.
Dautcourt, G. 1980. Physica Scripta 21, 714.
Davis, M. 1980. Physica Scripta 21, 717.
Davis, M. and Boynton, P.E. 1980. Ap.J., 237, 365.
Doroshkevich, A.G., Zel'dovich, Ya.B. and Sunyaev, R.A. 1977. Ast.Zh. 55, 913: Sov.Ast.J., 22, 523.
Fabbri et al. 1979. Proceedings Second Marcel Grossman Meeting. In press.
Fabbri, R., Guidi, I., Melchiorri, F. and Natale, V. 1980a. Phys.Rev.Lett., 44, 1563.
Fabbri, R., Guidi, I., Melchiorri, F. and Natale, V. 1980b. Phys.Rev.Lett., 45, 401.
Field, G.B. 1975. Stars and Stellar Systems, Vol. 9, Chap. 10. University of Chicago Press, Chicago.
Goldstein, S.J., Marscher, A.M. and Rood, R.T. 1976. Ap.J., 210, 321.
Goldstein, S.J., Turner, K.C. and Rood, R.T. 1979. Ap.J., 229, 455.
Gorenstein, M.V. and Smoot, G.F. 1981. Ap.J., 244, 361.
Gott, J.R. 1979 in Physical Cosmology Les Houches Session XXXII, 564.
Gott, J.R. and Rees, M.J. 1975. Astr.Astrophys. 45, 365. Ap.J., 194, 543.
Harrison, E.R. 1973. Ann.Rev.Astr.Astrophys., 11, 155.
Hogan, C.J. 1980. MNRAS, 192, 891.
Jones, B.T. 1976. Rev.Mod.Phys., 48, 107.
Kaiser, N. 1982. MNRAS to appear.
Ledden, J.E., Broderick, J.J. and Condon, J.J. 1980. A.J., J. 85, 780.
Lifschitz, E. 1946. J.Phys.USSR, 10, 116.
Martin, H.M., Partridge, R.B. and Rood, R.T. 1980. Ap.J.Lett., 240, L79.
Melchiorri, F., Melchiorri, B., Ceccarelli, G. and Pietranera, L. 1981. Preprint.

Negroponte, J., Rowan-Robinson, M. and Silk, J. 1981. Ap.J., 248, 38.
Ozernoi, L.M. 1978. IAU Symposium 79. Ed. M.S. Longair. Reidel Publ.Co. Dordrecht, Holland.
Ozernoi, L.M. and Chibisov, G.V. 1971. Sov.A.J., 14, 615.
Ozernoi, L.M. and Chibisov, G.V. 1972. Sov.A.J., 15, 923.
Parijskij, Yu.N. 1973a. Ap.J.Lett., 180, L47.
Parijskij, Yu.N. 1973b. Sov.A.J., 17, 291.
Parijskij, Yu,N., Petrov, Z.E. and Cherkov, L.N. 1977. Sov.A.J. Lett. 3. 263.
Partridge, R.B. 1980a. Physica Scripta 21, 624.
Partridge, R.B. 1980b. Ap.J., 235, 681.
Peebles, P.J.E. 1981a. The Large Scale Structure of the Universe. Princeton University Press, Princeton.
Peebles, P.J.E. 1981b. Ap.J.Lett., 243, L119.
Peebles, P.J.E. and Dicke, R.H. 1968. Ap.J., 154, 891.
Peebles, P.J.E. and Yu, J.T. 1970. Ap.J., 162, 815.
Penzias, A.A., Schraml,J. and Wilson, R.W. 1969. Ap.J.Lett., 157, L49.
Press, W.H. and Schechter, R. 1974. Ap.J., 187, 425.
Puget, J.L. and Heyvaerts, J. 1980. Astr.Astrophys., 83, L10.
Rees, M.J. 1971. Proceedings of the International School of Physics Enrico Fermi. Course 47. Ed. R.K.Sachs. Academic Press, New York.
Rees, M.J. 1978. Nature, 275, 35.
Rubin, V.C., Thonnard, N., Ford, W.K. and Roberts, M.S. 1976. Astr.J., 81, 719.
Sachs, R.K. and Wolfe, A.M. 1967. Ap.J., 147, 73.
Seielstad, G.A., Masson, C.R. and Berge, G.L. 1981. Ap.J., 244, 717.
Silk, J. 1968. Ap.J., 151, 459.
Silk, J. 1974. IAU Symposium 63. Ed. M.S. Longair,p. 173. Reidel Publ.Co. Dordrecht, Holland.
Silk, J. and Wilson, M.L. 1980. Physica Scripta, 21, 708.
Silk, J. and Wilson, M.L. 1981. Ap.J.Lett., 244, L37.
Sironi et al. 1981. Submitted to Paris Conference on Cosmic Rays.
Smoot, G.F. 1980. Physica Scripta, 21, 619.
Smoot, G.F., Gorenstein, M.V. and Muller, R.A. 1977. Phys.Rev. Lett., 39, 898.
Solheim, J.E., Mandolesi, N., Sironi, G. and Partridge, R.B. 1980. Physica Scripta 21, 629.
Sunyaev, R.A. 1978. IAU Symposium 79. Ed. M.S. Longair. Reidel Publ.Co. Dordrecht, Holland.
Sunyaev, R.A. and Zel'dovich, Ya.B. 1970. Astr.and Sp.Sci., 7, 3.
Weinberg, S. 1972. Gravitation and Cosmology. John Wiley Co. New York.
Wilkinson, D.T. and Partridge, R.B. 1967. Phys.Rev.Lett., 18, 557.

Wilson, M.L. and Silk, J. 1981. Ap.J., 243, 14.
Zel'dovich, Ya.B. 1972. MNRAS, 160, 1P.
Zel'dovich, Ya.B. 1978. IAU Symposium 79. Ed. M.S. Longair and J. Einasto, p. 409. Reidel Publ.Co. Dordrecht, Holland.

LARGE-SCALE FLUCTUATIONS IN THE MASS DISTRIBUTION AND THE MICROWAVE BACKGROUND: NATURE AND EVOLUTION*

P.J.E.Peebles

Joseph Henry Laboratories, Physics Department,
Princeton University, Princeton, N.J. 08544

1. INTRODUCTION

There is a striking contrast between the simplicity of the large-scale structure of the Universe and the complexity we encounter on smaller scales. Clusters of galaxies are on the dividing line and so are a natural focus for attempts to understand what this dichotomy might mean. I discuss here some scenarios for the evolution of departures from homogeneity with particular attention to what we learn from the large-scale clustering of galaxies and the large-scale fluctuations in the microwave background radiation.

It is well to recall another peculiar feature of cosmology: as we do not have a theory for the Big Bang there is no way to make ab initio calculations. We are left with a meaningful task: to discover whether each proposed scenario can start from initial conditions that seem not unreasonable, evolve according to accepted (or at least well-defined) laws of physics, and end up matching the observed character of the distributions of matter and radiation.

My interpretation of 'reasonable' initial conditions is based on the argument that, as the expanding Universe is unstable against the growth of departures from homogeneity and isotropy, the Universe must have been more nearly homogeneous and isotropic in the past than it is now (Peebles 1967, 1972, 1980a). This

*This research was supported in part by the National Science Foundation.

B. J. T. Jones and J. E. Jones (eds.), The Origin and Evolution of Galaxies, 143–169.

proposition is neither ineluctable nor universally accepted but it is one acceptable point of departure that provides a convenient framework for discussion of the above question. The proposition excludes primeval chaos (unless we can find some way gravity reverses its usual tendency to make matter pile up). It allows growing primeval irregularities of two linearly independent sorts, adiabatic and isothermal (Zel'dovich 1967). In the 'isothermal' case the primeval mass distribution and space curvature are homogeneous but the baryon number distribution (or the distribution of some new massive particle) is clumpy. (This was first discussed by Harrison 1967.) In the adiabatic case the local value of the entropy per baryon started out a universal constant. Press and Vishniac (1980) have noted that this might be considered the leading most conservative scenario because elementary particle physicists have some reason to believe that microscopic laws of physics fix the baryon number per unit of entropy. Ostriker and Cowie (1981) have made the excellent point that explosions or other effects of star formation in the intergalactic medium may have promoted development of clustering. As seeds are required for the explosions this may be regarded for our purposes as a variant of either the adiabatic or isothermal scenario.

Recent measurements of irregularities in the microwave background at $\theta = 6^o$ by Fabbri, Guidi, Melchiorri and Natale (1980) and of the quadrupole moment a_2 of the temperature distribution by Boughn et al. (1981) provide important new constraints on these scenarios. We must consider the possibilities that the temperature fluctuations originate (a) in our galaxy, (b) along the line-of-sight, or (c) in the primeval distribution. Case (b) seems highly unlikely (section 5 below). The main test for (a), yet to be done, will be the wavelength dependence of the effect: the temperature fluctuations are expected to be independent of wavelength under (c) but not under (a). If a_2 is not due to (a) it will be a serious problem for the Ostriker-Cowie scenario or any other where it is imagined that matter appears homogeneously distributed on the horizon and later is piled up into lumps. There is a similar problem in the primeval isothermal picture. It is shown in section 3 below that the amplitude of fluctuations wanted to make clusters of galaxies yields the temperature fluctuation $\delta T/T$ at $\theta = 6^o$ observed by Natale et al. (1980) in a natural enough way. However, the assumption that there are no primeval curvature fluctuations means a_2 is reproduced only by the postulate that the power spectrum of the primeval baryon number distribution has a particular form, $\mathcal{P} \sim$ constant at comoving wavelengths < 50 Mpc, $\mathcal{P} \propto k^{-3}$ at larger scales, which perhaps is artificial. The observed a_2 is produced in what seems to be a natural way if large-scale curvature fluctuations are allowed. This causes differences in redshifts of the radiation received in different directions and hence differences in temperature, an effect first discussed by Sachs and Wolfe (1967). It is an interesting and perhaps very attractive coincidence that if the fluctuations in the mass distribution on scales $< 30\ h^{-1}$ Mpc agree with the observed fluctuations in the galaxy distribution and if the mass autocorrelation function $\xi(r)$ vanishes at separations larger than that, then the Sachs-Wolfe effect reproduces the $\delta T/T$ observed by Fabbri et al. and the a_2 found by Boughn et al. (Peebles 1981a: section 4 below). This condition $\xi(r) = 0$ on large scales is obtained in the primeval adiabatic scenario by the postulate that the primeval power spectrum of density fluctuations is flat. Again we encounter a problem, that if neutrinos are massless or have a mass of a few tens of eV then the coherence length of the mass distribution at decoupling exceeds the coherence length of the present galaxy distribution, which seems unacceptable. Again the way out is to adjust the shape of the primeval spectrum, taking it to be flat only on large scales, varying as $\sim k^4$ on small scales to counter the strong dissipation during decoupling.

As the discussion depends on the calibration of mass fluctuations on scales $< 30\ h^{-1}$ Mpc inferred from the galaxy distribution I begin in the next section with a review of the status of that subject.

2. NATURE OF THE GALAXY DISTRIBUTION: CLUSTERING HIERARCHIES AND HOLES

The study of how galaxies are distributed is of interest in its own right, for galaxies are important objects in the scheme of things, and as a possible hint to the nature of the large-scale mass distribution. The n-point correlation functions used in the theory of non-ideal gases have proved to be useful measures of the galaxy distribution. (For reviews see Groth et al. 1977: Fall 1979: Peebles 1980a.) The main practical reason is that each reduced correlation function in the observed angular distribution of galaxies is a linear integral over the corresponding spatial function, so it is fairly easy to translate from one to the other: the usually vexatious problem of deciding whether two galaxies seen close together in the sky are close in space is dealt with in a straightforward statistical way. In catalogs with redshifts the projection problem is lessened but by no means removed, and I think it is likely that there will continue to be an important role for statistical measures of clustering in the 'field'.

The galaxy two-point spatial correlation function is defined by the joint probability of finding galaxies centered in the volume elements V_1 and V_2 at separation r,

$$\delta P = n[1+\xi(r)]\ \delta V_1 \delta V_2\ , \tag{1}$$

where n is the mean space number density of galaxies. As the galaxy distribution approximates a homogeneous isotropic random process, the two-point function ξ has been written as a function only of the magnitude of the separation. Of course, ξ may well vary with cosmic time but galaxy samples are not yet deep and large enough to have unambiguously detected the effect. We might also expect that ξ depends on absolute magnitude, but Soneira (1978a) found no evidence of it in the Zwicky sample. Davis and Geller (1976) found that ξ varies with morphological type in the expected direction: ellipticals are more strongly clustered than spiral galaxies. In the results stated below ξ should be understood to be the correlation function for bright galaxies of the sort found in samples selected by apparent magnitude.

The two-point function is well approximated by the power law

$$\xi = (r_o/r)^\gamma\ , \quad \gamma = 1.77 \pm 0.04, \tag{2}$$

over a remarkably large range of separations,

$$5\ h^{-1}\ \mathrm{kpc} < r < r_b \sim 10\ h^{-1}\ \mathrm{Mpc}, \tag{3}$$

where Hubble's constant is $H_o = 100\ h\ \mathrm{km\ s^{-1}\ Mpc^{-1}}$. It appears that ξ is slightly higher than the power law at $r \sim r_b$ and then breaks below the power law at $r > r_b$ (Groth and Peebles 1977: Fall and Tremaine 1977: Soneira and Peebles 1978: Gott and Turner 1979). Estimation of the clustering length r_o depends on the probability of detection of a galaxy as a function of its distance from us. This is not well known in the large deep samples, and where redshifts are known the samples are small and so may not be representative. My best present guess is

$$r_o = 5 \pm 0.5\ h^{-1}\ \mathrm{Mpc}. \tag{4}$$

One can compute from ξ the variance of the number of galaxies found within a randomly placed sphere with radius R (Peebles 1980a, section 36):

$$\begin{aligned} \delta N^2 &= \langle (N-\langle N\rangle^2)\rangle \\ &= \langle N\rangle + n^2 \textstyle\int_R dV_1 dV_2\ \xi\ (r_{12}). \end{aligned} \tag{5}$$

If $\xi(r)$ approaches zero more rapidly than r^{-3} at $r > r_b$ as the data suggest, then at $R \gg r_b$ we can approximate this as

$$(\delta N/N)^2 = K_3/V\ , \quad K_3 = \int d^3r\ \xi(r)\ . \tag{6}$$

This says that the large-scale fluctuations in the galaxy

distribution are the same as the fluctuations that would be produced by clumps randomly placed at number density

$$n_{eff} = K_3^{-1}. \tag{7}$$

If the large-scale mass distribution follows the galaxy distribution then the large-scale fluctuations in mass would be produced by a uniform random Poisson distribution of clumps with mass

$$M_{eff} = K_3 \rho_o , \tag{8}$$

where ρ_o is the mean mass density.

Estimation of the integral K_3 is discussed by Clutton-Brock and Peebles (1981). A reasonable working number is

$$K_3 = 9000 \; e^{\pm 0.25} \; h^{-3} \; Mpc^3 , \tag{9}$$

which makes the characteristic mass

$$M_{eff} = 2.5 \times 10^{15} \; \Omega_o \; h^{-1} \; M_o , \tag{10}$$

where the density parameter is $\Omega_o = \rho_o/\rho_c$, with ρ_c the density in an Einstein-de Sitter model with Hubble constant H_o.

The statistic $\xi(r)$ gives only a limited description of the clustering, and one way to add to the information is to estimate higher order correlation functions. The three- and four-point correlation functions both are known well enough (Groth and Peebles 1977: Fry and Peebles 1978) that we can see that they do vary in the way expected if on scales $\sim r_b$ galaxies are organized in a scale-free clustering hierarchy of the sort advocated on other grounds by de Vaucouleurs (1970), Mandelbrot (1977) and others.

An interesting test of the clustering hierarchy picture is to make a computer model galaxy distribution according to this pattern, project the distribution onto the model sky, and compare the visual impression of the resulting map to maps of the actual galaxy distribution. This was done using the Lick map as the standard by Soneira (1978b) and Soneira and Peebles (1978). In making the clustering hierarchies we used a dumbbell model: place a stick of length ℓ at random: place sticks of length ℓ/λ at random orientations and with their centers at either end of the first stick: repeat with four sticks of length ℓ/λ^2, and continue through L levels to 2^{L-1} sticks at the ends of which 2^L galaxies are placed. The construction is repeated at randomly chosen spots with L drawn from a distribution chosen to

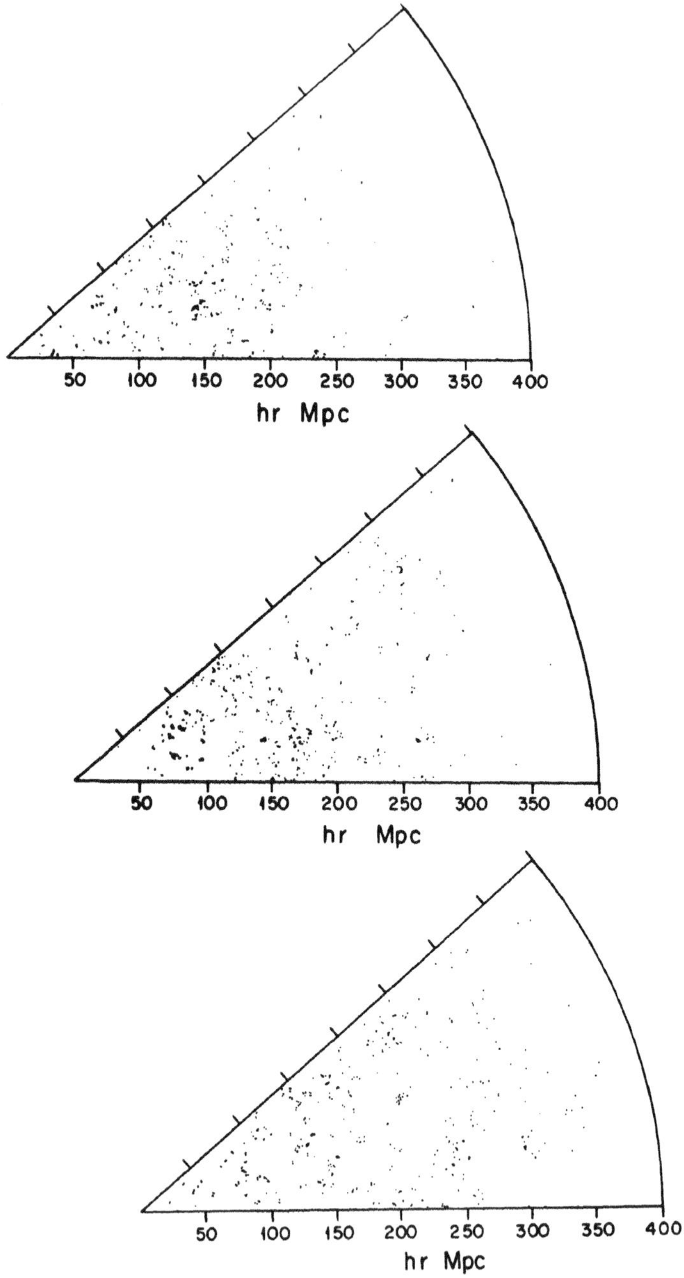
50
100
150
200
250
300
350
400
hr Mpc
50
100
150
200
250
300
350
400
hr Mpc
50
100
150
200
250
300
350
400
hr Mpc

produce the wanted mean galaxy number density n, the wanted value of K_3 (eq. [6]),

$$1+nK_3 = \langle 2^{2L}\rangle / \langle 2^L \rangle \,, \tag{11}$$

and like equations for the higher moments. The length ℓ determines the radius r_b of the break from the power law behaviour of $\xi(r)$. The power law index γ (eq. [2]) is reproduced by setting

$$\lambda^{(3-\gamma)} = 2 \quad , \quad \lambda \sim 1.76. \tag{12}$$

The visual texture of the map was found to be sensitive to two interesting features of the model. If were made much larger than the observed value of r_b, making the power law shape of $\xi(r)$ extend to much larger radii (at fixed r_o) than is observed, it made K_3 too large, and by thus placing too many galaxies in each clump it made the map look too patchy. Second, a rather broad distribution in L was needed not only to fit the n-point correlation functions but also to make the distinctive highlights seen in the Lick map.

Figure 1 shows the space distribution of the galaxies in a model constructed on this plan. All the parameters are the same as in Soneira and Peebles (1978) save that the limiting magnitude has been adjusted to make the mean surface density agree with the redshift sample of Kirshner et al. (1981). Each fan in the figure represents a slice of the model sky one degree thick and 40° wide. The three fans are placed one above the other at separations of 3.5°. Given that in the model $\xi(r)$ falls to unity at $r \sim 5\ h^{-1}$ Mpc and that ξ is negligibly small at $r > 20\ h^{-1}$ Mpc, it may seem surprising to see holes and ridges that can be traced over lengths ~ 100 Mpc, but that is only the result of statistical fluctuations. Each clump produced in the dumbbell construction has a diameter ~ 20 h^{-1} Mpc, and the mean space density of clumps is about equal to the reciprocal of the volume of a clump so that the clumps just about fill space. (That is because $r_b \sim r_o$.) However as each clump contains considerable empty space, and each clump must be placed at random (otherwise ξ would not vanish at $r > 20\ h^{-1}$ Mpc), the accidental fluctuations must produce substantial gaps here and there.

Chincarini and Rood (1979), Kirshner, Oemler and Schechter (1979), Kirshner et al. (1981) and others have noted that there are large holes in the actual galaxy distribution. The model results show that we should be cautious in seeking the physical interpretation of these holes. In the model they result from the conditions $\xi > 0$ at $r < 15\ h^{-1}$ Mpc, $\xi = 0$ at $r > 20\ h^{-1}$ Mpc: when galaxies are pulled up into clusters it makes holes, and if the cluster positions are uncorrelated on scales > 20 h^{-1} Mpc there have to be statistical fluctuations that here and there open up gaps much larger than 20 h^{-1} Mpc. Thus it seems to me that the existence of the holes does not mean there is reason to think $\xi(r)$ has been underestimated on large scales. One is free to assume that the holes were produced by a physical process operating in a coherent way over scales ~ 100 h^{-1} Mpc, but one must recognize that the process would have had to have operated in a peculiar way, leaving $|\xi| \ll 1$ on this scale.

In the following sections the integral K_3 of ξ is used as a measure of the mean square mass fluctuations on large scales.

3. PRIMEVAL ISOTHERMAL PERTURBATIONS

3.1 Evolution of irregularities

In this picture the primeval mass distribution, expansion rate and space curvature all agree with the Friedman-Lemaitre model but the baryon number is irregularly distributed. (For recent discussions see Grishchuk and Zel'dovich 1978: Doroshkevich, Zel'dovich and Sunyaev 1978, Wilson and Silk 1981 and Silk and Wilson 1981). The hypothesis of homogeneous space curvature is as ad hoc as any other, but still might be considered the least artificial choice if as it appears we are forced to make one. If everything were as uniformly distributed as possible at the moment of the Big Bang (quanta anticorrelated) it would make the Universe too smooth: the clumpy distribution of baryons (or some other massive component) is designed to remedy that. It violates the idea that microscopic physics at kT < 10^{15} GeV may fix a universal value for the entropy per baryon number but ways around that have been discussed (Senjanovic and Stecker 1980: Barrow and Turner 1981) and anyway it is doubtful that astrophysical speculations should be pinned on such a long extrapolation of elementary particle physics.

The hypothesis of uniform space curvature means that initial conditions have been set so the mass density of hypersurfaces of fixed cosmic (proper) time remains very nearly homogeneous, each part of the Universe evolving like an Einstein-de Sitter model (at $1 + z \gg |\Omega_0^{-1} - 1|$), as long as the pressure gradient force can be neglected. At kT ~ 10^9 eV, z_b ~ 10^{13}, the local excess baryon-antibaryon pairs combine leaving an irregular distribution of low pressure matter. (I shall assume that if there are residual patches of antimatter they are well enough separated from the patches of matter that the effect of subsequent annihilation on the dynamics may be neglected.) As the total density is uniform the fluctuations $\delta\rho_m$ in the matter mass density are balanced by opposite fluctuations $\delta\rho_r = -\delta\rho_m$ in the radiation mass density. At high redshift the mean density of the radiation is much higher than that of the matter so the fractional fluctuation in its density is much smaller. Assuming the neutrinos have negligible mass we have

$$\delta_r = \delta\rho_r/\rho_r \ , \quad \delta_m = \delta\rho_m/\rho_m \ , \ \delta_m = \delta_i + \frac{3}{4}\,\delta_r \ ,$$

$$\delta_m = \delta_i/(1 + \frac{3}{4}\frac{\rho_m}{\rho_r})$$
$$\delta_r = -(\rho_m/\rho_r)\ \delta_i/(1 + \frac{3}{4}\frac{\rho_m}{\rho_r}) \qquad (13)$$

The initial perturbation is δ_i. The last two equations just express the conditions that the total density is constant and the entropy per baryon conserved. The densities ρ_r and ρ_m are equal at redshift z_{eq},

$$1 + z_{eq} = 4.2 \times 10^4\ \Omega_o h^2 \ . \qquad (14)$$

At $z << z_{eq}$, δ_m and δ_r tend to approach

$$\delta_m \sim 0 \ , \quad \delta_r = -\ 4/3\ \delta_i \ . \qquad (15)$$

(The massive neutrino case is discussed later.) However, radiation pressure prevents this if the scale of the fluctuation is less than the Jeans length for the matter-radiation fluid (Peebles 1980a, eq. 92.47),

$$\lambda_x = 50(\Omega_o h^2)^{-1}\ \mathrm{Mpc} \ . \qquad (16)$$

Here and below comoving lengths have been expressed in units of proper length at the present epoch. A density fluctuation with comoving scale $\lambda < \lambda_x$ enters the horizon at redshift $z \sim z_{eq}(\lambda_x/\lambda)$, with amplitudes

$$\delta_r \sim -\delta_m\ \lambda/\lambda_x \ . \qquad (17)$$

The part δ_r then oscillates as a pressure wave until the primeval plasma combines to form hydrogen. Photon diffusion strongly damps the oscillations at wavelengths less than the Silk length (Peebles 1981b, Table 1).

$$\lambda_s \sim 2(\Omega_o h^2)^{-0.8}\ \mathrm{Mpc}. \qquad (18)$$

On scales less than this we are left at decoupling with an almost quiescent sea of radiation and the primeval distribution of matter. At $\lambda_s < \lambda < \lambda_x$ there is an appreciable part of the oscillating wave left over after decoupling but it has negligible effect on the residual distributions of matter and radiation: the matter perturbation adds to the static part with random phase and the radiation fluctuations are washed out by electron scattering during recombination (Peebles and Yu 1970, section 4.3). Thus the matter is left with the primeval spectrum of matter density irregularities truncated longward of $\sim \lambda_x$, and the power spectrum of the residual radiation energy density distribution vanishes at $\lambda < \lambda_x$ and follows the primeval baryon spectrum longward of that. As galaxy clustering has only been detected on scales $< 40\ h^{-1}$ Mpc $< \lambda_x$ the first effect might be used to build up the observed galaxy distribution. As λ_x subtends the angle

$$\theta_x \sim 40\ h^{-1} \text{ arc minutes} , \tag{19}$$

we can look for the second effect in angular fluctuations in the brightness of the microwave background radiation.

3.2 Fluctuations in the mass distribution and the microwave background

Let us suppose the primeval power spectrum of irregularities δ_i is

$$\mathcal{P}(k) = Ak^n , -3 < n < 1 . \tag{20}$$

The limits of integration assure convergence of the integral in equation (22) below. With $n \sim 0$ we get a reasonable initial spectrum from which to build the clustering hierarchy observed at $r < r_b$ (Peebles 1974). Now if we extrapolate this power law spectrum to wavelengths $> \lambda_x$ we get residual irregularities in the radiation distribution which we can compare to the measurements of microwave background radiation.

The mass autocorrelation function at decoupling is

$$\xi_p(x) = \int d^3k\, \mathcal{P}(k)\, e^{i\underline{k}\cdot\underline{x}} . \tag{21}$$

The variance of the mass found within a sphere of radius X is given in terms of ξ_p by equation (5):

$$\langle(\delta M/M)^2\rangle = 4\pi \int_0^\infty dk\, k^2\, \mathcal{P}(k)\, W(kX) ,$$

$$W(y) = 9\,(\sin y - y \cos y)^2/y^6 . \tag{22}$$

In the power law model in equation (20) for $\mathcal{P}(k)$ we find

$$(\delta M/M)_X = (4\pi A I_n)^{1/2}\, X^{-(3+n)/2} ,$$

$$I_n = \int_0^\infty dy\, y^{2+n}\, W(y) , \quad I_0 = 4.7 . \tag{23}$$

With $n > -3$ and $X \ll \lambda_x$ we can ignore the cutoff at λ_x. If $n > 1$ the variance $(\delta M/M)^2$ in the mass found within the sphere of radius X is determined by small-scale fluctuations in the mass distribution along the surface of the sphere so here $\delta M/M$ is not a useful measure of mass fluctuations on the scale X. However $n < 0$ is going to be wanted below so that need not be considered further.

Next, consider the distribution of the radiation at decoupling. The power spectrum of the radiation mass density is (eq. [15])

$$\mathcal{P}_r(k) = (4/3)^2\, \mathcal{P}(k)\ , \quad k < k_x = 2\pi/\lambda_x\ ; \qquad \simeq 0\ , \quad k > k_x \tag{24}$$

The rms value $\delta T/T$ of the background temperature difference seen along lines of sight (1) and (2) at separation θ depends on the autocorrelation function of the radiation distribution,

$$\left\langle \frac{(T_1-T_2)^2}{T} \right\rangle \equiv (\delta T/T)^2 = (\xi_r(0) - \xi_r(x_{12}))/8: \tag{25}$$

$$x_{12} = 2\theta/(H_o a_o \Omega_o)\ , \quad \theta < 1 \text{ radian.}$$

Using equations (21) and (24) we can write this as

$$(\delta T/T)^2 \sim \frac{2}{9}\int_0^{k_x} d^3k\, \mathcal{P}(k)\, \left(1 - \frac{\sin kx_{12}}{kx_{12}}\right) \tag{26}$$

and with equations (20) and (23) we get

$$\delta T/T \sim \left(\frac{2}{9(3+n)I_n}\right)^{1/2} (2\pi X/\lambda_x)^{(3+n)/2} (\delta M/M)_X, \quad x_{12} > \lambda_x. \tag{27}$$

Now we can find the expected value of $\delta T/T$ from the observed mass fluctuations on scales $< \lambda_x$ (under the assumption that the large-scale rms fluctuation in the galaxy distribution is a good measure of the fluctuation in the mass distribution). We have from equations (2) to (5)

$$(\delta M/M)^2 = 1.82(r_o/X)^{1.77},\ \delta M/M \sim 1.0,\ X = 7h^{-1}\text{ Mpc.} \tag{28}$$

The numerical factor in the first equation comes from Peebles and Groth (1976, eq. [28]). In the second equation $\delta M/M$ is evaluated at about the largest radius for which ξ is reliably known. Extrapolation of $\delta M/M$ at fixed comoving radius X from the present epoch back to decoupling is discussed by Peebles and Groth (1976). The result for the primeval fluctuations at decoupling is

$$\delta M/M_p \sim 1.7 \times 10^{-3}\, \Omega_o^{-1} \tag{29}$$

at $X = 7h^{-1}(1+z)^{-1}$ Mpc. This in equation (27) with $n \sim 0$, which is about what is wanted to form the clustering hierarchy observed at $r < r_b$, gives

$$\delta T/T \sim 1.6 \times 10^{-4}\ \Omega_o^{1/2} h^{3/2} \quad . \tag{30}$$

It is interesting that Fabbri et al. (1981) found

$$\delta T/T \sim 3 \times 10^{-5}\ ,\ \theta = 6^{o}, \tag{31}$$

which agrees with equation (30) for not unreasonable choices of Ω_o and h. However, we need a special adjustment to fit the quadrupole moment of the radiation distribution (Boughn et al. 1981), as follows

The ℓ,m multipole moment of the temperature distribution in the microwave background is

$$a_\ell{}^m = \int d\Omega\ Y_\ell{}^m\ T(\theta,\phi)/\langle T\rangle \quad . \tag{32}$$

The expected value is (see e.g. Peebles 1973, Appendices A and C)

$$a_\ell{}^2 = \langle |a_\ell{}^m|^2\rangle = \pi^2 \int_o^\infty k^2 dk\, \mathcal{P}_r(k)(j_\ell(kx))^2 \quad , \tag{33}$$

$$x = 2/(H_o a_o \Omega_o) \quad ,$$

where j_ℓ is the spherical Bessel function, and for the quadrupole moment it is

$$j_2 = \left(\frac{3}{y^3} - \frac{1}{y}\right) \sin y - \frac{3 \cos y}{y^2} \tag{34}$$

If the power law index for $\mathcal{P}(k)$ is $n < -1$ (eqs. [20] and [24]) then the main contribution to the integral in equation (33) is at $k \sim x^{-1} \ll \lambda_x{}^{-1}$ so we can approximate it as

$$a_2{}^2 = (16/9)\pi^2 A\ J_n\ x^{-(3+n)},$$
$$J_2 = \int_o^\infty dy\ y^{2+n}\ j_2\ (y)^2. \tag{35}$$

If $n > -1$ the integral is dominated by the short wavelength cutoff λ_x of $\mathcal{P}_r$. In this case a_ℓ is down from $\delta T/T$ by the factor $\sim (\Delta\Omega/4\pi)^{1/2}$, where $\Delta\Omega$ is the solid angle of the coherence length λ_x of the radiation distribution at decoupling. This makes a_2 much smaller than the Boughn et al. number. If $-1 > n > -3$ equations (23), (29) and (35) give

$$a_2 \sim 10^{-3}\Omega_o{}^{-1}\ (10^{-3}\Omega_o)^{(3+n)/2} \quad , \quad n < -1 \ . \tag{36}$$

Boughn et al. (1981) found

$$a_2 \sim 2 \times 10^{-4} \quad . \tag{37}$$

We see that to make the model agree with this we would have to take $n < -2$. That is not wanted at $\lambda < \lambda_x$ for it would mean the mass fluctuations at decoupling vary with scale no more rapidly than $X^{-1/2}$ (eq. [23]), so that if we adjusted the amplitude to make proto-objects at $X \sim 1$ Mpc form at moderately high redshift we would violate the limit $\delta M/M \sim 1$ now at $X \sim 7h^{-1}$ Mpc.

The major part of the mass of the Universe might be in massive neutrinos, mean mass density ρ_ν. In that case equations (13) and (15) for the density contrast in the radiation mass density at decoupling are replaced with

$$\delta_r = -4/3 \frac{\rho_b \delta_i}{\rho_b + \rho_\nu}, \tag{38}$$

where ρ_b is the mass density in baryons. If, as has been speculated, $\rho_b < 0.1 \rho_\nu$, then the temperature fluctuations in equations (30) and (37) are lowered by a factor of 10, which worsens the problems with the quadrupole moment. However, the isothermal model does offer an interesting picture for galaxy formation, as described next.

3.3 Massive neutrinos and galaxy formation

If massive neutrinos are invoked to account for the mass in clusters of galaxies it would be most economical to suppose they also provide the mass needed to make flat rotation curves in large isolated spiral galaxies. That requires that the neutrinos in the central part of the galaxy be fairly close to degenerate (Szalay and Marx 1976: Cowsik and McClelland 1973: Tremaine and Gunn 1979: Peebles 1980b), which is hard to arrange if galaxies formed by collapse of protogalaxies or protoclusters. It can happen in the primeval isothermal scenario because there may be primeval lumps of baryons (or some other material) around which the neutrinos collect as they cool. I will argue that this produces a not unreasonable mass distribution for a galaxy.

Let us recall several well-known features of a cosmology with massive neutrinos (see e.g. the proceedings of the 10th Texas Symposium on Relativistic Astrophysics, to be published in the Annals of the New York Academy of Sciences). For definiteness I take $\Omega_o = 1$ and I assume the mass of the Universe is dominated by one type of neutrino in just two spin states. The present mean number density of neutrinos is fixed by the neutrino temperature $\sim 2°$K (see eq. [50] below),

$$n_\nu \sim 109 \text{ cm}^{-3} \quad , \tag{39}$$

and the neutrino mass is therefore

$$m_\nu = \rho_o/n_\nu = 97 \text{ h}^2 \text{ eV} \quad . \tag{40}$$

In the absence of gravitational perturbations the neutrino momentum distribution is

$$dn \propto d^3p\ (e^{pc/kT_\nu} + 1)^{-1}, \tag{41}$$

and for $p < m_\nu c$ this gives the rms velocity

$$v = 3.60\ kT_\nu/m_\nu c = v_o(1+z) \quad , \ v_o = 1.85 \text{ h}^{-2}\text{km s}^{-1}, \tag{42}$$

where the present neutrino temperature is given by equations (50) and (51) below. As cosmic time in the model is

$$t = 2/(3H_o(1+z)^{3/2}) = 2.06 \times 10^{17} \text{ h}^{-1} (1+z)^{-3/2} \text{ sec}, \tag{43}$$

the characteristic Jeans length for the neutrino gas is

$$\lambda_j = vt = \lambda_o\ (1+z)^{-1/2} \quad , \ \lambda_o = 0.012 \text{ h}^{-3} \text{ Mpc} \ . \tag{44}$$

Now suppose there is at $z \sim 10^4$ an isolated mass concentration m_o (a clump of baryons, or a primeval black hole, or some other object). Since the neutrinos are moving rapidly at high redshift they are smoothly distributed to begin with but as they cool they collect around m_o. For the purpose of a crude analysis let us imagine we can follow fixed mass shells of the neutrinos. The shell that contains mass $m > m_o$ in neutrinos has radius

$$r = \left(\frac{3m}{4\pi\rho_o}\right)^{1/3} (1+z)^{-1} \quad , \tag{45}$$

at high redshift. At epoch $z_1(r)$ this radius agrees with the Jeans length λ_j (eq. [44]), and the radius is

$$r_1 = \lambda_o^2 \left(\frac{4\pi\rho_o}{3m}\right)^{1/3} \quad . \tag{46}$$

Thereafter we can neglect the effect of neutrino pressure on the motion of the mass shell, so its final radius is, by energy balance,

$$Gm_o/r_1 \sim Gm/(2r) \quad . \tag{47}$$

These equations give

$$m \sim \frac{m_o^{3/2} r^{3/2}}{\lambda_o^3} (6/\pi\rho_o)^{1/2} \ . \tag{48}$$

With $m \propto r^{3/2}$ the rotation speed for circular orbits varies as $v \propto r^{1/4}$, which is fairly close to what is observed (Rubin et al. 1980). This scaling relation, which is a variant of the one discussed by Gott (1975), applies at $m > m_0$. The shell at $m \sim m_0$ reaches maximum expansion with enough pressure to support itself so there is not much collapse and little entropy production. At $m \gg m_0$ the neutrinos gain considerable velocity during the collapse making the final mean occupation number $\ll 1$ well outside the core.

In a large spiral galaxy the rotation speed at $r = 10\ h^{-1}$ kpc typically is ~ 250 km s^{-1}. This fixes $m(r)$, and we get from equation (48) the required seed mass,

$$m_0 \sim 2 \times 10^9\ h^{-5}\ M_\odot \quad . \tag{49}$$

This is comparable to the mass in the central part of a large spiral because we were able to make the core radius of the neutrino distribution small, on the order of 1 kpc.

The seed masses m_0 are invoked ad hoc in this picture: conceivably they are the remnants of galaxies from the last phase of oscillation of the Universe.

4. PRIMEVAL ADIABATIC PERTURBATIONS

4.1 The mass distribution

In this scenario the primeval entropy per baryon is the same everywhere. In the variant that has been analysed in most detail (see e.g. Peebles and Yu 1970: Doroshkevich, Zel'dovich and Sunyaev 1978: Bonometto and Lucchin 1979: Silk and Wilson 1980: Press and Vishniac 1980: Peebles 1981b) the Universe contains matter (hydrogen and helium), radiation and zero mass neutrinos. The primeval power spectrum $\mathcal{P}(k)$ of density irregularities must be adjusted as follows to fit the observations (Peebles 1981b): $\mathcal{P}(k)$ must vary as $\sim k^4$ at comoving wavelengths ~ 1 to 30 Mpc to make the coherence length of the residual matter distribution after decoupling less than the present coherence length: at comoving wavelengths < 1 Mpc $\mathcal{P}(k)$ must be nearly flat to avoid large curvature fluctuations and relativistic collapse at high redshift: and as discussed below if the recently discovered irregularities in the microwave background are primeval, $\mathcal{P}(k)$ must again be nearly flat at $\lambda > 30\ h^{-1}$ Mpc. All this seems contrived and so argues against this scenario as well as the isothermal one.

In a variant of the scenario that has been much discussed recently the major part of the mass of the Universe now is in neutrinos with mass > a few tens of eV, or some other sort of weakly interacting particles. However we still encounter a problem as follows. At a redshift $z_\nu > 10^5$, cosmic time t_ν, the particle peculiar motions are just becoming nonrelativistic. The motions have smoothed out inhomogeneities in the particle distribution on comoving scales $< \lambda_\nu \sim ct_\nu/a(t_\nu)$. Longward of λ_ν the growth of inhomogeneities in the particle distribution is never hindered by the particle motions. Initially the radiation and baryon number distributions match that of the weakly interacting particles, but as inhomogeneities appear on the horizon the baryon and radiation part is stored as a pressure wave field with decreasing amplitude while the particle part is constant or keeps growing. At $z < z_\nu$ the comoving Jeans length for the weakly interacting particles decreases but the particle distribution is not appreciably perturbed by the irregular mass distribution in the radiation because the latter is oscillating. The result is that at decoupling the mass fluctuations in the free particle distribution can greatly exceed that in the baryons.

The smoothing length λ_ν has been computed by several authors (among others, Szalay and Marx 1976: Bisnovaty-Kogan and Novikov 1980: Bond, Efstathiou and Silk 1980: Wasserman 1980: Schramm and Steigman 1981: Klinkhamer and Norman 1981). The discussion here follows most closely that of Davis et al. (1981). It will be supposed that the Universe is dominated by neutrinos of mass m_ν, other neutrinos still present in appreciable abundances having masses $<< m_\nu$. The neutrino temperature extrapolated to the present by the equation $T_\nu \propto a(t)^{-1}$ is

$$T_\nu = \alpha T_o \ , \quad T_o = 2.7 \text{ K} \ . \tag{50}$$

In the standard cosmology the entropy dumped into the radiation by the annihilation of the electron-positron pairs gives

$$\alpha = (4/11)^{1/3} \ . \tag{51}$$

As pointed out by Davis et al., one could imagine that the decays of other particles made α smaller than that. The mean density of neutrinos, assuming only that spin states are present, is

$$n_\nu = 2 \int \frac{d^3p}{(2\pi h)^3} (e^{pc/kT} + 1)^{-1} = 300 \ \alpha^3 \ \text{cm}^{-3} \ . \tag{52}$$

For simplicity I assume that the mean mass density of the neutrinos exceeds that of the baryons, so

$$m_\nu = \rho_o/n_\nu = 35 \ \Omega_o h^2 \ \alpha^{-3} \ \text{eV} \ . \tag{53}$$

The rms neutrino momentum is (eq. [42])

$$p = p_0 \, a_0/a(t) \quad ,$$
$$p_0/m_\nu c = 3.60 \; kT_\nu/m_\nu c^2 = 8.37 \times 10^{-4} \, \alpha \, m_\nu^{-1} \quad , \tag{54}$$

where m_ν is expressed in electron volts. The neutrino peculiar velocity is

$$v = a(t) \; dx/da = p(m_\nu^2 + p^2/c^2)^{-1/2} \quad , \tag{55}$$

which gives coordinate displacement

$$x = \int_0^{t_0} dt \; v/a \quad . \tag{56}$$

On using the cosmological equation

$$(1/a \; da/dt)^2 = 8/3 \; \pi G(\rho_r + \rho_\nu),$$
$$\rho_r \sim a^{-4}, \quad \rho_\nu \sim a^{-3}, \tag{57}$$

which is accurate enough for our purpose, we find

$$\lambda_\nu = a_0 x = (\rho_0/m_\nu)(\tfrac{8}{3}\pi G\rho_r(t_0))^{-1/2} I(\alpha)$$
$$= 15 \; \alpha^4 (\Omega_0 h^2)^{-1} \; I(\alpha) \; \mathrm{Mpc},$$
$$I(\alpha) = \int_0^\infty dy[(1+y)(y^2+B)]^{-1/2}, \tag{58}$$
$$B = \left[(\rho_0/m_\nu c) \;\; (\rho_\nu/\rho_r)_0\right]^2 = 1.0 \; \alpha^8.$$

The integral varies only as the logarithm of α and for the standard value in equation (51) is

$$I = 3.53 \quad . \tag{59}$$

With the standard value for α and $h = 0.75$ we get $\lambda_\nu = 24 \; \Omega_0^{-1}$ Mpc. Even for $\Omega_0 = 1$ this exceeds the present clustering length for galaxies, $r_0 \sim 7$ Mpc (eq. [4]). As we expect gravity caused the clustering length to have grown since decoupling this is unacceptable. An interesting point made by Davis <u>et al</u>. is that for fixed Ω_0 the smoothing length is very sensitive to the neutrino temperature αT_0: if decays of elementary particles after decoupling of neutrinos and radiation had caused α to be half the standard value it would reduce λ_ν by a factor of about 16, removing the discrepancy with the clustering length. If we want to use these neutrinos to account for the dark mass in spiral galaxies we need pockets of low entropy neutrinos with comoving size $\sim 100 \; h^{-1}$ kpc, which we see requires only a further modest decrease in α. It would be equivalent to assume that the mass of the Universe now is dominated by any sort of massive weakly interacting particle whose primeval peculiar velocity now would have amounted to ~ 0.1 km s^{-1} (in the absence of gravitational perturbations) so as to make $\lambda_\nu < 100$ kpc.

There is however a serious problem that was pointed out by Simon White. Two important epochs in the scenario are the redshift z_{eq} of which the mass densities of radiation and of non-relativistic matter are equal (e.g. [14]) and the redshift z_r at which the peculiar motions of the weakly interacting particles become nonrelativistic and the comoving smoothing length λ_ν stops growing. If α has the conventional value, z_r is only slightly larger than z_{eq}, and reducing α reduces λ_ν and increases z_r/z_{eq}. If $z_r \gg z_{eq}$ it tilts the slope of the power spectrum for those Fourier components that enter the horizon at epochs between z_r and z_{eq}. The Fourier component with comoving wave number k enters the horizon when the proper wavelength is $\sim a(t)/k \sim t$. Prior to this epoch the amplitude of the component has been growing as $\delta \propto t \propto a(t)^2 \propto k^{-2}$. At t, pressure pushes matter and radiation out of the neutrino wave. From t to the epoch z_{eq} the Universe is radiation dominated so the amplitude δ_k of the neutrino wave is $\sim$ constant. Thus if the initial power spectrum of the density fluctuation is $\mathcal{P} \propto k^n$, it becomes $\mathcal{P} \propto k^{n-4}$ at $z < z_{eq}$ and $\lambda_\nu < \lambda <$ horizon at z_{eq}. If $n \sim 0$, as will be argued below is wanted to account for the anisotropy of the microwave background, then the spectrum $\propto k^{-4}$ at $\lambda < \lambda_{eq}$ makes negligible contribution to the variance of the density so the first generation of bound systems forms at λ_{eq}. This is just the size of the length λ_ν obtained when α has the conventional value (eq. [58]) and as before is uncomfortably large compared to the coherence length of the galaxy distribution. If $n \sim 4$ we avoid this problem but encounter two others. At large scales the power spectrum is too small to account for the large-scale fluctuations in the microwave background. At small scales the density fluctuations are diverging so rapidly as $k \to 0$ that they cause diverging space curvature fluctuations: we must postulate a break in the spectrum to $\mathcal{P} \sim$ constant at comoving wavelengths $\lambda < 1$ Mpc.

4.2 Large-scale fluctuations in the microwave background

In the primeval adiabatic scenario it is natural to expect that there are fluctuations in the mass distribution integrated over large scales (~ 1000 Mpc) as an extension of the fluctuations postulated on smaller scales to produce galaxies and clusters of galaxies. Large-scale fluctuations in the mass distribution cause appreciable gravitational potential differences which would be reflected in fluctuations in the redshift and hence the temperature of the microwave background.

Sachs and Wolfe (1967) first computed this effect. The analysis here follows Peebles (1981a). For further discussion see Grishchuk and Zel'dovich (1978) and Silk and Wilson (1981).

The Sachs-Wolfe relation can be written as (Peebles 1980a, section 93)

$$\delta T/T = \frac{G\rho_i a_i^2}{3} \int \frac{d^3x}{x} \left[\delta_i(\underline{x}_2-\underline{x}) - \delta_i(\underline{x}_1-\underline{x})\right]. \tag{60}$$

The mass density contrast is $\delta = \delta\rho/\rho$. The subscript i means that the density contrast, the mean mass density and the expansion parameter a are all evaluated at some starting redshift z_i chosen to satisfy the conditions listed below. The radiation temperature measured in two directions at separation θ differs by the fractional amount $\delta T/T$ (eq. [25]). The radiation detected along these two lines of sight was at coordinate positions $\underline{x}_1$ and $\underline{x}_2$ at epoch z_i, and the separation of these points is

$$x_{21} \sim 2\theta/(\Omega_o h_o a_o) \quad . \tag{61}$$

The Sachs-Wolfe relation assumes the following:

a) The optical depth for scattering since z_i is negligibly small.

b) Non-gravitational forces on the scale x_{21} may be neglected.

c) Linear perturbation theory is a good approximation. This is reasonable because we will be considering perturbations $\delta T/T \sim 10^{-4}$.

d) Initial temperature fluctuations are negligible. The initial hypersurface $t = t_i$ is placed so only the growing mode of the mass density perturbation is present, and it is assumed that $(\delta T/T)_i$ on this hypersurface may be neglected, which means

$$(\delta T/T)_i \ll \delta T/T \sim G\rho_i\delta_i x_{21}^2 a_i^2 \sim \delta_i(a_i x_{21}/t_i)^2 \quad . \tag{62}$$

In the primeval adiabatic picture we expect $(\delta T/T)_i \sim \delta_i$, so this assumption is reasonable if $a_i x_{21}$ > the horizon at decoupling, or $\theta > \theta_x$ (eq. [19]). The assumption is not valid in the primeval isothermal scenario because the initial conditions are set up to eliminate large-scale gradients in the gravitational potential.

e) $1 + z_i \gg |\Omega_o^{-1} - 1|$ so the expansion rate at z_i approximates the Einstein-de Sitter model.

f) The separation of the two lines of sight is smaller than the radius of curvature of space, so

$$\theta << \Omega_o/(2(1-\Omega_o)^{1/2}). \tag{63}$$

Eardley (1981) has shown how to compute $\delta T/T$ when the separation is larger than this.

We can rewrite the Sachs-Wolfe relation in terms of the present mass distribution by using

$$\delta_i = \delta_o D_i/D_o \quad , \tag{64}$$

where the subscript o means the present value and D is the growing solution to the equation

$$\frac{d^2D}{dt^2} + 2\frac{\dot{a}}{a}\frac{dD}{dt} = 4\pi G\rho D. \tag{65}$$

which in an open model gives (Groth and Peebles 1975)

$$D_o = 1 + 3/x + \frac{3(1+x)^{1/2}}{x^{3/2}}\ln\left[(1+x)^{1/2} - x^{1/2}\right],$$

$$x = \Omega_o^{-1} - 1, \quad D_i = 2/5\ (\Omega_o^{-1}-1)/(1+z_i) \ . \tag{66}$$

Then equation (60) is

$$\delta T/T = \frac{2G\rho_o(\Omega_o^{-1}-1)}{15D_o}\int\frac{d^3r}{r}\left[\delta(\underline{r}_2-\underline{r}) - \delta(\underline{r}_1-\underline{r})\right]. \tag{67}$$

where $\rho_o(1+\delta)$ is the present mass distribution as a function of proper position $\underline{r}$.

The mean square value of $\delta T/T$ is determined by the mass autocorrelation function

$$\xi(r) = \langle\delta(\underline{r}_1 - \underline{r})\ \delta(\underline{r}_1)\rangle \quad , \tag{68}$$

which gives

$$\langle(\delta T/T)^2\rangle = 2I\left[\frac{(1-\Omega_o)H_o^2}{20\ \pi D_o}\right]^2, \tag{69}$$

$$I = \int d^3r/r\ d^3t\ \xi(t)\ \left[1/(|\underline{r}-\underline{t}|) - 1/(|\underline{r}_{21}+\underline{r}-\underline{t}|)\right].$$

We can write the autocorrelation function in terms of the power spectrum of the mass distribution (eq. [21]) and then use the identity

$$\int d^3t\ \frac{e^{i\underline{k}\cdot\underline{t}}}{|\underline{r}-\underline{t}|} = 4\frac{e^{i\underline{k}\cdot\underline{r}}}{k^2} \tag{70}$$

to get

$$I = (4\pi)^3 \int_0^\infty \frac{dk\,\mathcal{P}(k)}{k^2} \left[1-\frac{\sin kr_{12}}{kr_{12}}.\right] \tag{71}$$

This can be compared to equation (26): the factor k^2 in the denominator shows the added weight given to the long wavelength part of the spectrum by the gravitational potential.

Let us consider first the assumption that the mass autocorrelation function is negligibly small at $r > 40\ h^{-1}$ Mpc so that the power spectrum levels off at the constant value $\mathcal{P}_0$ at $k \rightarrow 0$. Then equations (6) and (21) give

$$K_3 = \int \xi d^3r = 8\pi^3 \mathcal{P}_0 , \tag{72}$$

and with the identity

$$\int_0^\infty (dk/k^2)\,[1-(\sin kr)/kr] = \pi r/4 . \tag{73}$$

we find from equations (69) and (71)

$$\delta T/T = A\theta^{1/2},\ A^2 = \frac{(1-\Omega_0)^2}{50\pi\Omega_0 D_0{}^2}(H_0/c)^3\ K_3 . \tag{74}$$

If the galaxy distribution is a fair measure of the large-scale mass distribution then we have from equation (9) for K_3

$$\begin{aligned} \delta T/T &= 1.5 \times 10^{-5}\ \theta^{1/2} , & \Omega_0 &= 1.0 ; \\ &= 2.5 \times 10^{-5}\ \theta^{1/2} , & \Omega_0 &= 0.1 ; \\ &= 4.0 \times 10^{-5}\ \theta^{1/2} , & \Omega_0 &= 0.03 , \end{aligned} \tag{75}$$

where θ is in units of degrees. It is interesting and perhaps strongly encouraging that this result agrees with the measurement of $\delta T/T$ at $\theta = 6°$ by Fabbri et al. (1980) (eq. [31]) and also with the measurement of the quadrupole moment a_2 (eq. [32]) by Boughn et al. (1981). For the estimation of a_2 see Peebles (1981a).

In the primeval adiabatic scenarios with massive weakly interacting particles there may be appreciable mass fluctuations at decoupling on scales $<$ the matter-radiation Jeans length λ_x (eq. [16]) but we know Thomson scattering during recombination strongly suppresses fluctuations in the radiation distribution on scales $< \lambda_x$. Thus we would look for a cutoff in the $\theta^{1/2}$ law at $\theta \sim \theta_x$ (eq. [19]). For a more detailed discussion of $\delta T/T$ at small θ see Wilson and Silk (1981). The possible effect on the shape of $\delta T/T(0)$ by scattering by intergalactic plasma is discussed by Peebles (1981a).

Now let us consider what happens if the spectrum varies as some power of k at long wavelength,

$$\mathcal{P}(k) \simeq \mathcal{P}_o \quad , \quad k > k_c = 4.5/r_c \quad :$$

$$\mathcal{P}(k) \simeq \mathcal{P}_o(k/k_c)^n \quad , \quad k < k_c \tag{76}$$

$$r_c = 30\ h^{-1}\ \mathrm{Mpc} \quad , \quad \mathcal{P}_o = K_3/(8\pi^3) \quad .$$

The short wavelength part is chosen so the integral of ξ to r_c agrees with K_3 for galaxies (eq. [6]), which is measured to about the radius r_c. This radius subtends the angle

$$\theta_c \sim 20\ \Omega_o \text{ arc minutes} \quad . \tag{77}$$

If $-1 < n < 1$ then the integral I (eq. [71]) converges and we find

$$I = (8K_3/k_c)\ (2\theta k_c/H_o\Omega_o)^{1-n}\ L_n,$$
$$L_n = \int_0^\infty dy\ y^{n-2}\ (1-\sin y/y), \tag{78}$$

so that

$$\delta T/T \sim 1.5 \times 10^{-5}\ \theta^{1/2}\ (0.064\ \Omega_o/\theta)^{n/2}\ \Omega_o^{-0.2}\ (L_n/L_o)^{1/2} \tag{79}$$

where θ is in degrees.

We note that $\delta T/T \propto \theta^{(1-n)/2}$ (Peebles 1981a). If $n < 0$ the power spectrum increases with increasing wavelength, meaning that the large-scale mass fluctuations are larger than would be expected if the clusters observed at $r < r_c$ were randomly placed, and this increases $\delta T/T$. For example, if $n = -1/2$ it increases $\delta T/T$ at $\theta \sim 60°$ by a factor of about 6, which probably is excluded by the quadrupole moment measurement of Boughn et al. (1981). If $n > 0$ the power spectrum goes to zero at large wavelength: the large-scale mass fluctuations are smaller than random because the mass distribution is anticorrelated on large scales where $\xi(r)$ is negative and varies as $r^{-(3+n)}$. It would be difficult to detect this effect at $r \gg 30\ h^{-1}$ Mpc in the galaxy distribution. If it is present it reduces the large-scale gravitational potential fluctuations and hence $\delta T/T$. This autocorrelation was built into the primeval isothermal model discussed in the last section, so there we expect to see only the temperature fluctuations present at decoupling.

5. SUMMARY: CONSTRAINTS FROM THE MASS DISTRIBUTION AND THE MICROWAVE BACKGROUND

As noted in section 1 we must consider several possible sources of the fluctuations in the microwave background: in the Galaxy, along the line of sight, in the primeval entropy distribution, or in the primeval gravitational potential fluctuations. We may hope for a fairly convincing test of the first possibility because fluctuations produced in the Galaxy would not be expected to be independent of wavelength. It is easy to see that the second idea is unreasonable for a_2 as large as that found by Boughn et al. We note that in the present Universe appreciable structure is observed out to scales of perhaps 40 h^{-1} Mpc. That comoving length subtends the angle $\theta \sim 0.5°$ at redshift $z \sim 1$. Therefore we must expect that any perturbations to the microwave background temperature caused by galaxies or by whatever process made galaxies and clusters of galaxies must be uncorrelated to a good approximation at $\theta > \theta_o$. The auto-correlation function $w(\theta)$ of the background temperature thus would be expected to satisfy the equations

$$2w(0) \cong (\delta T/T)^2 \quad : \quad w(\theta) = 0 \quad , \quad \theta > \theta_o \sim 0.5° \quad , \tag{80}$$

where $\delta T/T$ is measured by Fabbri et al. (1981). The multipole moments (eq. [32]) of the temperature distribution would be (Peebles 1980a, eq. 46.16)

$$\begin{aligned} a_\ell{}^2 &= \langle |a_\ell{}^m|^2 \rangle = \int d\Omega\, w(\theta)\, P_\ell\,(\cos\theta) \\ &\cong \int d\Omega\, w(\theta) \lesssim \pi\theta_o{}^2\, w(0) \quad . \end{aligned} \tag{81}$$

The last two equations apply at $\pi/\ell \gtrsim \theta_o$. These equations with equation (31) give $a_2 \lesssim 3 \times 10\ h^{-7}$, three orders of magnitude below what Boughn et al. found (eq. [37]).

The Ostriker-Cowie (1981) picture, where galaxies are formed by explosions in the intergalactic medium, would not produce the observed a_2 because it postulates negligible inhomogeneities on the horizon (or at least, inhomogeneities much smaller than what is found by extrapolation of the present clustering back in time according to the gravitational instability picture). In the primeval isothermal perturbation picture the large-scale temperature fluctuations measured now have the amplitude that was present at the Big Bang. When the spectrum of fluctuations is adjusted to make a reasonable distribution of galaxies it makes temperature fluctuations $\delta T/T$ in accord with Fabbri et al. but then a_2 is well below the observation. This can be remedied by assuming that the primeval spectrum of fluctuations is roughly flat at $\lambda \lesssim \lambda_x$ (eq. [16]) and varies as $\mathcal{P} \propto k^{-3}$ at $\lambda \gtrsim \lambda_x$. It means the primeval fluctuations in the baryon number distribution are dominated by the largest allowed scale, which perhaps is artificial. In the primeval adiabatic picture we reproduce the observed temperature fluctuations at $\theta = 6^{\circ}$ and at $\theta \sim 1$ radian by assuming a flat primeval spectrum of density irregularities. It is an interesting coincidence that the wanted amplitude of the power spectrum, which measures the 'shot noise' of the mass distribution, agrees with what is found in the galaxy distribution at $r \lesssim 30\ h^{-1}$ Mpc. (Of course, this assumes that mass is clustered in the same fashion as the galaxies.) However, in standard scenarios with or without massive neutrinos the coherence length of the mass distribution is too long. As in the isothermal case we must postulate a kink in the primeval power spectrum to $\sim k^4$ at 1 Mpc $\lesssim \lambda \lesssim$ 30 Mpc.

A flat primeval spectrum is produced if at some starting epoch z_i each part of the Universe is separately prepared from a heat reservoir at the fixed temperature T_i (Peebles 1968). (It will be noted that this is not equivalent to the expectation that each part of the Universe relaxes to local statistical equilibrium for that can only rearrange mass on scales $\lesssim ct_i$ and so cannot produce a flat power spectrum at wavelengths $\gg ct_i$.) If the initial thermal equilibrium state is produced by independent applications of the heat reservoir in small parts so that the gravitational energy of each part can be neglected then the initial density fluctuations are given by the usual equation from thermodynamics,

$$\delta M/M = (kT^2 c_V/u^2 V)^{1/2} \tag{82}$$

where $\delta M/M$ is the rms mass density fluctuation in the proper volume V, u is the mean energy per unit volume at temperature T, and c_V is the heat capacity per unit volume. If the energy is in non-interacting quanta with $\varepsilon \simeq pc$ then $u = gaT^4$ where a is Stefan's constant and g counts particle types present, each spin state for each boson contributing 1/2 and each spin state for each fermion contributing 7/16. This gives

$$(\delta M/M)_{th} = (4k/g\ aT^3 V)^{1/2} \simeq (n_\gamma V)^{-1/2} , \tag{83}$$

where n_γ is the number density of particles. At the present temperature this thermal value for $\delta M/M$ is smaller than the value observed for the galaxy distribution (eq. [6]) by the factor

$$(\delta M/M)_{th}/(\delta M/M)_o = (4k/aT_o{}^3 K_3)^{1/2} \sim 10^{-40} \tag{84}$$

I presume that it is only a coincidence that this is the magic number of cosmology. Since $\delta M/M$ grows as $t \propto z^{-2}$ prior to z_{eq} and as $t^{2/3} \propto z^{-1}$ thereafter we get the present value $(\delta M/M)_o$

from initial thermal fluctuations if we start with $g \sim 1$ (the energy shared among a fairly small number of species) and at redshift

$$z_i \sim 10^{22} \quad , \quad kT_i \sim kT_o z_i \sim 10^{18} \text{ eV} \quad ,$$
$$t_i \sim 10^{-24} \text{ secs} \quad , \quad n_\gamma(i)\,(ct_i)^3 \sim 10^{27} \quad . \tag{85}$$

The last number is the mean number of particles within the horizon at z_i. This redshift is 10 orders of magnitude less than the redshift at the Planck time.

As the quantities in equation (85) are not particularly suggestive we might take note of another number. With a flat spectrum of mass density irregularities, space curvature fluctuations diverge as $\lambda^{1/2}$ and so the spectrum must be truncated at some maximum wavelength λ_m. If $\lambda_m >> cH_o^{-1}$ we would see a nearly homogeneous Universe with density parameter Ω_o perturbed from the global value (Grishchuk and Zel'dovich 1978). As the observed value of Ω_o is within an order of magnitude or so of unity the space curvature fluctuations on the scale λ_m cannot be much greater than 100%. As the fluctuation in curvature on our horizon is $\sim \delta T/T \sim 1$ part in 10^4 in the primeval adiabatic picture,

$$\lambda_m \lesssim (\delta T/T)^{-2}\, cH_o^{-1} \sim 3 \times 10^{11} \text{ Mpc} \quad . \tag{86}$$

The number of photons inside the length λ_m is

$$N_m \sim 10^{111} \quad , \tag{87}$$

roughly independent of time (if $g \sim 1$).

Though recent developments in our understanding of the large-scale distributions of matter and radiation have greatly improved the constraints on scenarios for evolution, to my mind they have not yet forced on us a natural choice of scenario: that awaits further stimulus from the measurements.

REFERENCES

Barrow, J.D. and Turner, M.S. 1981. Nature, 292, 35.
Bisnovaty-Kogan, G.S. and Novikov, I.D. 1980. Astron.Zh. 57, 899.
Bond, J.R., Efstathiou, G. and Silk, J. 1980. Phys.Rev.Letters, 45, 1980.
Bonometto, S.A. and Lucchin, F. 1979. MNRAS, 187, 611.

Boughn, S.P., Cheng, E.S. and Wilkinson, D.T. 1981. Ap.J., 243, L113.
Chincarini, G. and Rood, H.J. 1979. Ap.J., 230, 648.
Clutton-Brock, M. and Peebles, P.J.E. 1981. Astr.J., 86, 1115.
Cowsik, R. and McClelland, J. 1973. Ap.J., 180, 7.
Davis, M. and Geller, M.J. 1976. Ap.J., 208, 13.
Davis, M., Lecar, M., Pryor, C. and Witten, E. 1981. Preprint.
de Vaucouleurs, G. 1970. Science, 167, 1203.
Doroshkevich, A.G., Zel'dovich, Ya.B. and Sunyaev, R.A. 1978. Soviet Astron., 22, 523.
Eardley, D.M. 1981. (unpublished).
Fabbri, R., Guidi, I., Melchiorri, F. and Natale, V. 1980. Preprint.
Fall, S.M. 1979. Rev.Mod.Phys., 51, 21.
Fall, S.M. and Tremaine, S. 1977. Ap.J., 216, 682.
Fry, J.N. and Peebles, P.J.E. 1978. Ap.J., 221, 19.
Gott, J.R. 1975. Ap.J., 201, 296.
Gott, J.R. and Turner, E.L. 1979. Ap.J., 232, L79.
Grishchuk, L.P. and Zel'dovich, Ya.B. 1978. Astron.Zh., 55, 209: Soviet Astron., 22, 125.
Groth, E.J. and Peebles, P.J.E. 1975. Astr.Astrophys., 41, 143.
Groth, E.J. and Peebles, P.J.E. 1977. Ap.J., 217, 385.
Groth, E.J., Peebles, P.J.E., Seldner, M. and Soneira, R. 1977. Scientific American, 237, Nov., p. 76.
Harrison, E.R. 1967. Phys.Rev.Letters, 18, 1011.
Kirshner, R.P., Oemler, A. and Schechter, P.L. 1979. Astr.J., 84, 951.
Kirshner, R.P., Oemler, A., Schechter, P.L. and Schectman, S.A. 1981. Preprint.
Klinkhamer, F.R. and Norman, C.A. 1981. Ap.J., 243, L1.
Mandelbrot, B.B. 1977. Fractals: Form, Chance and Dimension (San Francisco: W.H. Freeman).
Ostriker, J.P. and Cowie, L.L. 1981. Ap.J., 243, L127.
Peebles, P.J.E. 1967. Ap.J., 147, 859.
Peebles, P.J.E. 1968. Nature, 220, 237.
Peebles, P.J.E. 1972. Comments on Astrophysics and Space Phys., 4, 53.
Peebles, P.J.E. 1973. Ap.J., 185, 413.
Peebles, P.J.E. 1974. Ap.J., 189, L51.
Peebles, P.J.E. 1980a. The Large-Scale Structure of the Universe (Princeton, N.J.: The Princeton University Press).
Peebles, P.J.E. 1980b. Physical Cosmology, (Les Houches XXXII), eds. R, Balian, J. Audouze and D.N. Schramm (Amsterdam: North Holland) p. 215.
Peebles, P.J.E. 1981a. Ap.J., 243, L119.
Peebles, P.J.E. 1981b. Ap.J., 248,
Peebles, P.J.E. and Groth, E.J. 1976. Astron.Astrophys., 53, 131.
Peebles, P.J.E. and Yu, J.T. 1970. Ap.J., 162, 815.
Press, W.H. and Vishniac, E.T. 1980. Ap.J., 236, 323.
Rubin, V.C., Ford, W.K. and Thonnard, N. 1980. Ap.J., 238, 471.

Sachs, R.K. and Wolfe, A.M. 1967. Ap.J., 147, 73.
Schramm, D.N. and Steigman, G. 1981. Ap.J., 243, 1.
Senjanović, G. and Stecker, F.W. 1980. Physics Letters, 96B, 285.
Silk, J. and Wilson, M.L. 1980. Physica Scripta, 21, 708.
Silk, J. and Wilson, M.L. 1981. Ap.J., 244, L37.
Soneira, R.M. 1978a. (unpublished).
Soneira, R.M. 1978b. Doctoral Dissertation, Princeton University.
Soneira, R.M. and Peebles, P.J.E. 1978. Astr.J., 83, 845.
Szalay, A.S. and Marx, G. 1976. Astron.Astrophys. 49, 437.
Tremaine, S. and Gunn, J.E. 1979. Phys.Rev.Letters, 42, 407.
Wasserman, I.M. 1980. Cornell University Preprint CRSR 760.
Wilson, M.L. and Silk, J. 1981. Ap.J., 243, 14.
Zel'dovich, Ya.B. 1967. Usp.Phys.Nauk., 89, 647: Soviet Phys.Usp., 9, 602.

EVOLUTION OF POTENTIAL PERTURBATIONS AFTER DECOUPLING (THE ADIABATIC SCENARIO)

S.F. Shandarin

Institute of Applied Mathematics, Moscow

Two hypotheses about type of initial perturbations (adiabatic and isothermal) give rise to different scenarios after decoupling at $z < 10^3$ (Jones 1980). This is caused by the difference of the spectra of inhomogeneities just after decoupling. The spectrum of primeval perturbations is assumed to possess no preferred scales.

The adiabatic spectrum after recombination does not contain short wave perturbations below a scale of about 10^{13} - $10^{15} M_\odot$ due to very strong damping resulting from interaction of matter with radiation before decoupling (see Bonometto, this volume). For this reason the mass of the first formed objects is about 10^{13} - 10^{15} $M_\odot$ in the adiabatic scenario. In 1970 Zeldovich showed that these objects must be very oblate ones - 'pancakes'. In this scheme the formation of galaxies was shown by Doroshkevich et al. (1978) to go on inside the pancakes.

Formation of the large scale structure in the neutrino dominated Universe is very similar to that in the adiabatic scenario in baryon dominated Universe. At the time when neutrinos become non-relativistic a preferred scale occurs in the spectrum of perturbations. Perturbations on the scales less than the horizon in that era are almost completely wiped out (Doroshkevich et al. 1980).

1. EVOLUTION OF PERTURBATIONS OF A GENERAL TYPE

The essential features of the evolution of inhomogeneities is determined by their random character. In other words the lack of any organisation in the distribution of density as well as in the perturbed velocity field provides rather strong restrictions on

B. J. T. Jones and J. E. Jones (eds.), The Origin and Evolution of Galaxies, 171–178.

the shape of structures arising at the non-linear stage. Thus considering the evolution of perturbations one needs to study the structures of general types.

Just to illustrate the main idea I shall discuss the simplest one dimensional case. But before doing this I would like to note that, as the smallest scale of perturbations is about 10^{13} -10^{15} $M_\odot$ which is very much larger than Jeans' mass after decoupling (10^6 $M_\odot$), it is reasonable to consider the model of cold (T = 0) medium. The influence of finite but small temperature I shall discuss later. In addition I shall assume that the medium is a collisionless one. Thus this model is applicable directly to a neutrino dominated Universe, while many of its features are similar to ones in the ordinary adiabatic scenario in the baryon dominated Universe.

A one-dimensional cold system is entirely described by the phase curve $v = v(x)$ plotted in the two-dimensional phase space. To find the density distribution one should project the phase curve onto the spatial axis x. The density in each point is proportional to a derivative $\partial v/\partial x$. At the points where the curve $v = v(x)$ has a vertical tangent the density distribution has a singularity. It is quite clear that this singularity is caused by the singular phase distribution of the cold sytem. If the temperature is not exactly zero but small, one gets regions of very high but finite density instead of singularities.

When perturbations are small $v = v(x)$ is a single valued function and $\partial v/\partial x$ is small. The evolution in the course of the time appears as changing of the phase curve. The phase curve can become rather complicated in shape (many-valued for instance) but self-gravitating systems under consideration are Lagrangian ones (for definition of Lagrangian systems as well as discussion of their properties see for example the book of Arnold 1980). Using only these general properties of evolution of Lagrangian systems one can find out the general types of density singularities arising in such a system. One need only to project a snapshot of the phase curve of a general type onto the spatial axis. As the phase curve is smooth it is possible to decompose it in Taylor's series in the vicinity of an arbitrary point. However, it is more convenient to deal with the function $x = x(v)$ (inverse to $v = v(x)$) since $v = v(x)$ can be many-valued near singularities in density:

$$x - x_0 \propto \sum a_n (v-v_0)^n, \qquad n = 1,2,3,..... \tag{1}$$

The density is determined by the lowest non zero term in the sum (1)

$$\rho(x - x_0) \propto \partial v/\partial x \propto (v-v_0)^{1-n} \propto (x-x_0)^{(1-n)/n} \tag{2}$$

If at the point in question $a_1 \neq 0$ there is not any singularity at all. If $a_1 = 0$ but $a_2 \neq 0$ there is a singularity in the density distribution and $\rho \propto (x - x_0)^{-1/2}$. If $a_1 = 0$ and $a_2 = 0$ also but $a_3 \neq 0$ one gets a steeper law for growth of density near singularity $\rho \propto (x - x_0)^{-2/3}$. The first type of singularity is realized at a general moment of time during the non-linear stage. However the second one is realized only at specific moments of time. The others do not arise at all in general case because they need too many conditions and thus need a particular organization of the initial conditions.

Now it is easy to understand what one ought to do to treat the 2 (or 3) dimensional case. Let us imagine a 2 (or 3) dimensional phase surface (hypersurface) of a general type enclosed in a 4 (or 6) dimensional phase space. Then we should project the surface onto a 2 (or 3) dimensional spatial plane (or space). To consider the evolution in time we simply add one more dimension and treat the problem in 5 (or 7) dimensional space. In both cases one can obtain both the laws of density growth and the patterns of the generic structures arising at the non-linear stage. A mathematical analysis of the problem in the spirit of the catastrophe theory was done by Arnold (1972, 1976, 1980). Some of the applications to cosmology were given in the paper of Arnold, Shandarin and Zeldovich (1981). Some patterns of 2-dimensional generic singularities are drawn in Figure 1.

This approach is a very general one and it can describe the evolution of a cold medium without any forces at all, the motion of a medium under the action of external forces, or in our case a medium under the influence of its self-gravitational field. The only demand is that the motion must be of a potential type. In each particular case the equations governing the evolution of phase curve (or surface or hypersurface) are different but in all of them the set of generic types of singularities is the same.

In the case of most interest to us, the evolution of perturbation in expanding Universe, the approximate non-linear solution proposed by Zeldovich (1970) was shown by Doroshkevich et al. (1973) to give a constructive method to investigate the generic singularities. This solution provides an exact description of the geometry of objects that appear first - the so called pancakes. The pancakes are one of a few generic singularities arising at the non-linear stage. It is tempting to identify some of the pancakes with the huge superclusters like the Local Supercluster.

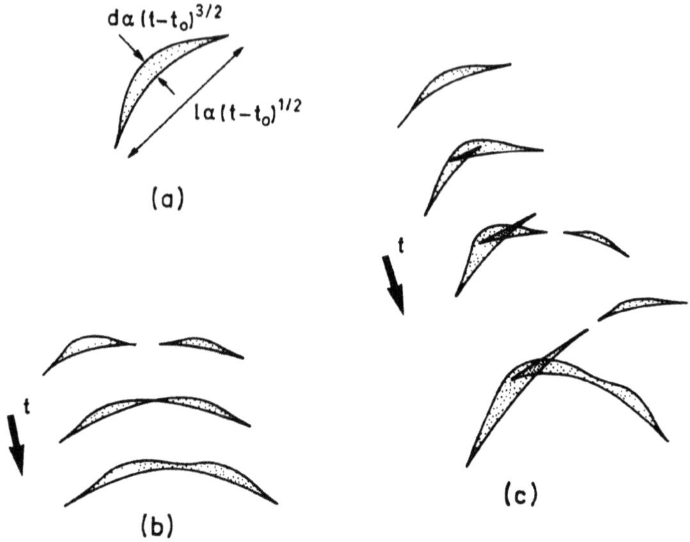

Figure 1. Examples of the general patterns arising in two-dimensional models. The dark regions are regions of three streams flows. a) 'pancake', b) and c) two types of merging of pancakes. Solid lines are caustics - lines of infinite density.

The approximate solution predicts correct asymptotic laws for density growth near singularities. In addition it agrees well with results of numerical simulations (Doroshkevich et al. 1973, 1980).

2. 3D NUMERICAL SIMULATIONS IN THE FRAME OF THE ADIABATIC SCENARIO

In this section we discuss some results of numerical simulations in the frame of adiabatic scenario.

Our model consists of 32^3 particles moving in their self-gravitating field in the expanding Universe model. The method of calculation is a multiple repetition of the following three principal steps:

i) evaluating the density at the mesh points, in our case the mesh is 32x32x32 and each particle was assigned a size equal to the size of a mesh cell,

ii) using the technique of the Fast Fourier Transforms and numerical differentiation, to obtain the acceleration of each particle.

iii) moving particles to new positions and calculating new values for the velocities.

The use of periodic boundary conditions simulated an infinite system.

The evident merit of the method is its high speed and the lack is strong smoothing of gravitational potential on the scale of the mesh (details are in the paper of Klypin and Shandarin 1981).

The quasihomogeneous distribution of density when particles were at the mesh points was perturbed in such a manner that only the growing mode was present. In addition the spectrum of initial perturbations did not contain perturbations with wavelength shorter than $8r_0$ where r_0 was the size of the mesh cell. Amplitudes and phases of waves were random and uncorrelated and the spectrum of density perturbations was flat.

To compare the results of simulations with observation the following parameters are used: the size of the mesh cell $r_0 = 5\ h^{-1}$ (where h is the Hubble parameter in units of 100 km/sec/Mpc), the particles have the same size, the cutoff in the spectrum of the initial perturbations was at $\lambda_{min} = 40\ h^{-1}$ Mpc ($M_{min} = 4/3\ \lambda^3{}_{min}\ \rho \sim 2.10^{16}\ h^{-1}\ M_\odot$), the longest wave was equal to the total size of the region $\lambda_{max} = 160\ h^{-1}$ Mpc ($M_{max} = 4/3\ \lambda^3{}_{max}\ \rho \sim 1.2\ 10^{18}\ h^{-1}\ M_\odot$), the mean density parameter $\Omega = 1$. The numerical simulation commenced at $z \sim 5.2$, when density perturbations at the scale r_0 was $(\delta\rho/\rho)_{r_0} \sim 0.2$. This choice of λ_{min} and the amplitude of the perturbations provides the best fit of the model with observations at $z = 0$.

The main parameter of comparison of the model with observations is the spatial two-point correlation function (Peebles 1980). In Figure 2 the model correlation function (points) is plotted together with one inferred from observation (straight line). The deviation observed at small scales is believed to be due to the poor accuracy of our simulations at scales less than about $r_0 \sim 5\ h^{-1}$ Mpc.

Our simulations show that during the non linear stage the distribution of particles exhibits rather complicated structures. To illustrate this a surface of constant density ($\rho \sim 2.5\ \rho$) is shown in Figure 3.

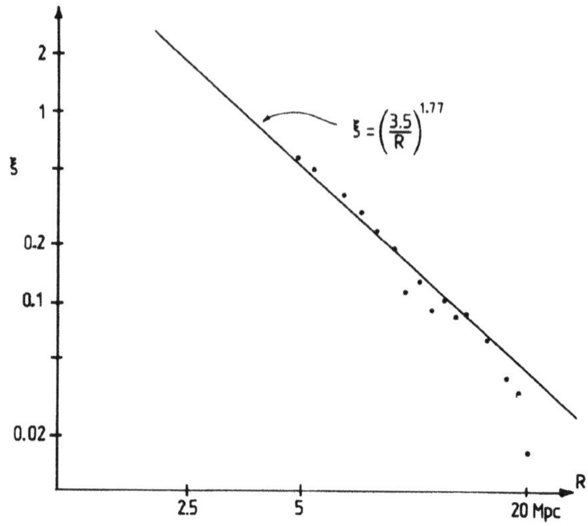

Figure 2. Spatial two point correlation function. Points indicate the model correlation function.

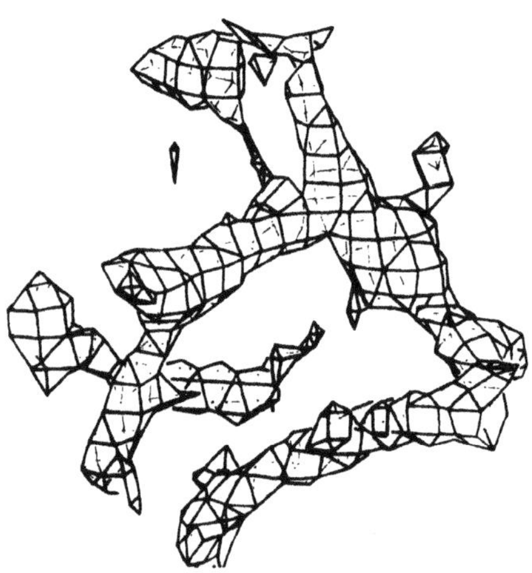

Figure 3. An example of the structures arising in 3D numerical simulations of the adiabatic scenario. The surface is a surface of constant density $\rho \sim 2.5\ \rho$.

In accordance with general theory of singularities (Arnold et al. 1981) the most pronounced patterns (i.e. having the steepest growth of density) possess a rather simple compact shape. They are naturally identified with rich clusters. The prolate condensations connecting rich clusters are less pronounced (Figure 3) and the least pronounced ones are oblate structures.

3. DISCUSSION

The model of the formation of the large scale structure presented here describes directly the evolution of the collisionless component of the density. Thus it is directly applicable to the process of large scale structure formation in a neutrino dominated Universe and it is reasonable to suppose that the formation of the large scale structure in a baryon-dominated Universe can equally be described by such a model. Of course this is not true for galaxy scales.

This short discussion shows that in the adiabatic scenario it is possible to obtain a clustering correlation function in reasonable agreement with the observed one (Figure 2) as well as very prolate structures (Figure 3) which are similar to some observed superclusters.

The mean density parameter $\Omega = 1$ was taken as an example only. This value is somewhat preferable in the neutrino dominated Universe. Note that in the dense models ($\Omega \sim 1$) the lifetime of the large scale structure is not too long. After a while the bridges between rich clusters are destroyed and the distribution of particles becomes quite similar to that commonly arises in simulations of the isothermal scenario. The present model satisfies to the rather conservative demand that the first galaxies formed at $z > 2$ (see for instance Peebles 1981), however the question needs more careful study.

REFERENCES

Arnold, V.I. 1972. Functional Anal.Appl., 6, 254.
Arnold, V.I. 1976. Commun.Pure Appl.Math., 29, 557.
Arnold, V.I. 1980. Mathematical Methods of Classical Mechanics, Springer Verlag, New York, Heidelberg, Berlin.
Arnold, V.I. 1981. Trudy Seminara im I.G. Petrovski, 8.
Arnold, V.I., Shandarin, S.F. and Zeldovich, Ya,B. 1981. Geophys.Astrophys.Fluid Dynamics (to be published).
Doroshkevich, A.G., Khlopov, M.Yu., Sunyaev, R.A., Szalay, A.S. and Zeldovich, Ya.B. 1980. Proc. Xth Texas Symp. on Relativistic Astrophys., Baltimore.

Doroshkevich, A.G., Ryabenki, V.S. and Shandarin, S.F. 1973. Astrophysics 9, 144.

Doroshkevich, A.G., Shandarin, S.F. and Saar, E. 1978. MNRAS, 184, 643.

Doroshkevich, A.G., Kotok, E.V., Novikov, I.D., Polyudov, A.N. Shandarin, S.F. and Sigov, Yu.S. 1980. MNRAS 192, 321.

Jones, B.J.T. 1980. Phil.Trans.R.Soc.Lond., A296, 289.

Klypin, A.A. and Shandarin, S.F. 1981. Preprint of Inst.Appl. Math.Acad. of Sciences USSR.

Peebles, P.J.E. 1980. The Large Scale Structure of the Universe, Princeton University Press, Princeton.

Peebles, P.J.E. 1981. Astrophys.J. (to be published).

Zeldovich, Ya.B. 1970. Astr.Astrophys., 5, 84.

PRIMORDIAL STARS - THE PRECURSORS TO GALAXY FORMATION

Janet E. Jones

Institut d'Astrophysique, 98 bis, Bd Arago, 75014 Paris

1. INTRODUCTION

We have already heard that a knowledge of how galaxies form gives us information about the early universe. It is also true that a knowledge of how and when the first stars formed gives us information about how galaxies formed, and hence about the early universe. The theory of galaxy formation has undergone a big advance in the last decade. Previously, the work was largely theoretical, concerned only with deriving the masses and angular momenta of proto-galactic structures. Now, a lot more attention is given to questions of the metallicity, morphological type, dynamics and evolution of galaxies: questions far more closely tied to observation than before. It is important to remember that ultimately, everything we know rests on observation.

The lack of direct observational data has greatly inhibited our analysis of the post-recombination, pre-galactic universe. However, we are now becoming aware of a number of observations and deductions that have vital bearing on this epoch. It is important to emphasize that no single observation is by itself compelling evidence for any particular model. However, taken together, the different pieces of evidence collectively provide some vital constraints on what sequence of events may have occurred, from which we can construct a general picture of that epoch. I shall summarize this evidence now, and indicate the conclusions it points to.

First, I shall define some important terms, which tend to be used ambiguously in the literature.

B. J. T. Jones and J. E. Jones (eds.), The Origin and Evolution of Galaxies, 179–187.

The epoch of consideration is that between decoupling and the existence of recognizable galaxies, that is, approximately

$1000 > Z > 2$

My attention is centred on the very first generation of stars: let me therefore define stellar populations as follows:

Population I	recently formed stars.
Population II	old stars, formed during or after collapse.
Population III	stars formed prior to the collapse of the Galaxy.
Population IV	sites of nucleosynthesis in zero-metal gas (= 'primordial stars').

I have introduced the group 'Population IV' in order to avoid confusion between 'primordial' and 'pregalactic' stars. Population III is the name commonly used nowadays to denote pre-galactic stars, but not all of these are necessarily primordial - the first generation. Note that there is still some ambiguity even with these definitions: it may be impossible to distinguish observationally between Population II and Population III. Some halo stars described by observationalists as Population II may well have been formed prior to the collapse of the Galaxy, that is, be Population III.

The evidence may be divided into four categories, which I shall discuss in turn.

a. Metallicity - the G-dwarf problem.
b. Dynamics - violent relaxation and triaxiality.
c. Cosmic Microwave Radiation - distortions.
4. Sundry other.

2. METALLICITY

A good discussion of the chemical evolution of the Galaxy and the G-dwarf problem may be found in Pagel and Patchett (1975). There is also an excellent brief summary in Audouze and Vauclair (1980).

The only elements emerging out of the Big Bang were hydrogen, helium, deuterium and perhaps some lithium 7. All the heavier elements that make up the present cosmic metal abundance are thought to be the products of nucleosynthesis in stars and supernovae. The significance of G-dwarfs is that these are stars whose projected lifetimes are longer than the present age of the universe, and whose evolution has not yet changed their surface composition. A G-dwarf formed from zero or near zero metal gas would therefore be seen to have zero or near zero metallicity. The G-dwarf problem is simply the observation that although current chemical evolution models can explain quite well most of the stellar abundances, there are too few low-metallicity G-dwarfs, and indeed we see no zero-metal stars at all.

The four main arguments offered to explain this problem are:
(i) Variable initial mass function: this supposes that in the earliest phases of galactic evolution, the mass function was weighted towards the formation of high mass stars. Only a negligible number of zero metal, low mass stars were therefore produced.
(ii) Prompt initial enrichment: this supposed that by some unspecified means there was an initial injection of heavy elements after which evolution proceeded in the normal way.
(iii) Metal enhanced star formation: this supposes that the star formation efficiency was very low in low metal gas, so that stars tended to form only where metals were already present.
(iv) Inflow of partially processed material: this supposes that gas flows in from the outer parts of the Galaxy, bringing with it some metal enrichment and becoming further enriched as it falls in. Infall models do not fit all the data very well, and seem to demand some form of prompt initial enrichment (possibly in the halo) to explain the G-dwarf data.

Although there are different mathematical descriptions pertaining to each of these models, which variously improve the fit to the data, the principle behind each 'explanation' is actually identical: for some reason, the first stars to form from primordial gas were massive, and lower mass stars tended to form only after some enrichment had taken place. Notwithstanding the plausibility of each model, there is no firm reason offered as to why that should be so.

The metallicity data also provide another important constraint in this problem: only a small number of primordial massive stars can have been formed. We know this from the fact that the oldest observed stars in our Galaxy, the halo stars, are very metal poor ([Fe/H] ~ -2 to -1). If a substantial proportion of primordial gas had turned into high mass stars before the formation of second generation stars then too much enrichment would have occurred to be compatible with the metallicity of the halo. The limit is quite tight: to satisfy the observed enrichment of halo stars we need only one massive star (~ 35 $M_\odot$) per 10^5 or 10^6 $M_\odot$ or primordial gas. This limit assumes a uniform mixing of enriched gas, so it may vary quite a bit from this. Nevertheless, a high formation efficiency of population IV stars is clearly ruled out.

3. DYNAMICS

Attempts to model the collapse of galaxies from an initial protogalactic cloud of gas and/or stars indicate that at least a substantial number of stars must have formed before or at the very start of collapse. The problem is most severe for elliptical galaxies: the presence of even a small amount of gas leads to the formation of a disk (Larson 1976). The argument is strengthened by the observation of triaxiality in elliptical systems: only in dissipationless collapse is triaxiality preserved. The timescale for gravitational collapse and violent relaxation of stellar systems is $\sim 10^8$ years, and dynamical arguments show that the halo stars of our own Galaxy must have been formed before this process was complete (Eggen, Lynden-Bell and Sandage 1962). This places quite a severe time constraint on the formation of Population IV stars and the subsequent mixing of their heavy element enrichment:

(i) the oldest observed stars have metals (Population II or III)

(ii) the oldest observed stars formed before the Galaxy had collapsed (Population III?)

therefore (iii) the first metals formed before the Galaxy had collapsed (Population IV).

The distribution of metallicities among the halo stars provides a constraint on the density distribution of their Population IV precursors: there is a high degree of uniformity in the (low) metallicity of halo stars (Bond 1981). This indicates that the density distribution of Population IV stars was not too even: it is difficult to distribute the metals if their production is too isolated. In particular, they cannot have been distributed as sparsely as galaxies themselves. The timescale for mixing on these length scales is about 10^8 years, the collapse or rotation timescale of a galaxy. The earliest metals must have been reasonably well mixed in much less than this time, and hence the distribution of Population IV stars was more evenly spread than galaxies. This argues against the production of Population IV stars in large but isolated pockets around the universe, as has been suggested by some authors (Carr 1981).

4. COSMIC MICROWAVE RADIATION

Recent observations of the cosmic microwave radiation (CMR) by Woody and Richards (1979) reveal the presence of distortions longward of the peak. It has been suggested that these distortions are due to emission by dust at pre-galactic epochs (Rowan-Robinson *et al*. 1979: Puget and Heyvaerts 1980). Such a model demands two relevant input parameters. The dust itself must have been produced by a generation of primordial stars. From the amount of dust required this need involve only a small fraction of the material in the universe, but the enrichment must have taken place at early epochs (no later than $Z \sim 10$, and possibly much earlier). Then, the radiation producing the distortions must also have been produced before $Z \sim 10$, but the amount of energy required means that a large amount of material in the universe was involved in star formation. Rowan-Robinson et al. and Puget and Heyvaerts suggest that both dust and radiation were produced by the same burst of star formation. However, by the arguments given above, it is clear that two different generations would be needed: an earlier dust-forming generation involving only a few massive stars, and a subsequent radiation-producing generation involving very many normal stars. This is because the number of low mass Population IV stars must be negligibly small, to avoid the G-dwarf problem, while the number of high mass Population IV stars must also be low in order not to contradict the observed low metallicity of the halo stars. Thus the CMR distortions, and their presumed explanation, not only indicate the existence of pregalactic Population IV stars, but also suggest that large numbers of normal stars (Population III) had already formed before the collapse of galaxies.

5. SUNDRY OTHER

A number of other arguments and questions relate to how and when the first stars formed.

(i) Observations of element abundances in the most metal-poor stars suggest that these may have come from only a single precursor generation of massive stars (Truran 1981). In particular the N and O anomalies suggest that the first metals were made in the supernovae explosions of stars in the mass range (Barbuy 1981):

$$10\ M_{\odot} < M < 100\ M_{\odot}$$

(ii) There is a problem concerning dwarf spheroidal galaxies: why do we see them only in groups and clusters, and how did they form? Evidently, the process of star formation was surprisingly efficient, under conditions apparently less favourable than those in our Galactic disk today. This is part of a far more general and basic problem, that of fragmentation. We do not understand what triggers star formation nor determines its efficiency.

(iii) Why do spheroidal galaxies have no visible edge? If star formation were confined only to within galaxies we might expect to see such an edge. On the other hand, if stars exist also in intergalactic space, as they would if many stars were pregalactic, then galaxies would merge smoothly into their background. Their haloes would be of low contrast and very difficult to detect (Jones 1982 in preparation).

6. GENERAL DISCUSSION

It is a consequence of cosmology that the first heavy elements were made in primordial (Population IV) stars. The arguments given above also indicate that

(i) Population IV stars were pregalactic:

(ii) they were massive, and

(iii) relatively few such stars were ever formed.

The approximate number consistent with the metallicity data corresponds to about one massive star per 10^5 $M_\odot$ of primordial gas. Furthermore, their spatial distribution was much less isolated than that of protogalaxies. These general conclusions suggest a simple model for the evolution of the universe between Z = 1000 and 10.

Sometime during this epoch, primordial Jeans mass clouds form and begin to collapse. Their collapse time is given by:

$$t_{coll} = \frac{\pi}{(\Omega h^2)^{1/2}} \cdot \frac{10^{10}}{Z_{rec}^{3/2}} \cdot \left(\frac{1}{3} \frac{\delta\rho}{\rho} - \frac{\Omega^{-1}-1}{Z_{rec}} \right)^{3/2} \text{ years}$$

where $\delta\rho/\rho$ is the density conttast at recombination (Jones 1977). In practice, this is only a crude estimate since (a) the lumps are non-spherical and hence will tend to collapse faster (Palmer 1981), (b) they are non-uniform and hence any 'peak' spots will collapse faster than the rest of the material, and (c) fragmentation may occur and hence several cusps of very high density may develop within the cloud. The analysis of Wood (1980) suggests that only a few (~ 3) such fragments form in the global structure. This kind of break-up is qualitatively distinct from the suggested small-scale fragmentation proposed by Low and Lynden-Bell (1976) and Rees (1976) which is based on thermodynamic arguments and depends critically on the ability of the gas to cool. In the crude model presented here, the presumption is made that small-scale fragmentation does not occur in primordial material but requires the presence of heavy elements to act as efficient cooling agents. Indeed, it is sufficient to argue simply that the timescale for cooling and fragmentation in the absence of heavy elements is longer than the timescale for the formation and evolution of one or a few massive stars.

Under these assumptions, a Jeans mass cloud having mass $M_J < 10^5 - 10^6\ M_\odot$ will begin to collapse sometime after recombination and the density in the centre (or possibly a few such centres) will rise to a very high value. A collapse process of this kind has been considered by B. Jones and S. Weber (unpublished). The presumption is that the density and temperature in the cusp(s) will rise until hydrogen burning ignites, after which accretion onto the star will be reduced but not terminated by the radiation of the star. The density spike that ignites is cusp-like, not flat-topped, so the initial star involves a relatively small amount of material. However the star continues to accrete material and grow, until the accretion timescale is equal to its evolution timescale. At that point the star will rapidly pass along its evolutionary sequence and explode as a supernova before any more material can accrete. This places the final mass of the star somewhere in the range

$$16\ M_\odot < M < 100\ M_\odot$$

Any star larger than 100 $M_\odot$ would evolve too quickly ($< 10^6$ years) for a greater amount of material to be accreted. Any star smaller than 16 $M_\odot$ would survive for long enough ($> 10^7$ years) to be able to acquire more material and hence grow larger. Such stars are discussed by Weaver and Woosley (1980).

A model of this kind is also needed to satisfy the mixing constraints and avoid the likelihood of gross inhomogeneities in the metallicities of the oldest (2nd generation) observed stars. The typical energies release in the supernova explosions of massive ($\sim 30\ M_\odot$) stars is

$$E_{sup} \sim 10^{52} \text{ ergs}$$

which are larger than the binding energy of the clouds:

$$E_J \sim \frac{G\ M_J^2}{\lambda_J} \sim 10^{50} \text{ ergs for a } 10^5\ M_\odot \text{ cloud.}$$

Of course, because of 'local' variations in the density, shape and rotation of the $10^{5-6}\ M_\odot$ clouds, there will be local variations in the Population IV nucleosynthesis depending on what mass of stars form locally. This might be a source of some of the abundance variations observed with globular clusters (Freeman and Norris 1981).

The formation of these Population IV stars may occur over quite a long period, inhomogeneously throughout the universe. However, wherever they occur, the metal enrichment they produce provides efficient cooling agents that can trigger gas fragmentation and widespread star formation. Such a scenario is suggested by the CMR distortions, if these are to due to emission by dust at pregalactic epochs.

7. SUMMARY

Although the overall model presented here is crude, most of its crucial features are a consequence of the observational constraints outlined above. These features will probably remain, even though the details of the picture undergo further modification and elucidation.

The key points are:

(i) The first metals were produced in primordial or Population IV stars. These stars were:

- nearly all massive (G-dwarf problem)
- relatively few in number: 1 per 10^5 $M_\odot$ of gas (metallicity of halo stars)
- fairly uniformly distributed on galactic scales (mixing problems)
- non-uniform on scales of globular clusters (abundance variations)
- pregalactic (dynamics of galaxy collapse and distortions to CMR)

(ii) Efficient star formation, and the formation of low mass stars did not occur until after this initial enrichment had taken place.

(iii) The initial enrichment may have occurred at different times from place to place throughout the universe.

(iv) Many second and later generation stars may have formed before the galaxies collapsed.

REFERENCES

Audouze, J. and Vauclair, S. 1980. An Introduction to Nuclear Astrophysics. D. Reidel Publ. Co., Dordrecht, Holland.

Barbuy, B. 1981. Astr.Astrophys., 101, 365.

Bond, H.E. 1981. Ap.J., 248, 606.

Carr, B.J. 1981. MNRAS, 195, 669.

Eggen, O.J., Lynden-Bell, D. and Sandage, A.R. 1962. Ap.J., 136, 748.

Freeman, K.C. and Norris, J. 1981. Ann.Rev.Astr.Astrophys., 19, 319.

Jones, B.J.T. 1977. MNRAS, 180, 151.

Larson, R.B. 1976. MNRAS, 176, 31.

Low, C. and Lynden-Bell, D. 1976. MNRAS, 176, 367.
Pagel, B.E.J. and Patchett, B.E. 1975. MNRAS, 182, 13.
Palmer, P. 1981. MNRAS, 197, 721.
Puget, J.L. and Heyvaerts, J. 1980. Astr.Astrophys., 83, L10.
Rees, M.J. 1976. MNRAS, 176, 483.
Rowan-Robinson, M., Negroponte, J. and Silk, J. 1979. Nature 281, 635.
Truran, J.W. 1981. Astr.Astrophys., 97, 391.
Weaver, T.A. and Woosley, S.E. 1980. Ann.N.Y.Acad.Sci., 336, 335. (Proc. 9th Texas Symp. on Relativistic Astrophysics, ed. J. Ehlers, J.J. Perry and M. Walker).
Wood, D. 1981. MNRAS, 194, 201.
Woody, D.P. and Richards, P.L. 1979. Phys.Rev.Lett., 42, 925.

VERY MASSIVE OBJECTS AND PREGALACTIC STARS*

B.J. Carr

Institute of Astronomy, Madingley Road,
Cambridge, UK.

1. INTRODUCTION

A 'Very Massive Object', henceforth referred to as a VMO, is defined to be a star which goes pair unstable during its oxygen burning phase. Stars below about 10^2 $M_\odot$ remain stable until their iron core phase, when they may explode as type II supernovae, possibly leaving black hole or neutron star remnants. Stars above about 10^5 $M_\odot$ go unstable to general relativistic effects before they begin nuclear burning (Fowler 1966) and either collapse directly to black holes or, for sufficiently high metallicity, explode (Fricke 1973). VMO's would therefore exist in the intermediate mass range 10^2 - 10^5 $M_\odot$. Hitherto little attention has been paid to VMO's because it has not been clear that there are any plausible circumstances in which they might form. In this talk, however, I will present arguments for believing that there may have been a population of pregalactic VMO's. Furthermore, recent evidence suggests that VMO's may be forming even in the present epoch (Andriesse et al. 1978; Cassinelli et al. 1981). This has prompted several people, most recently Arnett, Bond and Carr (1981), henceforth referred to as ABC, to study the evolution of VMO's in some detail.

I begin by summarizing the results of these studies. During its hydrogen burning phase, a VMO is radiation-dominated and therefore unstable to nuclear energized pulsations (Stothers and Simon 1970). However, the amplitude of these pulsations is

* Report of work done collaborating with W.D. Arnett and J.R. Bond. See Arnett et al. (1981).

B. J. T. Jones and J. E. Jones (eds.), The Origin and Evolution of Galaxies, 189–195.

limited by shock formation to about 10% in radius at the centre (Talbot 1971), so the star should certainly survive this phase even though it may lose an appreciable fraction of its mass (Papaloizou 1973). The pulsationally induced mass loss may continue during a VMO's helium burning phase, after which one ends up with a large radiation-dominated isentropic oxygen core. This core goes pair unstable, overshooting the stable oxygen-burning configuration and igniting it into a nuclear statistical quasi-equilibrium mix of heavy elements. The released nuclear energy creates thermal pressure on the overlying unburnt oxygen. If the weight of this overlying region is not too large, the pressure pulse results in complete disruption: if the weight is large enough, the core continues collapsing, photodissociating iron peak nuclei into alphas, becoming further unstable and finally forming a black hole. Within the framework of simplified polytropic models, ABC calculate that the mass of the oxygen core required for complete collapse is about 130 $M_{\odot}$. The associated helium core mass is 150 $M_{\odot}$ and the mass of the precursor hydrogen star M_C must be in the range 300-500 $M_{\odot}$, the large uncertainty reflecting our ignorance of how much mass loss occurs during the hydrogen burning phase. The important qualitative conclusion is that sufficiently large VMO's collapse to black holes after copious mass loss and that smaller VMO's disrupt entirely, producing in the process considerable enrichment. This conclusion is supported by detailed numerical studies, as reviewed in ABC.

I will now present arguments for believing that VMO's may have formed in the pregalactic era, thereby explaining various cosmological conundrums.

2. THE MASS OF PREGALACTIC STARS AND THEIR REMNANTS

The reason one might expect pregalactic stars to form is that the existence of galaxies implies that the early Universe must have contained density fluctuations. Providing the fluctuations were isothermal, the form of the fluctuations required on a galactic scale and above, if extrapolated to smaller scales, would be of order unity on a scale 10^6-10^8 $M_{\odot}$ (Peebles 1974: Fall 1979). Thus bound regions could form well before galaxies and one would expect these regions to fragment into stars. Calculating the characteristic mass of the fragments is complicated but both the initial absence of metals and the influence of the background radiation would tend to make the stars considerably more massive than those which form in the present epoch (Silk 1980). In particular, calculations of Kashlinsky and Rees (1981) suggest masses of order 100 $M_{\odot}$. There are also observational reasons for believing that the mass spectrum falls off less steeply with decreasing metallicity and, for $Z < 10^{-3}$ $Z_{\odot}$, it may be shallow enough for most of the mass to be in the largest stars (Terlevich 1982). These considerations suggest that, if pregalactic stars do form, they could well consist predominantly of stars in the VMO range.

The black hole remnants of pregalactic stars would be natural candidates for explaining the missing mass in galactic halos and clusters (White and Rees 1978). However, the considerations above indicate that only VMO's larger than $M_c \sim 300\ M_\odot$ and supermassive objects could be expected to undergo complete gravitational collapse. Smaller stars might also leave black hole remnants but only a small fraction of their mass would be involved, so it would be difficult to explain how most of the Universe could get into these holes. The most important observational constraint on the sort of holes which could provide the missing mass in the galactic halo stems from considering the tidal disruption of loose star clusters by the holes as they pass through the disc. These considerations suggest that the holes which dominate the mass of the halo need to be smaller than $10^5\ M_\odot$ (Carr 1978). This means that the missing mass in the halo can comprise black holes only if the holes derive from VMO's rather than supermassive objects.

3. NUCLEOSYNTHESIS AND METAL ENRICHMENT

An important consequence of pregalactic stars would be their nucleosynthetic effects. Calculations of Weaver and Woosley (1980) suggest that the mass fraction of a star in the mass range $15 < M/M_\odot < 100$ returned as metals should be

$$Z_{ej} = 0.5-(M/6.3\ M_\odot)^{-1},$$

i.e. Z_{ej} should lie between 0.2 and 0.5. Since there exist Population II stars with metallicity as low as 10^{-5}, we infer that either the fraction of the Universe going into pregalactic stars was tiny (less than $\sim 10^{-5}$) or that most of the stars were more massive than $M_c \sim 300\ M_\odot$ so that they collapsed to black holes together with their nucleosynthetic products. This argument therefore supports the suggestion that the first pregalactic stars were VMO's. Of course, if <u>all</u> the stars were larger than M_c, one would not expect any pregalactic enrichment whereas from some points of view a burst of pregalactic enrichment would be desirable (e.g. to explain the G-dwarf problem, see Truran and Cameron 1971). Also any model which depends on grains to thermalize the 3K background, either in total (Rees 1978) or in part (Rowan-Robinson <u>et al</u>. 1979; Puget and Heyvaerts 1980), requires a pregalactic enrichment of at least 10^{-5}.

However, it is not difficult to conceive of ways in which one can produce a small amount of enrichment even though most of the stars are in the collapsing mass range. For example, one might have a mass spectrum with $M_{min} < M_c < M_{max}$ which is weighted to the more massive end, so that only a small fraction of the Universe goes into stars smaller than M_c. Alternatively, one might envisage a small number of stars having sufficiently large rotation that the value of M_c is increased above M_{min} even though most stars have M_c below M_{min}. Both these situations are rather contrived.

A somewhat more attractive scenario, in which an enrichment of 10^{-5} arises fairly naturally, is as follows: as the fraction of the Universe going into stars increases, the amount of radiation generated by them will increase and eventually the Universe will be completely reionized. Simple energetic arguments (Hartquist and Cameron 1977) suggest that this happens when the fraction of the Universe gone into stars is of order 10^{-5}: more detailed arguments (ABC), accounting for the effects of the 3K background, give a somewhat larger fraction. Once the Universe is ionized, the formation of further bound regions and stars may be suppressed by the Compton drag of the 3K background until a redshift of order 140 (Hogan 1979). When the next generation of stars do form, they may be more massive than the first generation because the Jeans mass has been boosted as a result of the ionization. This could explain why the second generation of stars were processed into black hole remnants without producing any further enrichment.

Yet another scheme for generating small Z has been proposed by Jones (1981): she envisages a 30 $M_{\odot}$ star forming at the centre of each 10^6 $M_{\odot}$ bound region and then exploding, thereby generating a small enrichment and dispersing the rest of the region before further stars can form.

One might also invoke pregalactic VMO's to explain various abundance features. For example, the well-known oxygen anomaly (Sneden et al. 1979), that metal poor stars have an (O/Fe) ratio which is about 3 times solar, suggests that O and Fe must have had different nucleosynthetic histories. Twarog (1980) has inferred that, whereas 90% of the Fe in the disc must have been produced during the disc's lifetime, only two-thirds of the O was so produced. Since VMO's explode during their oxygen burning phase, they would naturally produce a large amount of oxygen and the (O/Fe) ratio generated could plausibly be as large as 3 times solar.

Another property of VMO's is that, unlike ordinary stars, they might produce nitrogen as a primary element due to convection effects. It is usually assumed that N is a secondary element, so that (N/H) ~ $(Fe/H)^2$. However, recent observations of very metal poor stars (Edmunds and Pagel 1978; Barbuy 1981) suggest that (N/H) ~ (Fe/H), in which case one must seek a way of generating nitrogen directly. Finally, of course, it should be pointed out that VMO's could be prolific generators of helium. A VMO might return 20-50% of its mass to the background medium as helium during its pre-oxygen-core phase (Talbot and Arnett 1971) and, in the mass range $M > M_c$, there would be no danger of overproducing heavy elements at the same time. Massive VMO's would therefore be natural candidates for generating the observed 20-25% helium abundance, even though this is usually assumed to be of cosmological origin. This is significant because, if one also wants the VMO's or their remnants to generate the 3K background, the early Universe would have been cold, in which case there may have been no cosmological helium production at all (Kaufman 1970).

4. BACKGROUND RADIATION

Another important consequence of pregalactic stars is that they would have generated a lot of radiation during their nuclear burning phase. Background light limits therefore place interesting constraints on their mass spectrum and formation redshift. Roughly speaking, one percent of each star's rest mass energy may be converted into radiation, so if the stars burn at a redshift z_*, the present radiation density generated by them would be $\Omega_R \sim 10^{-2}\ \Omega_*(1+z_*)^{-1}$ where Ω_R and Ω_* specify the radiation and stellar densities in units of the critical density. Since the radiation density over all wavebands (with the possible exception of the IR band) cannot exceed $\Omega_R \sim 10^{-4}$, we infer a limit $z_* > 10^2\Omega_*$. Thus, if the black hole remnants of these stars are to provide the missing mass in halos (which requires $\Omega^* \sim 0.1$), their precursors certainly need to form before $z = 10$. Furthermore, in order to burn their nuclear fuel before $z = 10$, they must have a mass of at least 10 $M_\odot$. This sort of limit has been discussed in more detail by ABC, Thorstensen and Partridge (1975) and Eichler and Solinger (1973).

More positive evidence for the effects of pregalactic starlight may come from certain distortions observed in the spectrum of the 3K background (Woody and Richards 1979): there appears to be an excess of energy, together with a distinctive dip, shortward of the peak. Rowan-Robinson et al. (1979) have suggested that these features can be explained if 25% of the 3K background density is radiation generated by pregalactic stars and thermalized by grains, the grains also being generated by the stars. This model requires that the 25% component of the 3K background be generated at a redshift of order 100 and that the stars have a density Ω_* of at least 0.1. The stars can burn their nuclear fuel by $z_* \sim 100$ only if they have a mass exceeding about 30 $M_\odot$.

Puget and Heyvaerts (1980) have suggested another way in which pregalactic starlight could explain the Woody-Richards distortion: their scheme requires that the stars generate about 30% of the 3K background and that they do so at a somewhat later epoch. Of course, if pregalactic stars generate such a large fraction of the 3K background, it is not implausible that they may have generated all of it. Pregalactic stars would be expected to span a range of masses and Carr (1981) has suggested that the initial component of the background may have been generated by the most massive ones. These would have completed their evolution earliest, before the Rowan-Robinson et al. or Puget and Heyvaerts stars, and perhaps early enough for the radiation to be thermalized by free-free processes rather than grains. In this scenario, energetic and thermalization criteria require that the first stars produce their light at $z \sim 10^3$. This corresponds to a time of order 10^6y which, remarkably, is just the characteristic lifetime of stars more massive than $10^2\ M_\odot$. Thus pregalactic stars could generate the entire 3K background providing they are in the VMO mass range. These same stars could give rise to black hole remnants which could today constitute the missing mass in halos and clusters.

REFERENCES

Andriesse, C.D., Donn, B.D. and Viotti, R. 1978. MNRAS, 185, 771.

Arnett, W.D., Bond, J.R. and Carr, B.J. 1981. Preprint.

Barbuy, B. 1981. Ann.Phys.Fr., 6.

Carr, B.J. 1978. Comm.Astrophys., 7, 161.

Carr, B.J. 1981. MNRAS, 195, 669.

Cassinelli, J.P., Mathis, J.S. and Savage, B.D. 1981. Preprint.

Edmunds, M.G. and Pagel, B.E.J. 1978. MNRAS, 185, 77P.

Eichler, D. and Solinger, A. 1973. Ap.J., 203, 1.

Fall, S.M. 1979. Rev.Mod.Phys., 51, 21.

Fowler, W.A. 1966. Proc.Enrico Fermi International School of Physics (Course XXXV) Academic Press, New York.

Fricke, K.J. 1973. Ap.J., 183, 941.

Hartquist, T.W. and Cameron, A.G.W. 1977. Astrophys.Space Sci., 48, 145.

Hogan, C. 1979. MNRAS, 188, 781.

Jones, J.E. 1981. Preprint.

Kashlinsky, A. and Rees, M.J. 1981. Preprint.

Kaufman , M. 1970. Ap.J., 160, 459.

Papaloizou, J.C.B. 1973. MNRAS, 162, 169.

Puget, J.L. and Heyvaerts, J. 1980. Astr.Astrophys., 83, L10.

Peebles, P.J.E. 1974. Ap.J.Lett., 189, L51.

Rees, M.J. 1978. Nature, 275, 35.

Rowan-Robinson, M., Negroponte, J. and Silk, J. 1979. Nature, 281, 635.

Silk, J. 1980. In Star Formation, 10th Saas Fee Lecture Course (Geneva, ed. L. Martinet and A. Maeder).

Sneden, C., Lambert, D.L. and Whitaker, R.W. 1979. Ap.J., 234, 964.
Stothers, R. and Simon, N.R. 1970. Ap.J., 160, 1019.
Talbot, R.J. 1971. Ap.J., 165, 121.
Talbot, R.J. and Arnett, W.D. 1971. Nature Phys.Sci., 229, 150.
Terlevich, R.J. 1982. Ph.D. Thesis, University of Cambridge.
Thorstensen, J.R. and Partridge, R.B. 1975. Ap.J., 200, 527.
Truran, J.W. and Cameron, A.G.W. 1971. Astrophys.Space Sci., 14, 179.
Twarog, B.A. 1980. Ph.D. Thesis, Yale University.
Weaver, T.A. and Woosley, S.E. 1980. Ann.N.Y.Acad.Sci., 336, 335.
White, S.D.M. and Rees, M.J. 1978. MNRAS, 183, 341.
Woody, D. and Richards, P.L. 1979. Phys.Rev.Lett., 42, 925.

THE CHEMICAL EVOLUTION OF GALAXIES

C. Chiosi* and Bernard Jones+

Istituto di Fisica Galileo Galilei, Padova 35100, Italy
+Institute of Astronomy, Madingley Road, Cambridge

1. INTRODUCTION

I shall review in these lectures some recent developments in the theoretical interpretation of the distribution and abundances of elements in stars and in the interstellar medium. Nucleosynthesis in stars is responsible for the production of heavy elements in the Galaxy and so the observed abundances form a record of past events of nuclear production. Coupled with dynamical information, galactic chemistry thus provides an important tool for studying the origin and early evolution of galaxies. We shall further recognise that different nuclei are products of different processes, and this will give more detailed insight into how things happened. Thus the isotopes of H and He are indicators of conditions in the Big Bang, while isotopes of elements produced and expelled from metal poor stars are clearly tracers of early galactic evolution. Isotopes produced or destroyed in metal rich stars tell us about more recent nuclear production.

There are a number of important recent reviews on the subject of abundances. The 22nd Liege Colloquium (Boury et al. 1979: referred to hereafter as BGR) contains numerous short articles on every aspect of the subject, and the review of Pagel and Edmunds (1981) is a complete survey and critique of abundance determinations. The book by Audouze and Vauclair (1980) provides an excellent introductory overview of the subject.

B. J. T. Jones and J. E. Jones (eds.), The Origin and Evolution of Galaxies, 197–226.

2. THE SOLAR VICINITY

2.1 Standard abundances

The solar system provides the richest source of data, and the solar vicinity within, say, a kiloparsec of the Sun provides more information where accurate spectroscopic determinations are possible. There is a Standard Abundance distribution (Cameron 1980: Meyer 1979) relative to which comparisons are generally made. Table 1 gives an overview of the abundances of He, O, C, N, Ne, S, Fe in various sites. It can be seen that there is little difference between the Sun, young stars and the local, present day, interstellar medium. O and N both appear to be underabundant by a factor ~ 2 in the Orion Nebula and the Cygnus loop: this may be due to condensation on grains (Meyer 1979). The isotope ratios of C, N, O, Si, S in the local interstellar medium differ from solar system ratios by less than a factor 2 (Wannier 1980). Interstellar D/H, while on average consistent with solar system values shows considerable fluctuations.

Table 1

Site	He/H	O/H	O/C	O/N	O/Ne	O/S	Fe/H
Photosphere		-3.1	0.2	0.9			-4.4
Prominences	-1.2						
Corona				0.9	1.1		
B Stars	-1.0	-3.1	0.5	1.0	0.9		-4.5
Orion Nebula	-0.2	-3.4	0.1	0.9	0.7	1.3	-5.3
Cyg Loop		-3.5	0.2	0.7		1.8	
PN (disk)		-3.3					

Note: the table entry for X/Y is the value of $\log_{10}$ (X/Y).
: PN (disk) denotes disk planetary nebulae.

2.2 The metallicity distribution of the solar neighbourhood

The metallicity distribution of long lived stars has played an important role in shaping our ideas about the chemical evolution of the Galaxy. The recognition of the 'G-dwarf' problem by van den Bergh (1962) and Schmidt (1963) and the discovery of the link between kinematics and chemistry by Eggen, Lynden-Bell and Sandage (1962) are landmarks of the subject. The metallicity distribution of G dwarfs in the Galactic disk is well approximated by a log normal distribution: that is, the values of

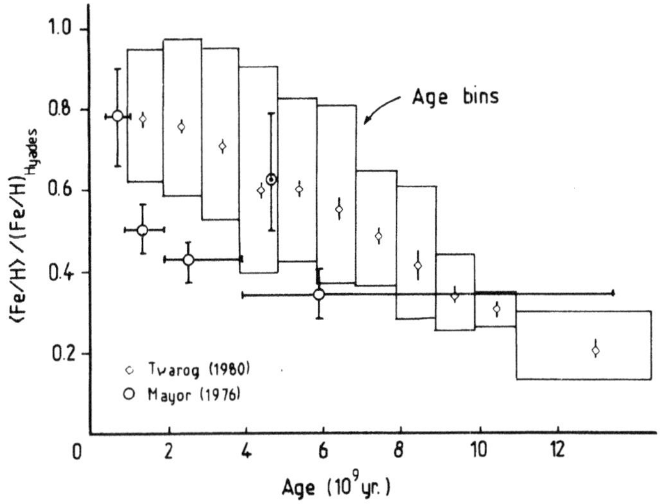

Figure 1. Metallicity as a function of age for F stars. The data are from Mayor (1976) and Twarog (1980), and the abundance is given relative to the Hyades.

[Fe/H][1] are Gaussian distributed with mean <[Fe/H]> ~ -0.3 relative to the Hyades (i.e. ~ 0.20 relative to the Sun) and a dispersion of ~ 0.25 (Pagel and Patchett 1975). The total absence of disk stars with [Fe/H] < -1.0 is the G dwarf 'problem'. Twarog (1981) has shown that the metallicity distributions of G and K giants and F dwarfs of the disk are similar to that of the G dwarfs.

The metallicity distributions of halo subdwarfs has recently been derived by Bond (1981) from the data of Carney (1979). These are very metal-poor in comparison with the disk G dwarfs, having a mean <[Fe/H]> ~ -2.0. Bond (1980) remarks that the halo subdwarfs have a similar metallicity distribution to the globular clusters, and that there are very few disk G dwarfs that are as metal poor as the most metal rich halo subdwarf in Carney's list. There is also a 'G dwarf problem' for the halo in that there appears to be an extraordinary lack of halo giants with [Fe/H] < -3.0.

---

1 We use the standard notation that abundance is measured logarithmically relative to the solar value: $[X] = \log X - \log X_\odot$.

2.3 The age-metallicity relationship and the star formation history

The relationship between age and metallicity is shown in Figure 1 which is taken from the work of Mayor (1976) and Twarog (1980). The data is based on studies of F stars: these have the advantage that age determination for individual stars is possible, while their evolution is slow enough that it is possible to find very old stars in the sample. Moreover, the determination of their temperatures and metallicities is quite reliable. Clearly the younger stars are the more metal rich, and star formation in the disk (as viewed in terms of the solar neighbourhood) has been an on-going process during the past life of the Galaxy.

The star-formation rate over the past 10^{10} years can be estimated by counting stars as a function of age. The counts have to be corrected to account for effects of stellar evolution (for example stars spend only a certain fraction of their lives as F stars, and this fraction depends on metallicity) and for effects of galactic evolution (the scale height of the galaxy has changed with time). Twarog (1980) finds that the star formation rate has remained constant over the past 4×10^9 years, but was at most a factor 2 ~ 3 times higher between 4 and 8×10^9 years ago. Beyond that point, the star formation rate seems to have fallen off. Indeed, if the star formation rate has remained fairly constant over the lifetime of the Galactic disk, the metallicity distributions of the F stars is entirely consistent with the G dwarf data. (See also Tinsley and Danly 1980.)

3. ABUNDANCE GRADIENTS IN THE GALAXY

3.1 Large scale trends in the stellar disk

When we look beyond the solar neighbourhood we find evidence from stars (Janes 1975, 1979: Mayor 1976), HII regions (Talent and Dufour 1979) and planetary nebulae (D'Odorico *et al*. 1976) for a metallicity gradient of a factor

$$d[M/H]/dR \sim 0.05 \text{ dex kpc}^{-1}.$$

The gradient is similar for O, N and metals except in the case of planetary nebulae where, owing to complications arising from the way in which self-enrichment works in different types of planetary, there is a steeper N-gradient than O gradient. Supergiants, perhaps because of mass loss and mixing, are also a little anomalous in that they show larger abundances and gradients than other objects (Luck and Bond 1980).

There is also a gradient of metallicity perpendicular to the disk (Trefzger 1980) with a gradient of ~ -0.6 dex kpc^{-1} near the plane. Trefzger shows lines of constant metallicity which appear to run more or less parallel to the Galactic plane.

3.2 The diffuse interstellar medium

The situation in the interstellar medium is complicated because of depletion and fractionation processes: the review of Wannier (1980) covers this in great detail. Fractionation, the separation of isotopes by chemistry, happens because the small mass difference can cause different chemical behaviour for isotopes. It is not an important process for elements like S, O, N, Si, but in some circumstances it can be important for H and C. The $^{12}C/^{13}C$ ratio has the value 60 ± 8 throughout the disk at radii R > 3 kpc. There is no evidence for any gradient though there is an overabundance of C^{13} by a factor 1.5 relative to the solar system. The $^{18}O/^{17}O$ ratio is also roughly constant throughout the disk, with a mean value ~ 3.2 which is significantly different from the solar neighborhood value of 5.5. There is also a lack of spatial variation among the isotope ratios of ^{28}Si, ^{29}Si and ^{30}Si and of ^{32}S, ^{33}S, and ^{34}S. The Galactic centre is generally different from the disk: the $^{12}C/^{13}C$ ratio is different there from the disk, there is an overabundance of ^{18}O by a factor 2, and an underabundance of ^{15}N by the same factor. The element deuterium, which is so important for the Big Bang nucleosynthesis is very difficult to measure, though it appears to be depleted towards the Galactic centre (Penzias et al. 1977: Audouze et al. 1976: see also Laurent et al. 1979 in BGR).

3.3 The Galactic halo

The situation with regard to the metallicity distribution in the Galactic halo can be summarised by saying that the metal-rich globular clusters of integrated spectral type G are usually confined to the inner halo (R < 10 kpc), whereas the metal-poor globular clusters of type F are found at all distances. Any attempt at a more precise statement becomes somewhat controversial. Thus Searle and Zinn (1978) contend that while the mean metallicity of the inner halo globular clusters is certainly greater than those in the outer halo, the distribution for R > 8 kpc is independent of distance. Harris and Canterna (1979) on the other hand argue that both the mean and the maximum 'metallicity (Z)' decrease continuously out to radii of 40 kpc. Caputo (1981) recalibrated the Z values for about 100 globular clusters and came to the conclusion that while the maximum Z at a given radius decreases the radius, the mean Z decreases only within 20 kpc, remaining constant outside of that radius.

The difficulty comes from two directions. Firstly there is the crucial problem of calibrating [Fe/H] in terms of some observable quantity. For globular clusters this rests largely on 47 Tuc and M71, for which different methods yield metallicities differing by factors of up to four. Secondly, there is the problem of the so-called 'second parameter'. If globular clusters were a one parameter family, the ratio of the number of blue horizontal branch stars to red ones would be a function of that one parameter, Z. With theoretical models for horizontal branch stars we can evaluate this and it is clear that some metal-poor globular clusters do not behave as expected in regard to their horizontal branch morphology. Hence there must be some other factor, the second parameter, playing a role and several possibilities have been suggested: age, He/H ratio, or the ratio of C, N, O relative to Fe in clusters with the same [Fe/H]. Whichever explanation is correct, the second parameter is largest among the distant globular clusters and dwarf spheroidal galaxies and decreases with decreasing R. It is absent in the metal poor globular clusters within 5 kpc of the Galactic centre.

Searle and Zinn (1978) identify the second parameter with age, and argue that its dependence on position is evidence that metal enrichment proceeded more rapidly in the central parts of the Galaxy. This is what would be expected if the gas density rose rapidly in the core of the collapsing Galaxy. Over the radius range $0 < R < 40$ kpc, the mean age would decline by about 3×10^9 years whereas the scatter in ages at a given radius increases from less than 10^9 years to 2×10^9 years going from the $0 < R < 5$ kpc region to the 12 kpc $< R <$ 40 kpc region. If the He abundance were the second parameter, its value would vary from $Y = 0.23$ in the central regions to $Y = 0.19$ in the outer regions.

4. EXTERNAL GALAXIES

4.1 Metallicity indicators

Because of their distance, direct spectroscopic analysis of individual stars can only be undertaken for the very brightest giants in the nearest galaxies. In the Magellanic Clouds this is a reasonable way of doing things, and, although the data are somewhat poorer, there have been direct low dispersion observations of giants in the Draco and Ursa Minor dwarf galaxies. On any larger scale it is necessary to resort to using the systematics of the colour magnitude diagram, the colour-period relation for Cepheids or the colour distribution of supergiant stars. Emission line observations of HII regions, supernova remnants and planetary nebulae in external galaxies (when such objects exist) in principle provide a good method in determining the abundance of their present-day interstellar medium (though planetary nebulae present a few difficulties). The results are not, however, entirely uncontroversial (see for example section 4.5). Elliptical and SO galaxies have virtually no HII regions, nor can they be resolved into individual stars. In those cases we have to use the integrated light from the stellar populations that make up the system.

4.2 Irregular and blue compact galaxies

Emission lines in these galaxies are easily observed and they provide a reasonably large sample of objects in which the abundances of H, He, O, S, Ne can be determined. The ability to study the element He is particularly important since we can see how He/H changes with metallicity, and by extrapolating to zero metallicity infer the primordial He abundance. The slope of the helium abundance:metallicity relationship is denoted by $\Delta Y/\Delta Z$, resulting in $Y = 0.23$ for the primordial Helium abundance. Talent (1980) from a large sample derives $\Delta Y/\Delta Z = 2.83 \pm 2.69$ and settles for $Y = 0.21$, whereas Kunth (1981) finds no evidence for a ΔY, ΔZ correlation and argues that $Y < 0.23$.

Such low values of Y as $Y = 0.21$ would seriously constrain the standard Big Bang nucleosynthesis.

4.3 Elliptical and SO galaxies

There are two phenomena that must be distinguished: the variation of colour or line strength across a given galaxy and the variation of mean colours or mean line strengths from one galaxy to another expressed as a function of, say, absolute magnitude. The variation of colour within a galaxy is such that the colours generally redden towards the centre. This is consistent with the data from line strength studies and indicates that the central regions of ellipticals are more metal rich than the outer regions, with the metallicities of the central regions of the intrinsically brightest ellipticals being two or three times solar.

It has long been known that there is a strong correlation of (U-V), (V-K) and (U-R) colours with absolute visual magnitude M_V. The slope of the (U-V):M_V relation is about $\Delta(U-V)/\Delta M_V = +0.10$, and this would imply a factor ~ 10 change in metallicity between $M_V \sim -23$ and $M_V \sim -15$. SO and E galaxies have identical colour magnitude diagrams, and there is no apparent difference between field and cluster galaxies. Despite the apparent trends of optical colours with M_V, the infrared colour (H-B) does not appear to correlate with M_B (Wyse 1982). Some line strength indices correlate very well with absolute magnitude (Faber, 1973), and at a given absolute magnitude the Mg-index correlates with velocity dispersion (Terlevich et al. 1981). Hence we are seeing in elliptical galaxies an intimate link between chemistry and dynamics which is telling us about the formation of these systems (see for example Struck-Marcell 1981).

4.4 Dwarf spheroidal systems

These are galaxies with absolute magnitudes roughly in the range $M_V > -15$: they are generally satellites of galaxies and may be regarded as elliptical galaxies in their own right, quite distinct from the globular clusters which are far more centrally concentrated. The Fornax dwarf even has its own retinue of globular clusters, all of which are metal poor. As would be expected from the general trend of decreasing metallicity with luminosity seen in the normal ellipticals, the dwarf spheroidals are metal poor. The metallicities are difficult to determine and are generally uncertain, Table 2 is taken from Pagel and Edmunds (1981).

Table 2

Draco	$M_V = -8.6$	[Fe/H] = −1.8 to −2.1
Ursa Minor	−8.8	−2.1
Sculptor	−11.7	−1.5 to −2.2
Fornax	−13.6	−0.5

4.5 Spiral galaxies

Because of the presence of a young stellar population in spiral galaxies, colour indices are not useful metallicity indicators unless the galaxy has a very prominent bulge. The spheroidal bulges of the early type spirals S0/a, Sa, Sb show a colour magnitude relation similar to that for ellipticals (Visvanathan and Griersmith 1977: Griersmith 1980): $\Delta(U-B)/\Delta M_V = 0.13$, and this too may be a metallicity effect. Furthermore, there is a decrease of line strength index away from the nucleus.

The data on the disks of spiral galaxies comes from emission line studies, and there is now quite a large number of these for which abundances are reasonably well determined. Many, but not all, galaxies show gradients of metallicity and these are always in the sense of metallicity decreasing outwards. Moreover, neither the metallicity nor its gradient has any obvious dependence on either morphological type or galactic mass.

Of particular interest are the nitrogen and oxygen abundances in galaxies (Alloin et al. 1979). There is no gradient of N/O across galaxies (though it may decrease a little in the inner regions), yet there is a marked correlation of N/O with O/H in the sense that $N/O \propto O/H$ when all galaxies are taken together

(Pagel and Edmunds 1981). This generally supports the idea that nitrogen is a secondary product of stellar and galactic evolution.

5. NUCLEOSYNTHESIS IN STARS AND MASS EJECTION

5.1 Characteristic mass ranges

The heavy elements that we see today were formed in stars and ejected into the interstellar medium as a consequence of stellar evolution. A central problem in galactic evolution is thus the need to understand, in some detail, how much of each element a star of given initial mass synthesises during its lifetime and what portion of the material is ultimately ejected. We are therefore inevitably led to consider questions of the mixing of elements of stellar interiors, the details of mass loss, and the effects of mass loss on the star's evolution. The details depend on both the mass of the star and its chemical composition, so while it is convenient to discuss the behaviour of specific mass ranges, these are to be regarded simply as useful guidelines. Thus the division between what are called 'intermediate mass stars' and 'massive stars' depends on whether or not carbon burning occurs in degenerate matter. The dividing line lies somewhere in the range 7-10 $M_\odot$ and depends on initial chemical composition (Becker and Iben 1979).

The 'intermediate mass' range can be roughly taken as $1\ M_\odot < m < 8\ M_\odot$ (Renzini and Voli 1981). There is a subdivision of this range occurring at $m_W \sim 4$-$5\ M_\odot$. Below m_W the star loses its H-rich envelope before the degenerate C-O core can reach the Chandrasekhar limit of 1.4 $M_\odot$: these stars presumably become planetary nebulae and leave white dwarf remnants. Above m_W the C-O core would reach 1.4 $M_\odot$ and carbon would ignite under degenerate conditions were it not for the fact that such stars become violently pulsation-unstable and eject their envelopes (Tuckman et al. 1979). These stars also have a planetary nebula phase and leave white dwarf remnants. Romanishin and Angel (1980) present evidence that stars below ~ 8 $M_\odot$ leave white dwarf remnants, and are not supernova progenitors.

'Massive stars' occupy the range 8-15 $M_\odot$ and are thought to evolve to the classical type II supernova (Wheeler 1981). These stars almost certainly form neutron star remnants, but the mechanism whereby neutron star formation causes an explosion is not really understood. While these stars can burn carbon non-degenerately, they are not massive enough to form iron cores. Their evolution is very complicated and sensitively mass dependent, but from the present point of view what is important is that because of their structure they contribute relatively little to the synthesis of heavy elements, blowing off mainly

H + He (see section 5.3).

From the point of view of chemical enrichment of the interstellar medium it is the 'very massive stars', m > 15 $M_\odot$, that are important. These stars form iron cores which collapse to make neutron stars or black holes. The early evolution of these stars is complicated by problems such as overshooting and strong mass-loss via stellar winds. However, neutrino cooling is important in promoting degeneracy and postponing the ignition of successive fuels: this leads to an onion skin structure which is relatively simple to analyse.

5.2 Massive stars: onion skin structure

The recent model computations of Weaver and Woosley (1980) show the evolution of models of 15 $M_\odot$ and 25 $M_\odot$. They neglect mass-loss, rotation and magnetic fields, and start with a solar mix of elements. Elements in the mass range from oxygen to iron are ejected with abundance ratios close to those observed in the solar system. The model nitrogen abundance is deficient, but in any case we know N can be synthesised in stars of lower mass, and the C/O ratio is 0.24 in comparison with the solar value of 0.4. The mass fraction Z_{ej} of high Z ashes in H stars with masses in the range 15 $M_\odot$ < m < 50 $M_\odot$ is given approximately by Z_{ej} = 0.5 - 6.3/m. Multiplying this by the stellar birth function of Miller and Scalo (1979) gives the contribution to nucleosynthesis of stars having a specified mass. This function is strongly peaked at 18.5 $M_\odot$ and has a median value at 29 $M_\odot$: thus this mass range is the crucial one for the enrichment of the interstellar medium.

The chemical evolution of the interstellar medium cannot, however, be due entirely to such stars. The nitrogen and C/O discrepancies were commented on above. There is also direct evidence from population II stars and planetary nebulae (Snedden et al. 1979: Kaler 1980) that oxygen has a different history than either C or Fe, though this may be simply a consequence of the fact that the stars that produce the oxygen have different lifetimes than the stars that produce the iron.

5.3 Mass loss and mixing in intermediate mass stars

While the massive stars are responsible for most of the processing and recycling of heavy elements in the interstellar medium, they cannot explain all the abundances. The intermediate mass stars lose considerable amounts of material and owing to a variety of mixing processes in these stars this material will be enriched (Renzini and Voli 1981). These latter stars are particularly important as possible sources of ^{14}N and ^{12}C (Alloin et al. 1979). There are three important aspects of the evolution of intermediate stars: (1) convective 'dredge up' when the convective envelope penetrates inwards mixing processed material up to the surface, (2) nuclear burning in the deepest layers of the convective envelope, and (3) mass loss by stellar winds.

There are three dredge-up phases (Iben and Truran 1978), the first of which occurs when the stars reach the bottom of the red giant branch for the first time. The convection reaches zones where ^{12}C partly burns to ^{13}C and ^{14}N, thus depleting ^{12}C and enhancing ^{13}C and ^{14}N. In stars more massive that 3-5 $M_\odot$ (depending on composition) there is a second dredge-up following the ignition of the He-burning shell. The convection penetrates to the He-burning envelope bring ^{14}N and He to the surface and depleting ^{12}C, ^{13}C and ^{16}O while the C^{12}/CC^{13} ratio remains unchanged. The third dredge-up phase occurs during evolution along the asymptotic giant branch and is a sequence of mixing events following each helium shell flash. The convective envelope then penetrates the H-He discontinuity, thus bringing ^4He and ^{12}C to the surface. This phase ends with the ejection of what is thought to be a planetary nebula, signalling the end of the asymptotic giant branch evolution. The composition of ejected gas depends on when this planetary nebula-like mass loss occurs since the longer the third dredge-up phase lasts the greater will be the chemical change.

It is possible that during the asymptotic giant branch phase the temperature at the bottom of the convective envelope becomes high enough for CNO burning. The products are brought to the surface by the convection. This process has the graphic name of 'hot bottom burning', and enhances ^{13}C and ^{14}N while depleting ^{12}C and ^{16}O.

Mass loss during the asymptotic giant branch evolution is an important factor in determining the change of chemical composition during the third dredge-up. On the one hand the mass loss controls the decrease of stellar mass and thus determines the mass of the envelope with which the dredged up products are mixed. In addition, the temperature at the base of the convective zone is sensitive to both the mass of the stellar core and the total stellar mass (as well as the mixing length): this strongly affects the CNO hot bottom burning. The issue is a very important one since the third dredge-up and hot-bottom burning combine to produce <u>primary</u> ^{13}C and ^{14}N as opposed to the first and second dredge-ups which produce <u>secondary</u> ^{13}C and ^{14}N (they are formed from ^{12}C and ^{16}C already in the star). There is increasingly stronger evidence (Edmunds and Pagel 1978: Alloin <u>et al</u>. 1979) that some ^{14}N must be of primary origin, and this emphasises the importance of this aspect of stellar evolution. It should be noted that quite apart from the uncertainties in the mass loss and the mixing length the dredge up phenomenon has only been studied with population I abundances: a low metallicity may significantly reduce the amount of mixing since that depends on the opacity.

5.4 Chemical yield

To construct simple models of galactic chemical evolution (see section 6) it is convenient to define averaged quantities like the star formation rate (SFR), $\psi(t)$, and the initial mass function (IMF), $\phi(m)$, and the chemical yield y. It is then possible, as we shall see later, to model the evolution of the amount of gas and its rate of enrichment in rather broad terms. The initial mass function $\phi(m)dm$ describes the number of stars in the mass interval m to m+dm and will be normalised so that

$$\int_0^\infty m\ \phi(m)\ dm = 1 \tag{1}$$

It is often supposed that $\phi(m)$ is zero outside of some range $m_L < m < m_u$ and for simplicity the model

$$\phi(m) \propto m^{-(1+\kappa)} \tag{2}$$

is frequently used together with the assumption that $\phi(m)$ does not depend on time, metallicity or anything else. There is no empirical justification for this and indeed the work of Burki (1977, 1978) and Boisse et al. (1980) may be interpreted as evidence for variations in the IMF. Given that one adopts a simple invariant $\phi(m)$, for historical reasons, a favourite value for κ is 1.35 corresponding to the Salpeter (1955) initial mass function: the situation is reviewed by Lequeux (1979), and by Miller and Scalo (1979) who opt for κ = 1.9 in the range 2-100 $M_\odot$ if a power law is insisted on.

Models also need the local star formation rate $\psi(t)$ expressed as the total mass of stars formed locally per unit time, and is frequently expressed as mass per kpc^2 per unit time. The rate per unit area enters simply because the samples from which ψ is determined are sufficiently deep in space that their scale heights are small in comparison with their radial extent. We shall discuss various models for $\psi(t)$ in section 6.

For the purposes of simple models it is common to make an approximation called the 'instantaneous recycling approximation' wherein the stars are divided into two mass ranges: $m < m_1$ which effectively live forever and once formed play no role in the chemical enrichment process and $m > m_1$ which are supposed to die at the time of their birth. The value of m_1 will be roughly 1 $M_\odot$. If a star of mass m leaves a remnant of mass m_r, we can define the quantity

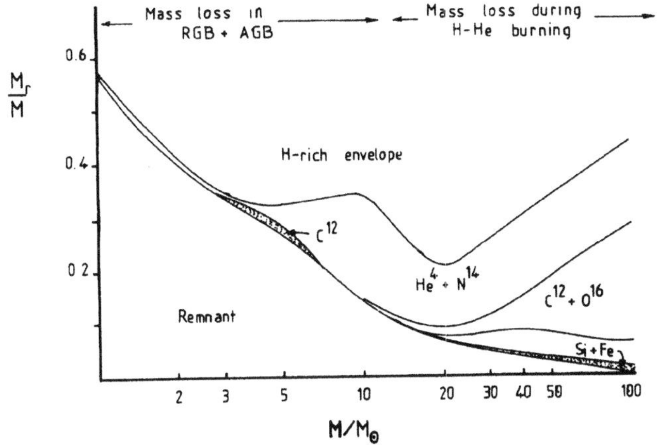

Figure 2. Fractional mass of newly synthesised elements and collapsed remnants as a function of the initial mass of the star.

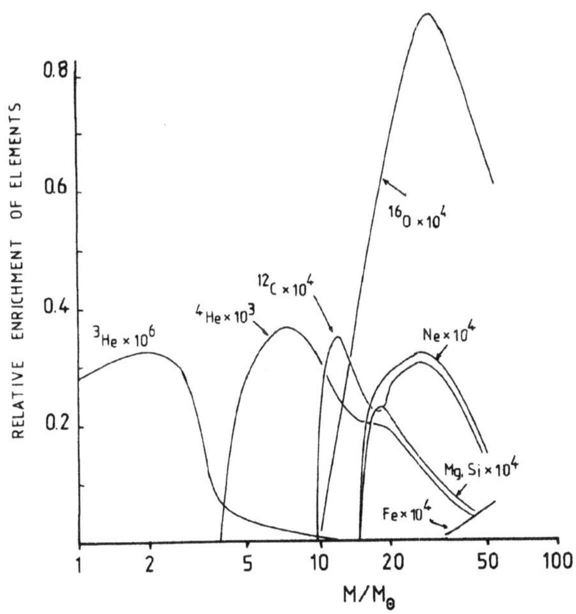

Figure 3. The relative contributions of stars of different masses to the enrichment of various elements.

$$r = \int_{m_1}^{\infty} (m - m_r)\, \phi(m)\, dm \tag{3}$$

which represents the 'returned fraction' of the IMF. (Compare with equation (1).) Its value is thought to lie in the range 0.1 - 0.3. The rate of ejection of unprocessed material having metallicity Z(t) is just $rZ(t)\psi(t)$: these heavy elements were in the star at the time of its birth.

If we now denote by P_{im} the mass fraction of a star of mass m that is ejected as newly synthesised primary element 'i', the yield of the element i is

$$y_i = \frac{1}{1-r} \int_{m_1}^{\infty} m P_{im}\, \phi(m)\, dm. \tag{4}$$

y_i is just the mass of element i newly ejected into the interstellar medium per unit net mass turned into stars. (Note that a fraction 1-r of stars formed live forever and hence the net stellar birthrate is just $(1-r)\psi$).

It is the task of stellar evolution theory to provide the P_{im} factors: for more details see Talbot and Arnett (1973) and Chiosi and Matteucci (1982). We can most simply summarise the information available at present by reference to Figure 2 which shows the fractional mass of newly synthesised elements and collapsed remnants as a function of the initial mass of the star. The diagram should be regarded as highly schematic in view of the considerable uncertainties that go into its construction. Figure 3 shows the product of the distributions of Figure 2 with a $\bar{\kappa} = 1.55$ initial mass function (Audouze and Vauclair 1980) and hence shows the relative contributions of stars of different masses to the enrichment in various elements. As commented earlier (section 5.3) it is stars of mass $> 15\ M_\odot$ that are primarily responsible for the enrichment of the heavier elements. Again it must be stressed that this diagram is only schematic, and may even fail to take account of some important but as yet not identified source of heavy elements. It is for example quite conceivable that iron peak elements could be created in intermediate mass stars as the result of the C-detonation of a C-O degenerate core. Type I supernovae evidently produce significant quantities of iron peak elements, but we have little understanding of what their progenitors are, whether they be intermediate mass stars or slowly accreting white dwarfs (see Nomoto 1981 and Wheeler 1980, 1981 for a discussion of the possibilities).

6. STAR FORMATION

6.1 Laws for the star formation rate

The rate at which stars form in a galaxy is the most important single factor in galactic evolution. The overall shape of a galaxy is thought to depend on the ratio of the timescale for star formation to the collapse and dissipation timescales. The colours and luminosity depend on the age distribution of the observed stars which in turn depends on the star formation rate in comparison with the gas flow rate, and also on the gas mass.

Models for galactic evolution need a 'law' of star formation which tells us the rate at which stars are formed in varying circumstances. The simplest of these laws is the Schmidt Law

$$\dot{\rho}_* = \nu \rho_g^s , \quad s > 1 \tag{5}$$

relating the rate of star formation ρ_* to the local gas density ρ_g via two universal constants ν and s. If $s > 1$ the star formation rate per unit gas mass would be higher closer to the galactic plane where the gas is denser and this would 'explain' the relative distribution of gas and stars in the direction perpendicular to the plane. This provides a means for estimating ν and s. Another way of getting at these parameters is to look for correlations between the gas density and the density of young objects in external galaxies. Several attempts have been made along these lines (Guibert 1978: Lequeux 1979a), but there is always the problem of determining ρ_g. Whereas the distribution of atomic hydrogen is easy to determine, the molecular hydrogen content can only be inferred indirectly from CO observations (Lequeux 1981). A further complication comes from the fact that many authors use a form of the Schmidt law that relates the rate of star formation per unit area to the gas surface density σ_g:

$$\sigma_* = \mu \sigma_g^s . \tag{6}$$

(5) and (6) are not equivalent since $\sigma \propto \rho z$ where z is the scale height of the relevant component.

When one considers sites of star formation in the Galaxy, it is evident that 'laws' like (5) and (6) can only represent global averages. There must be many parameters determining ρ_*, but the hope is that these are themselves related to ρ_g (the magnetic field or turbulent velocity could well be) and that the combined effect of all parameters is a law like the Schmidt Law. It would be easily possible to model the star formation process in what appears to be more realistic detail. Thus we might suppose that gas compression in spiral density waves triggers star formation in clouds by making them unstable (Roberts 1969). Such an idea has been much debated: Woodward (1976), using numerical simulations, showed that this process would be rather inefficient, but Silk (1980) has argued for a slightly different picture where the increased density of clouds in arm regions enhances their coalescence creating bigger and bigger clouds which eventually become Jeans-unstable and collapse (see also Chieze and Lazareff 1980).

The observation that star formation in one region of a galaxy can trigger off star formation in a neighbouring region (see for example Herbst and Assousa 1977) has been taken as the basis of a star formation model that produces remarkable looking global spiral structure from local phenomena (Mueller and Arnett 1976: Gerola and Seiden 1978: Seiden and Gerola 1979: Seiden et al. 1979). In the Gerola-Seiden models there is a probability of inducing star formation in a neighbouring region, P_{st}, which depends on the gas density. It turns out that although we have no proper understanding of P_{st}, it must lie in a rather narrow range of values to give reasonable looking spiral galaxies. The authors further argue that since values of P_{st} lying outside this range yield either no star forming activity or else a violent explosion of star formation, the nonlinear nature of the star forming process actually gives this stabilising value of P_{st}. It should be noted, however, that this model is only a model for producing spiral structure: as yet there is no conclusion about galactic chemistry. The main parameter is the differential rotation $d\Omega/dR$, and this model explains why irregular galaxies which have almost solid body rotation form stars slowly in comparison with most spirals (Gerola et al. 1980). The model is physically reasonable in that it represents the kind of 'contagious' star formation that is often seen in our Galaxy where one region of star formation induces stars to form in neighbouring regions.

Of course, whatever the model, we have little or no understanding of what is the mass distribution of the stars that form. It is quite conceivable that high mass and low mass stars form in different environments under quite different circumstances. Some of the various physical processes which may be relevant, such as the role of fragmentation, have been discussed at length (Rees 1976: Silk 1980) while others, such as the role of magnetic fields or initial conditions, have been largely neglected. We are reduced to more or less ad hoc simplification and parametrisation for this important stage of galactic evolution.

6.2 Simple analytic models

We shall present here a simple but effective model, which

considering our ignorance of the star forming process goes a long way to emphasising some key issues of galactic chemical evolution. We need a considerable amount of notation to get started so the equations look quite complicated until they are simplified. The total mass of the system will be M, and this may increase as a result of gaseous accretion at a rate f(t) which has to be modelled on the basis of some idea for building up galaxies:

$$\dot{M} = f \tag{7}$$

Gas, having total mass M_g is consumed by forming stars at a rate ψ, but is also replenished at rate E by the ejecta from evolving stars

$$\dot{M}_g = - \psi + E + f. \tag{8}$$

The metal content of the gas is ZM_g, Z being the metallicity, and this changes because metals fall into new stars at a rate Z , and get ejected from evolving stars at a rate E_Z:

$$(ZM_g)^{\bullet} = -Z\psi + E_Z + Z_f f. \tag{9}$$

The term Z_f is there to account for the fact that the accreted gas could be enriched.

The quantities E(t) and $E_Z(t)$ are somewhat complicated since they depend on the number of stars dying at t. These stars are born a time τ_m prior to t, where τ_m is the lifetime of a star of mass m. The number of such stars formed was proportional to the mass function $\phi(m)$ and the star formation rate at time $t-\tau_m$, and these stars when they die leave remnants of mass m_r. The material thrown out when they die can be regarded as consisting of a fraction $(1-P_{Zm})$ of the star's mass which is ejected but which has not been processed and therefore has the same metallicity as the gas from which the star was born, $Z(t-\tau_m)$, and a fraction P_{Zm} of gas that is both ejected and turned completely into heavy elements. Thus

$$E(t) = \int_{m_t}^{\infty} (m-m_r)\, \psi(t-\tau_m)\, \phi(m)\,dm \tag{10}$$

$$E_Z(t) = \int_{m_t}^{\infty} [(m-m_r-mP_{Zm})Z(t-\tau_m)+mP_{Zm}]\, \psi(t-\tau_m)\, \phi(m)\,dm \tag{11}$$

It is useful to recall the two quantities r and y which take up the ϕ dependence (see equations (3) and (4)).

$$r = \int_{m_1}^{\infty} (m - m_r)\ \phi(m)\,dm$$
$$y = \frac{1}{1-r} \int_{m_1}^{\infty} m P_{Zm}\ \phi(m)\,dm, \tag{12}$$

where the yield y here refers to all heavy elements rather than just one species.

These equations have been discussed by numerous authors since the pioneering paper of Talbot and Arnett (1971): among these we should mention the comprehensive review by Tinsley (1981) and the recent paper by Chiosi and Matteucci (1982) where the physics and chemistry going to y (taking account of several atomic species) is reviewed and updated.

Fortunately, for the present purposes it is possible to simplify these equations considerably by invoking the 'instantaneous recycling approximation' which is the limit $\tau_m \ll t$ (see section 5.4 and for example Lynden-Bell 1975). Then

$$E(t) = r\psi \tag{13}$$

$$E_Z(t) = rZ\psi + y(1-r)[1-Z] \tag{14}$$

If $Z < 1$ (a thoroughly reasonable approximation for the current epoch and all past epochs)

$$E_Z(t) = rZ(t)\psi + y(1-r)\psi \tag{14'}$$

and so finally we have

$$\dot{M}_g = -(1-r)\psi + f \tag{15}$$

$$M_g\dot{Z} = y(1-r)\psi + (Z_f - Z)f \tag{16}$$

If $\phi(m)$ does not vary with time, r and y are constants.

The mass of stars at time t is

$$M_*(t) = (1-r)\int_0^t \psi(t)dt = (1-r)\Psi\, t, \tag{17}$$

where the last equality merely defines Ψ. The mean metallicity of these stars is

$$Z_* = y - \left(\frac{\mu}{1-\mu}\right)Z, \quad \mu = \frac{M_g}{M} \tag{18}$$

and $Z_* \longrightarrow y$ as $\mu \rightarrow 0$. (19)

This simply means that when there is no gas left all the metals ever made and ejected (yM_g) are incorporated into later generations of stars.

There are two simple models of interest. The first, the closed model has $f = 0$, $M_{go} = M$, $M_{*o} = 0$, $Z_o = Z_f = 0$, and the solution can be expressed in the form

$$Z = y \ln \mu^{-1} = y \ln (1+\frac{M_*}{M_g}) \quad \text{if } Z \ll 1. \tag{20}$$

If Z gets large we have $Z = 1-\mu^y$. This model can be safely applied to irregular galaxies and blue compact objects (Lequeux et al. 1979): however it does not give acceptable results for the solar neighbourhood, as we shall see below.

The second simple model assumes that the rate of inflow of gas plus the stellar gas loss balances the star formation rate. Then M_g = constant and $\psi = f + r\psi$. It is then easy to show that

$$Z = y(1 - e^{-\gamma}), \quad \gamma = \mu^{-1} - 1 \tag{21}$$

Hence as $\mu \longrightarrow 0$, $\gamma \rightarrow \infty$ and $Z \longrightarrow y$.

Models of considerable complexity can be similarly constructed, and it is possible to give a few general trends that seem to hold even when instantaneous recycling is not a valid approximation, but break down if $\phi(m)$ is time dependent.

i) $Z \propto y$: any primary element whose stellar production parameters are independent of the composition will have abundances in proportion with their yield. Since y can be calculated independently of ψ and f, nucleosynthesis theories can be tested by comparing the theoretical yields with observed abundances.

ii) Z/y depends chiefly on current properties of the system and is insensitive to past history.

iii) Z/y depends weakly on μ and f/ψ. This implies that regions of galaxies where the gas fraction differs by orders of magnitude may have very similar abundances.

6.3 Application to the solar vicinity

The major difficulty in applying such models of chemical evolution to the solar vicinity comes in trying to explain the observed distribution of G-M dwarfs. It is easy to show in the simple model (Pagel and Patchett 1975) that the ratio of the number of stars of metallicity < Z to the ratio of stars with metallicity less than Z_1, the present day metallicity, is

$$\frac{S(Z)}{S(Z_1)} = \frac{M_*}{M_{*1}} = \frac{1 - \mu}{1 - \mu_1} = \frac{1 \; e^{-Z/y}}{1 \; e^{Z_1/y}} \tag{22}$$

Thus the differential distribution follows an exponential distribution on the basis of this model. As we commented earlier, the observed distribution is log normal with somewhat arbitrary truncations and a maximum around $Z/y \sim 1$, and hence the model predicts too many metal-poor stars. This is referred to as the 'G-dwarf' problem. Although increasingly larger numbers of very metal poor stars are being found, it is believed that the deficit of metal poor stars in the solar neighbourhood is not an effect of observational selection (Bond 1980). There are a number of more of less ad hoc ways out of the dilemma.

Variable mass function, $\phi(m)$. Because the yield y depends on the mass function (equation 12), an easy way around the G dwarf problem is to postulate a mass function that gives a greater y in the past.

Prompt initial enrichment (PIE). Truran and Cameron (1971) suggested that an outburst of massive star formation either before the Galaxy formed, or during its formation, would explain not only the metallicity distribution but also the fact that we see no zero metal stars (Bond 1980).

Metal enhanced star formation (MESF). Talbot and Arnett (1973) argued that star formation would be more efficient in regions of already enhanced metallicity as a consequence of more efficient cooling of the gas. For this process alone to solve the G dwarf problem would require a very low star formation rate in metal poor regions, together with large inhomogeneities. Nevertheless, the model can certainly be made to work rather well.

Inflow of unprocessed material. It is unlikely that the region called the solar neighbourhood has evolved at constant mass ($f = 0$). In particular the disk could have formed by gradual accretion on a timescale comparable with that for star formation (Larson 1972: Sciama 1972). An analytic solution was given by Larson (1972) for this model in the case $f = \psi$. This gives a strong peak at $Z = y$ and rather too few stars with $Z < Z_1$, though this is evidently a consequence of the rather naive assumption $f = \psi$. Lynden-Bell (1975) gives a 'Best Accretion Model' which is analytic and fits the data quite well.

6.4 The dynamical collapse model

From a study of the age distribution of globular and open

clusters, Demarque and McClure (1977) and Dermarque (1979) concluded that the formation of the Galactic disk was a long and drawn out process. As mentioned above, Lynden-Bell (1975) pointed out that such an accretion model could also provide a solution for the G dwarf problem. Lynden-Bell's model assumes a particular relation between the quantity of gas and the number of stars, and so the chemical history derived is independent of any assumptions about accretion. The essence of the model is that at first there is a rapid rise in metallicity until a mass M_0 of disk is turned into metals. Provided $M_0 < M_\infty$, the final mass of the Galaxy, then at the end of this phase $Z \sim 3/8$ y. Thereafter, as accretion proceeds, the disk grows and gets more metal rich, reaching $Z \sim 3/4y$ when half of the stars have been formed. The model gives few stars of low Z and most of the stars within a narrow range of metallicities.

More general accretion models have been presented by Chiosi (1980) and Chiosi and Matteucci (1982). Instead of Lynden-Bell's ansatz relating gas and star content, they assume an accretion law that depends exponentially on time:$M \propto \exp(-t/\tau)$ (Hunt 1975), and adopt a star formation law which takes account of the balance between the increased gravitational attraction of the disk and the increase in its scale height due to energy input from short lived stars (Talbot and Arnett 1975). This kind of model simulates the fully dynamical models of Larson (1976) and Tinsley and Larson (1978), and because the back reaction of the newly formed stars on the gas is taken into account in the star formation, it bears some resemblence to the ideas underlying the models of Gerola and Seiden referred to earlier. The model can be quite sophisticated, accounting for spatial variations of accretion rate and the production of different elements. The results for the solar neighbourhood from this model are very good with an accretion timescale $\tau \sim 3$ to 4×10^9 years.

Dynamics and chemical evolution were looked at simultaneously by Tinsley and Larson (1978): they used numerical models of disk galaxies and obtained results for the radial distribution of metallicity similar to those just described. Vader and de Jong (1979, 1981) computed both the kinematic and chemical evolution of the solar neighbourhood, getting good looking results for an accretion timescale of $\sim 5 \times 10^9$ years. An outstanding problem for models attempting to predict the radial distribution of metals in the disk is the question of radial flow of gas. Such radial flow results not only from the transfer of angular momentum by viscosity, but also from spiral and bar-like disturbances in the disk. It is difficult to evaluate the effect of noncircular gas motion quantitatively because of the lack of suitable dynamical models. However it is clear that rapid inflow will soon obliterate any gradients, while slow inflow will cause large gradients.

7. THE GALACTIC HALO

7.1 The simple model

The chemical evolution of the Galactic halo poses special problems because it is very metal poor and yet totally converted into stars. The radial abundance gradient is a further problem. There are two ways around the first problem: the yield y could be lower (by a factor 10), in which case the ratio of the mass of the luminous halo to disk would be very large, or mass could be removed from the star forming process either by winds or by forming a disk. Hartwick (1976) has constructed simple models based on the last hypothesis and finds that it does indeed provide a way of getting the right mean metallicity for the disk.

7.2 The halo Z-gradient

Several processes have been suggested to produce the metallicity gradient in the halo. The oldest of these is Larson's (1969) dissipative collapse model. In this model, because of dissipative processes, the gas sinks inwards building up a central concentration of gas and stars, and at the same time a metallicity gradient. In the outer regions where dissipation is not effective, there is little or no Z gradient. The situation is thus reminiscent of the globular cluster Z distribution where the lack of gradient beyond ~ 20 kpc suggests that collapse began with everything in free fall, dissipation becoming important only at radii around 20 kpc.

An alternative scheme is to attribute the Z gradient to a change in the initial mass function with radius (or metallicity), since this changes the yield y. It could be that either the slope changes, or the lower mass limit for star formation changes. Either way, the preponderance of lower mass stars would change with radius and we would get an increase of the mass to light ratio with increasing radius. Discontinuities in the metallicity distribution may, however, be hard to understand this way. Variations of the IMF have been studied by Chiosi and Matteucci (1982).

Zinn (1980) has described qualitatively a model to account for both the lack of a sizeable metal gradient among the outermost globular clusters and their age dispersion. The model consists of an ensemble of clouds of ~ 10^6 $M_\odot$ which are possible sites of star formation and globular cluster formation. At small radii ($R < 8$ kpc) frequent collisions and mergers among clouds produce larger clouds of higher Z, until eventualy one large cloud is formed. This cloud collapses along its axis of rotation and evolves into a disk that grows as more material is accreted. At radii 8 < R < 40 kpc, collisions are less frequent, but star formation nevertheless goes on. Those low angular momentum clouds fall to the central regions of the protogalaxy where they are rapidly disrupted. In this zone, the age spread of objects must be on the order of the free fall time. Those clouds of high angular momentum remain intact for a long time and the stars and star clusters that form within them have a relatively larger age spread. As the disk grows these clouds are disrupted, but the constituent stars and clusters remain in the same zone. At radii in excess of 40-50 kpc collisions between clouds are rarer, thus stars form in these clouds later than elsewhere since collisions are rare. Again some of these systems will be disrupted if they wander into the central regions or near the disk.

7.3 Mass-metallicity relation for elliptical galaxies

It is evident that if the gas is removed from a star forming galaxy, both star formation and metals production stop. Larson (1974) suggest that supernova explosions would eventually serve to remove the gas: with the hypothesis that smaller galaxies are less tightly bound we arrive at a model where the large galaxies are the most metal rich. If E is the thermal energy imparted to the medium when a unit mass of stars if formed, the gas is driven out when

$$k\frac{GMM_g}{R} \sim EM_s, \quad M_g = M - M_s \tag{23}$$

where M_g and M_s are the mass in gas and stars, and k is a geometrical constant. If we further suppose that there is a mass-radius relation for protogalaxies of the form

$$M \propto R^{\alpha}, \tag{24}$$

the escape condition is

$$M^{1-1/\alpha}\,(M - M_s) \propto M_s \tag{25}$$

The mass of metals ejected is yM_s, and so at early stages when $M_s << M$ we have

$$Z_s \propto y\,\frac{M_s}{M_g} \sim y\,\frac{M_s}{M} \propto M_s^{(\alpha-1)/(2\alpha-1)} \tag{26}$$

where the last expression follows from (25). Since we observe $Z_s \propto M_s^{0.25}$, we need $\alpha = 1.5$ at least on those small enough scales where $M_s << M$ at the time of mass loss. The dynamical consequences of such mass loss has been discussed by Saito (1979) who suggested this mechanism for the origin of the loosely bound dwarf elliptical galaxies.

Tinsley and Larson (1979) and Struck-Marcell (1981) considered a model wherein elliptical galaxies are built up from merging subsystems. The mass-metallicity relation is governed by the efficiency of star formation as a function of the mass of the system. Thus the mass of stars formed when the total mass is M and the mass in gas is M_g is taken as

$$\Delta M_s = M_g(M/M_e)^p \qquad (27)$$

where M_e and p are constants. While $M_s << M$ this yields a relation

$$Z_s \propto M_s^{p/(1+p)} \qquad (28)$$

and hence the value p = 1.3 is adopted for the model. Struck-Marcell (1981) shows how, with a suitable energy dissipation law this kind of model can also produce the $L \propto \sigma^4$ law, so despite the great uncertainties at present surrounding this model it is certainly promising.

8. DYNAMICS AND CHEMISTRY

8.1 The general problem

What is it that determines the morphological type of a galaxy, and how is this related to the chemical properties of galaxies? Since galaxy morphology is a fundamental aspect of galaxies this is an important question, but one that is not easy to answer despite a large amount of data. (See for example Sandage 1961 and Kormendy and Norman 1979). There is a suggestion that the bulge to disk ratio and the gas content lie at the base of this problem. Both these quantities certainly vary along the Hubble sequence, though what causes what is somewhat debatable. The situation is not made any easier by our being unable to give an answer to the question 'which came first - stars or galaxies'? One idea has stars forming first, and hierarchical gravitational clustering building up both galaxies and galaxy clusters (White and Rees 1978), whereas another has vast gaseous clouds collapsing, cooling and fragmenting to form galactic gas clouds, which in turn fragment to form stars (see Shandarin's lecture). If stars formed first, it is important to know their mass spectrum for this determines whether the subsequent evolution is gas dynamics or stellar dynamics dominated. If we wish to be pessimistic we might go on to ask about the role of complicating factors like angular momentum and removal of gas by winds or ram pressure stripping.

Evidently, the only approach is to postulate as an initial condition a protogalaxy of a given mass and angular momentum and examine some simple cases: a purely gas dynamical model (Larson 1969) or a purely stellar dynamical model (Gott 1975, Miller and Smith 1980). The gas dynamical models inevitably involve a need for formulae describing the rate of star formation and dissipative processes. The stellar models contain no gas at any time, and so do not look at chemical evolution. Moreover, they must have sufficient numbers of particles to properly mimic the relaxation of the system.

8.2 Stellar dynamical models

The purely stellar dynamical models are aimed at explaining the origin of certain aspects of elliptical galaxies. In the past most emphasis was laid on understanding the run of density with radius and the variations of ellipticity of the light distribution (see King 1978 for a collection of data). The smooth density profile was set up as a consequence of violent relaxation (Lynden-Bell 1967), but as Gott (1973) realised in his earliest models, that in itself was not enough to explain the $r^{-\alpha}$ with $\alpha \sim 2.7$ to 3 fall-off in density. The model profiles were too steep, and Gott (1975) later proposed models which continually accreted stars. This remedied the profile situation in a plausible way. At that time the question of the ellipticity was resolved by appeal to rotational flattening and redistribution of the angular momentum of the stellar orbits during the collapse phase (Gott and Thuan 1976). We now know, however, that rotation in elliptical galaxies is too insignificant to have an important flattening effect (Illingworth 1977: Binney 1978).

The situation regarding the origin of the density profile has not really changed except that we now have a variety of non-symmetric simulations in which to observe what happens. Miller and Smith (1980) have performed calculations involving ~ 50000 bodies collapsing collisionlessly from quite irregular conditions to form smooth and reasonable-looking relaxed systems. White (1981 and this volume) has considered the origin of the profile shape as a consequence of mergers between stellar systems: a Hubble-type profile emerges though the roles of violent relaxation and two body relaxation are not clear cut since White used a relatively small number of particles. The question of the origin of the shape is particularly important since it cannot be due to rotation. Binney and Aarseth (1978) showed that anisotropic initial shapes would lead to flattened systems now, even in nonrotating galaxies, since there has not been enough time for relaxation processes to make the galaxy rounder.

The low rotation speed of elliptical galaxies indicates that the luminous material did not undergo a dissipative gaseous collapse, and thus gives grounds for believing that the stellar dynamical models are the relevant ones. Why the luminous system formed dissipationlessly is not clear (it certainly contradicts the model proposed by White and Rees 1976): the solution of this dilemma inevitably lies in the formation of the first generation of stars, and hence in some form of the gaseous models. Moreover, we are also left with the question of the formation of disks (which is certainly a dissipative gaseous process, see Efstathiou and Jones 1980) and the formation of bulges, which unlike ellipticals are rotationally supported.

8.3 Gas-dynamical models

As mentioned above, we have no idea of the physical state of the pregalactic gas: how lumpy was it, did it have turbulent motions, what was its viscosity, how does it turn into stars, etc? The collapse was almost certainly not spherical, but should we do the gas dynamics in a pre-existing dark halo of the kind postulated by White and Rees (1978), or should we try to explain everything, including the dark halo, and start simply with an irregular distribution of gas? The only way of proceeding is to try and see what happens, and then see what is learned from these models.

This work has been done mostly by Larson in a series of articles dating back over 13 years. The first models (Larson 1969) were purely spherical and incorporated no chemical evolution. However, they did succeed with a suitable star forming law in producing a reasonable run of density with radius. Chemical evolution and modification of some of the hypotheses in this early model were incorporated in Larson (1974). These latter models showed a chemical abundance gradient and a change in the nature of the stellar orbits from radially elongated ones in the outer regions to isotropic ones in the central region where turbulent motions succeeded in stopping the free fall collapse. In these models, the timescales for star formation, τ_{SFR}, for free fall, τ_{ff}, and for viscous dissipation, τ_{vis}, are all of the same order, and the resultant 'light profile' (Figure 3 of Larson 1974) is remarkable.

The effects of rotation are discussed in Larson (1976), together with subtleties regarding the effect of velocity dispersion and tidal forces on the star formation rate. Were it not for the fact that $\tau_{vis} \sim \tau_{ff}$ the gas would simply settle into a thin disk. However, since $\tau_{vis} \sim \tau_{ff}$ the gas at the centre loses angular momentum at roughly the same rate as it loses energy and so the central regions roughly preserve their ellipticity as the collapse proceeds. Thus the variation of isophote ellipticity with radius in real galaxies may be due to differences in the details of the viscous transfer process in different galaxies. The metallicity profile is not radially symmetric but reflects the fact that in rotating systems the gas collapse occurs preferentially along the rotation axis. Thus the models produce a strong Z gradient perpendicular to the plane, and a mild one in the plane.

A potential difficulty with the Larson model is the need for a slow dissipation τ_{diss} and a large viscosity. The fact that $\tau_{vis} \sim \tau_{diss}$ implies a low Reynolds number (O(1)). Without large dissipation it is difficult to form a reasonable disk. Furthermore, the star formation rate cannot be too high early on since we do not wish too many stars to form before the gas settles to a disk. This allows us to speculate that the disk to bulge ratio in galaxies may be a consequence of the value of the star formation rate. Precisely why the star formation rate should vary in this way is not known. Two effects have been suggested, but the models indicate that they are less important than originally suspected. Variations of angular momentum per unit mass is one possibility, but this simply flattens the spheroidal component as the angular momentum increases. A second suggestion (Gott and Thuan 1976) suggested that the variations in the pregalactic density would work. However, the effect is not very marked. One suggestion that worked was due to Larson (1976) and that is that tidal forces would inhibit small scale instabilities. This was modelled by the relation

$$\rho_* = A\rho_g{}^S \left[1 + B \frac{\langle\rho(z)\rangle}{\rho_g(z)} \right]^{-1} \tag{29}$$

where A and B are suitable constants and other quantities are as stated before. The expression (and the evaluation of the constants A,B) is somewhat ad hoc, but the model performs quite impressively.

9. CONCLUDING REMARKS

It is difficult to conclude about a subject that is as rapidly changing as this. Rapid advances are being made on both the observational and theoretical fronts. However, whereas a clearer picture is emerging from the samples of stars studied by Twarog and others, there are areas where confusion reigns (such as the question of the metallicity scale of the globular clusters). On the extragalactic front we are rapidly learning more about our Galaxy in relation to others, though there are major difficulties in determining the metallicity of systems of stars which have yet to be overcome. The major impediment to theoretical progress is an almost total ignorance of how stars form. Indeed it is perhaps remarkable that our naive models go so far towards producing reasonable looking models at all. A depressing thought is that even if we knew about star formation here and now in our own galaxy, the conclusions may not necessarily apply in different chemical and dynamical environments. At least, our better understanding of the dynamical aspects of galaxy formation gives us a framework within which to follow the evolution of stars and gas.

References

Alloin, D., Collin-Souffrin, S., Joly, M. and Vigroux, L. 1979. Astron.Astr., 78, 200.
Audouze, J. and Vauclair, S. 1980. An Introduction to Nuclear Astrophysics, D. Reidel Publ. Co., Dordrecht, Holland.
Becker, S.A. and Iben, I. Jr. 1979. Ap.J., 232, 831.
Binney, J. 1978. MNRAS, 183, 501.
Binney, J. and Aarseth, S.J. 1978. MNRAS, 185, 227.
Boisse, P., Gispert, R., Coron, N., Wynbergen, J.J., Serra, G., Ryter, C. and Puget, J.L. 1981. Astron.Astrophys., 94, 265.
Bond, H.E. 1981. Ap.J., 248, 606.
Boury, A., Grevesse, N. and Remy-Battian, L. 1979. Les elements et leurs isotopes dans l'univers, Liege Colloque No. 22, June 1978.
Burki, G. 1977. Astron.Astrophys., 57, 135.
Burki, G. 1978. Astron.Astrophys., 62, 159.
Cameron, A.G.W. 1982. To be published in Festschrift for Willy Fowler's 70th birthday.
Caputo, F. 1982. To be published.
Carney, B.W. 1979. Ap.J., 233, 877.
Chieze, J.P. and Lazareff, B. 1980. Astron.Astrophys., 91, 290.
Chiosi, C. 1980. Astron.Astrophys., 83, 206.
Chiosi, C. and Matteucci, C.M. 1982. To be published.
Demarque, P. 1980. In Star Clusters, IAU Symposium No. 85, ed. J.E. Hesser, p. 281.
Demarque, P. and McClure, R.D. 1977. Ap.J., 213, 716.
D'Odorico, S., Peimbert, M. and Sabbadin, F. 1976. Astron. Astrophys., 47, 341.
Efstathiou, G. and Jones, B.J.T. 1980. Comments Astrophys., 8, 169.
Eggen, O.J., Lynden-Bell, D. and Sandage, A.R. 1962. Ap.J., 136, 748.
Faber, S.M. 1973. Ap.J., 179, 731.
Gerola, H. and Seiden, P.E. 1978. Ap.J., 223, 129.
Gerola, H., Seiden, P.E. and Schulman, L.S. 1980. Ap.J., 242, 517.
Gott, J.R. 1973. Ap.J., 186, 481.
Gott, J.R. 1975. Ap.J., 201, 296.
Gott, J.R. and Thuan, T.X. 1976. Ap.J., 204, 649.
Griersmith, D. 1980. Astr.J., 85, 1135.
Guibert, J., Lequeux, J. and Viallefond, F. 1978. Astron. Astrophys., 68, 1.
Harris, W.. and Canterna, R. 1979. Ap.J.Lett., 231, L19.
Hartwick, F.D.A. 1976. Ap.J., 209, 418.

Herbst, W. and Assousa, G.E. 1977. Ap.J., 217, 473.
Hunt, R. 1975. MNRAS, 173, 465.
Iben, I. and Truran, J.W. 1978. Ap.J., 220, 980.
Illingworth, G. 1977. Ap.J.Lett., 218, L43.
Janes, K.A. 1975. Ap.J.Suppl., 29, 161.
Janes, K.A. 1979. Ap.J.Suppl., 39, 135.
Kaler, J.B. 1980. Ap.J., 239, 78.
King, I.R. 1978. Ap.J., 222, 1.
Kormendy, J. and Norman, C.A. 1979. Ap.J., 233, 539.
Kunth, D. 1981. Thesis presented at University of Paris VII.
Larson, R.B. 1969. MNRAS, 145, 405.
Larson, R.B. 1972. MNRAS, 157, 121.
Larson, R.B. 1974. MNRAS, 166, 585.
Larson, R.B. 1975. MNRAS, 173, 671.
Larson, R.B. 1976. MNRAS, 176, 31.
Laurent, C., Vidal-Madjar, A., Audouze, J., Lequeux, J., Vigroux, L., Combes, M. and Encrenaz, T. 1979. In Les elements et leurs isotopes dans l'univers, Liege Colloque No. 22, June 1978, p. 515.
Lequeux, J. 1979a. Astr.Astrophys., 71, 1.
Lequeux, J. 1979b. Astr.Astrophys., 80, 35.
Lequeux, J. 1981. Comments Astrophys., 9, 117.
Lequeux, J., Peimbert, M., Rayo, J.F., Serrano, A. and Torres-Peimbert, S. 1979. Astron.Astrophys., 80, 155.
Luck, R.E. and Bond, H.E. 1980. Ap.J., 241, 218.
Lynden-Bell, D. 1967. MNRAS, 136, 101.
Lynden-Bell, D. 1975. Vistas in Astronomy, 19, 299.
Mayor, M. 1976. Astr.Astrophys., 48, 301.
Meyer, J.P. 1979. In 16th International Cosmic Ray Conference held at Kyoto Japan.
Miller, G.E. and Scalo, J.M. 1979. Ap.J.Suppl., 41, 513.
Miller, R.H. and Smith, B.F. 1980. Ap.J., 235, 793.
Mueller, M.W. and Arnett, W.D. 1976. Ap.J., 210, 670.
Nomoto, K. 1981. In Fundamental Problems in the Theory of Stellar Evolution, IAU Symp. No. 93, ed. D. Sugimoto, D.Q. Lamb and D.N. Schramm, Reidel, Dordrecht, Holland, p. 295.
Pagel, B.E.J. and Edmunds, M.G. 1981. Ann.Rev.Astron.Astrophys. 19, 77.
Pagel, B.E.J. and Patchett, B.E. 1975. MNRAS, 172, 13.
Penzias, A.A. 1979. Ap.J., 228, 430.
Rees, M.J. 1976. MNRAS, 176, 483.
Renzini, A. and Voli, M. 1981. Astron.Astrophys., 94, 175.
Roberts, W.W. 1969. Ap.J., 158, 123.
Romanishin, W. and Angel, J.R.P. 1980. Ap.J., 235, 992.
Saito, M. 1979. P.A.S.J., 31, 193.
Salpeter, E.E. 1955. Ap.J., 121, 161.
Sandage, A.R. 1961. The Hubble Atlas of Galaxies, Carnegie Institute of Washington.
Schmidt, M. 1963. Ap.J., 137, 758.
Sciama, D.W. 1972. Nature 240, 456.

Searle, L. and Zinn, R. 1978. Ap.J., 225, 357.
Seiden, P.E. and Gerola, H. 1979. Ap.J., 233, 56.
Seiden, P.E., Schulman, L.S. and Gerola, H. 1979. Ap.J., 232, 702.
Silk, J. 1980. in Star Formation, Saas-Fee, ed. A. Maeder and L. Martinet, p. 131.
Sneden, C., Lambert, D.L. and Whitaker, R.W. 1979. Ap.J., 234, 964.
Struck-Marcell, C. 1981. MNRAS, 197, 487.
Talbot, R.J. and Arnett, W.D. 1971. Ap.J., 170, 409.
Talbot, R.J. and Arnett, W.D. 1973. Ap.J., 186, 51 and 69.
Talbot, R.J. and Arnett, W.D. 1975. Ap.J., 197, 551.
Talent, D.L. and Dufour, R.J. 1979. Ap.J., 233, 888.
Terlevich, R., Davies, R.L., Faber, S.M. and Burstein, D. 1981. MNRAS, 196, 381.
Tinsley, B.M. 1980. Fund.Cosmic Phys., 5, 287.
Tinsley, B.M. and Danly, L. 1980. Ap.J., 242, 435.
Tinsley, B.M. and Larson, R.B. 1978. Ap.J., 221, 554.
Tinsley, B.M. and Larson, R.B. 1979. MNRAS, 186, 503.
Trefzger, Ch.F. 1981. Astron.Astrophys., 95, 184.
Truran, J.W. and Cameron, A.G.W. 1971. Astrophys.Sp.Sci., 14, 179.
Tuchman, Y., Sack, N. and Barkat, Z. 1979. Ap.J., 234, 217.
Twarog, B.A. 1980. Ap.J., 242, 242.
Twarog, B.A. 1981. Astr.J., 86, 386.
Vader, J.P. and de Jong, T. 1979. In Les elements et leurs isotopes dans l'univers, Liege Colloque No. 22, June 1978.
Vader, J.P. and de Jong, T. 1981. Astron.Astrophys., 100, 124.
Van den Bergh, S. 1962. Astr.J., 67, 486.
Visvanathan, N. and Griersmith, D. 1977. Astron.Astrophys., 59, 317.
Wannier, P.G. 1980. Ann.Rev.Astron.Astrophys., 18, 399.
Weaver, T.A. and Woosley, S.E. 1980. In 9th Texas Symposium on Relativistic Astrophysics, ed. J. Ehlers, J.J. Perry and M. Walker, p. 335.
Wheeler, J.C. 1980. In Type I Supernovae, Proc. of Texas Workshop, ed. J.C. Wheeler, p. 185.
Wheeler, J.C. 1981. In Fundamental Problems in the Theory of Stellar Evolution, IAU Symp. No. 93, ed. D. Sugimoto, D.Q. Lamb and D.N. Schramm, Reidel, Dordrecht, Holland, p. 285.
White, S.D.M. and Rees, M.J. 1978. MNRAS, 183, 341.
Woodward, P.R. 1976. Ap.J., 207, 484.
Wyse, R.F. 1982. MNRAS, 199, 1P.
Zinn, R. 1980. Ap.J., 241, 602.

STELLAR DYNAMICAL PROCESSES IN THE EVOLUTION OF GALAXIES AND GALAXY CLUSTERS

Simon D. M. White

Astronomy Department and Space Sciences Laboratory,
University of California, Berkeley

1. INTRODUCTION

Many important effects which influence the observed structure of galaxies and of galaxy clusters are purely stellar dynamical in nature and can be viewed as different manifestations of relaxation processes in inhomogeneous non-equilibrium systems. Three general types of relaxation process may be distinguished. The first, kinematic phase mixing, is the stellar dynamical analogue of Landau damping and results in the effective loss of much information about initial conditions in systems which have been in existence for many characteristic dynamical times. It is the supposition that phase-mixing can be highly efficient within the lifetime of a galaxy which leads to the traditional assumption that the phase space distribution of stars in an elliptical galaxy must satisfy Jeans' theorem and be a function only of the isolating integrals of motion.

Violent relaxation, as described by Lynden-Bell (1967), is the dynamical aspect of phase mixing and involves the scattering of particle orbits by the time-dependent gravitational field which results from coherent changes in density as a system evolves. It is of great importance during galaxy formation and during the formation and evolution of galaxy clusters. Finally, collisional or two-body relaxation is the stellar dynamical equivalent of thermal relaxation in a gas. As a result of the long range nature of the gravitational force and of the non-existence of bounded maximum entropy states for finite isolated self-gravitating systems, the effects of collisional relaxation are rather different in galaxies and galaxy clusters than in a thermalizing gas.

B. J. T. Jones and J. E. Jones (eds.), The Origin and Evolution of Galaxies, 227–245.

Because the N-body equations of motion are time-reversible, no information is really lost during the evolution of a stellar system: effective information loss occurs in two ways, however. Both violent relaxation and phase mixing can be described using the Vlasov equation

$$\frac{\partial f}{\partial t} + \underline{v} . \frac{\partial f}{\partial \underline{r}} - \underline{\nabla \phi} . \frac{\partial f}{\partial \underline{v}} = 0, \tag{1}$$

for the phase space density distribution $f(\underline{r},\underline{v},t)$: violent relaxation requires the simultaneous solution of Poisson's equation while pure phase mixing may not. Since this equation is also time-reversible, all information on the initial distribution function of a collisionless system is contained in the detailed $f(\underline{r},\underline{v},t)$ at later times: effective information loss occurs because of coarse graining - in practice $f(\underline{r},\underline{v},t)$ is only known with limited resolution and evolution transfers much initial information to progressively smaller scales in phase space where it is no longer accessible.

When collisional relaxation is important the Vlasov equation no longer applies and the previous history of any given system can only be reconstructed if the actual positions and velocities of all its constituent particles are known rather than just their phase space density: the collisional Boltzman equation is not time-reversible. Relevant information is thus lost in constructing $f(\underline{r},\underline{v},t)$ as well as in coarse-graining it.

The following article gives simplified analytic formulae for the effects of relaxation processes in a number of situations which are of importance for galaxy and cluster evolution. Unfortunately an analytic approach is successful in only a rather limited number of problems: many problems of astrophysical interest can be treated adequately only by direct computer simulations. Such simulations are subject to technical limitations which can be quite severe and the latter part of this article gives a comparative discussion of different computational approaches followed by a detailed description of their application to the problem of galaxy collisions. A final section discusses some aspects of the computer simulation of the formation of galaxy clusters.

2. ANALYTIC FORMULAE FOR 'ORDER-OF-MAGNITUDE' ESTIMATES

For point particles the N-body equations immediately yield the virial theorem for a system in equilibrium

$$\sum m_i v_i^2 - \sum_{i>j}\sum G m_i m_j / r_{ij} = 2T + W \simeq 0. \tag{2}$$

Note that this form of the theorem holds only for point particles. In numerical work it is often convenient to use a 'softened' potential,

$$\Phi_{ij} = -Gm_i m_j/(r_{ij}^2+\varepsilon_{ij}^2)^{1/2}, \tag{3}$$

in order to avoid singularities when particles approach one another. For such a potential the virial theorem takes the form:

$$2T + W + \sum_{i>j}\sum Gm_i m_j \varepsilon_{ij}^2/(r_{ij}^2+\varepsilon_{ij}^2)^{3/2} \approx 0. \tag{4}$$

It is useful to define characteristic scales for gravitating systems as follows:

$$M = \sum_i m_i; \quad V^2 = 2T/M; \quad R_g = -GM^2/2W; \quad t_d = 2R_g/V, \tag{5}$$

where M, V, R_g and t_d are then the total mass, the dispersion velocity, the gravitational radius and the dynamical or crossing time respectively. For systems which resemble star clusters, galaxies, or galaxy clusters R_g is normally close to R_h, the radius containing half of the system's mass, while t_d is of order one third the typical orbital time of particles near the half-mass radius. From the virial theorem $V^2 \simeq GM/2R_g$.

Chandrasekhar (1942) showed that a star moving through an infinite uniform background of stars, with velocity dispersion V and density ρ, will deviate strongly from its initial direction of motion as a result of encounters with other objects after a time

$$t_r = \frac{V^3}{17.7\ G^2 m\rho\ \ln(R_{max}/R_{min})} \tag{6}$$

where m is the mass of the stars, $R_{min} = 2Gm/V^2$ and R_{max} is the maximum distance at which encounters can perturb a star's motion. More recent theoretical work and comparison with numerical simulations show $0.6R_g$ to be the appropriate value of R_{max} in equations such as (6) (Henon 1975). The discrete nature of stars is responsible for collisional relaxation and appears in this equation through the m in the denominator: if stars in a given system were divided into smaller and smaller objects t_r would increase like m^{-1}. In any real system t_r is a strong function of position because of its variation as ρ^{-1}. As a result 2-body relaxation processes usually operate much more rapidly near the centre than near the edge of stellar systems. In most galaxies relaxation effects due to individual stars are quite negligible. A characteristic relaxation time is conventionally defined for inhomogeneous systems using their overall velocity dispersion and the mean density within their half mass radius (Spitzer and Hart 1971). This leads to

$$t_r/t_d = 0.06N/\ln(0.15N). \tag{7}$$

The long relaxation time of galaxies is a result of the enormous number of stars they contain. A clear and much more extensive discussion of collisional relaxation processes is given by Henon (1973b), who shows how they may be modelled by a Fokker-Planck diffusion equation in which a dynamical friction term counter-balances a diffusive spreading term and leads to the formal possibility of a Maxwellian equilibrium velocity distribution.

If we consider the orbit of a heavy particle through a system of much lighter particles, the effect of encounters is quasi-deterministic rather than diffusive. This is because the orbital speed of the massive object is similar to that of the background particles and so its kinetic energy is much greater. In such a situation the dynamical friction term greatly exceeds the diffusive term in the Fokker-Planck representation of the relaxation process and the heavy particle experiences an effective mean force,

$$\underline{F} = -4\pi G m^2 \rho(v) \ln(R_{max}/R_{min}) \underline{v}/v^3, \tag{8}$$

where m is the mass of the heavy particle, $\underline{v}$ its velocity, and $\rho(v)$ the density of background particles with speed less than v (Chandrasekhar 1942). For small v ($\ll$ V), $\rho(v) \propto v^3$, and $F \propto v$ as with a normal frictional force: for large v ($\gg$ V) $F \propto v^{-2}$ and so becomes ineffective. The m^2 dependence of $\underline{F}$ means that massive particles decelerate at a rate proportional to their mass, and indeed comparison of (6) and (8) shows that a massive particle slows down in a timescale equal to the relaxation time the system would have if it were made entirely of particles of mass m. Notice that the mass of the background particles does not appear in (8) but only their density: this means, for example, that the effect of dynamical friction on the orbit of a massive galaxy in a cluster is the same whether the 'missing' mass is made of neutrinos, faint stars or 10^6 $M_\odot$ black holes.

As a simple model for the effects of dynamical friction in real systems, consider a heavy particle, mass m, moving in a circular orbit of radius r within a system which can be described as a singular isothermal sphere of one-dimensional velocity dispersion σ. Such a system has

$$\rho(r) = \sigma^2/2\pi G r^2, \quad M(r) = 2\sigma^2 r/G;$$

although it has infinite extent, total mass and central density, it is not a bad model for intermediate regions of many real stellar systems. The evolution of the orbit of the heavy particle is easily calculated from its energy loss rate (given by the product of (8) with the circular velocity) and its radial distance as,

$$r(t)/r(t_o) = (1-(t-t_o)/T)^{1/2},$$
$$T = \frac{0.19\ t_{orb,i}}{\ln(R_{max}/R_{min})}\ \frac{M(r_i)}{m}, \qquad (9)$$

where $t_{orb,i}$ is the initial orbital period. The number of initial periods required for a heavy object to reach the centre thus depends primarily on the ratio of its mass to the mass enclosed within its orbit. Notice that in centrally concentrated systems the effect of dynamical friction is not to slow particles down (in the above example the orbital velocity is constant) but rather to cause them to move towards the centre of the system in which they are embedded. Equation (9) shows the result that evolution speeds up as particles move inwards, and thus that the rate at which objects reach the centre is determined mainly by the rate at which they begin to move in from distant regions, rather than by processes operating at the system centre.

Dynamical friction models have been used to explain the formation of galactic nuclei by infall of globular clusters (Tremaine et al. 1975), the accretion of dwarf satellite galaxies by their parent objects (Tremaine 1976) and the formation of cD galaxies by the piling up of galaxies at the centre of a cluster (Ostriker and Tremaine 1975). White (1976a) showed that the weak degree of luminosity segregation in the Coma cluster implies that only a small fraction of the cluster mass can be attached to the galaxies, otherwise dynamical friction would have brought all the luminous galaxies to the cluster centre. Surprisingly enough, although the dynamical friction formula is derived for linear orbits through an infinite uniform medium, direct numerical tests show it to work quite well for circular orbits in a spherical, highly inhomogeneous system (White 1978).

Although galaxies may be island universes which at some level can be considered as independent systems, their isolation is not sufficient to ensure that they avoid each other's influence for their entire lifetime. Indeed, if the mass distribution of galaxies is extended to large radii by massive dark halos, close gravitational encounters become a common and important part of galactic life. Such encounters lead to tidal stripping of the outer parts of galaxies, to the conversion of relative orbital kinetic energy into internal thermal motions within each object, and to the capture of colliding pairs onto bound orbits from which they later coalesce into a single system.

The simplest description of tidal truncation is via the classical Roche theory. Consider, for definiteness, a globular cluster with mass profile $m_c(r)$ on a circular orbit of radius R_0 about a galaxy with mass profile $M_g(R)$. Equating the galactic tidal force on a star at radius r within the cluster to the cluster restoring force leads to a tidal radius r_t given by

$$m_c(r_t)r_t^{-3} = (2-\frac{d\ln M_g}{d\ln R})M_g(R_0)R_0^{-3} \tag{10}$$

Thus the cluster is limited at a radius such that its mean density is similar to the mean density of the galaxy within its orbit. This statement is equivalent to placing the tidal limit at the radius where the orbital period of a star in the cluster becomes equal to that of the cluster about the galactic centre. At larger radii cluster and star follow independent orbits about the galaxy and are separated by phase mixing: at smaller radii the orbital frequency of the star is sufficiently high that its orbit preserves an adiabatic invariant in the changing galactic field and so is shielded from long term evolution.

This kind of process appears to limit the extent of globular clusters in real galaxies (King 1962) and ensures that in rich galaxy clusters the total mass cannot be attached to individual galaxies in halos with the kind of velocity dispersion inferred from galactic rotation curves. Stripped material from the outer parts of galaxies may make up much of the outer envelope of cD galaxies in rich clusters (Gallagher and Ostriker 1972).

In rich clusters another closely related truncation mechanism plays an important role. This is the impulsive injection of energy into the outer layers of galaxies by rapid tidal encounters between individual objects. A star, distance r from the centre of a galaxy in the direction of the pericentre of a perturber of mass M_p which passes by rapidly at velocity V_p and distance d, experiences an impulsive velocity change

$$\Delta v \sim r\,\frac{GM_p}{d^3}\,\frac{d}{V_p}$$

Averaged over all stars at the same position the change in velocity dispersion is

$$\Delta v^2 = \overline{(\underline{v}_i+\Delta\underline{v})^2} - \overline{v_i^2} = |\Delta\underline{v}|^2 .$$

A more careful calculation averaging over all objects at radius r in the target galaxy was carried out by Spitzer (1958) and gives a mean change in energy/unit mass of

$$\Delta E = \frac{4}{3}\;\frac{G^2M_p^2r^2}{d^4V_p^2}\,. \tag{11}$$

Estimating the initial energy/unit mass of these stars by $E \sim -GM_g(r)/2r$ leads to

$$\Delta E/E \sim \frac{M_p d^{-3}}{M_g(r) r^{-3}} \quad \frac{V_{para}{}^2}{V_p{}^2}, \tag{12}$$

where V_{para} is the relative velocity of the two objects for a parabolic encounter at distance d. Since stripping corresponds to $\Delta E/E \sim 1$, (12) reduces to the Roche limit (equation 10) for objects in a bound orbit (corresponding to $V_{para}/V_p \sim 1$).

For strongly hyperbolic encounters, stripping in a single encounter is less efficient than suggested by the Roche criterion but a series of such encounters can strip galaxies to radii smaller than the Roche limit because the large collision velocities require that stars must have smaller orbital periods in order to be protected from tidal damage by their adiabatic invariants. Simulations of galaxy collisions in rich clusters have been carried out by Richstone (1976) and show that massive halos are indeed stripped to relatively small radii by successive encounters. More recent work by Farouki and Shapiro (1981) studies the effects of collisions on disk galaxies in clusters, and Dekel <u>et al.</u> (1980) have considered the tidal effects of slower collisions.

In a slow encounter the mean total energy change/unit mass is given approximately by averaging (11) and setting $V_p = V_{para}$. This energy change must occur at the expense of the relative orbital energy of the two protagonists. Thus in a weakly hyperbolic encounter between identical objects the two galaxies will become bound to each other if their initial relative velocity at large separation satisfies

$$MV_\infty{}^2/4 < 8\, G^2M^3\overline{r^2}/3V^2{}_{para}\, d^4, \quad d^2 > \overline{r^2}\,.$$

Setting $V_{int}^2 = GM/2R_g$, $V^2{}_{para} = 4GM/d$ this inequality becomes

$$V_\infty^2/V_{int}{}^2 < \frac{16}{3} \quad \frac{R_g\overline{r^2}}{d^3}\,. \tag{13}$$

If two stellar systems undergo a nearly overlapping encounter with a relative velocity at infinity comparable to their internal velocity dispersion, the loss of orbital energy is sufficient to throw them into a bound orbit. Subsequent collisions should cause an even more efficient loss of orbital energy and the two galaxies are thus expected to coalesce rapidly into a single object. This possibility was first discussed in the present context by Alladin (1965) and has been amplified by him in more recent work (Alladin *et al.* 1975).

Notice that since the amounts of orbital energy dissipated in such collisions can be comparable to the binding energy of the individual galaxies, one might expect a significant amount of material to become unbound during the coalescence process and to be lost to infinity. Unfortunately the impulsive approximation on which (13) is based is clearly very uncertain for the slow collisions which lead to strong tidal interactions and subsequent mergers. Although it is convenient as a basis for calculating the cumulative effect of collisions (Knobloch 1978) and the parameter dependence of merger cross-sections (White 1979a), it is not even evident whether it gives the correct scaling for such processes (*cf* the discussion in Richstone 1976, White 1979a and Dekel *et al.* 1980).

In practice numerical simulations are required to gain an understanding of the effects of collisions in many situations of interest and it is to simulation techniques that we now turn. The size of the effect predicted by equations (12) and (13) ensures that tidal disruption and merging will be of paramount importance in any dissipationless clustering hierarchy of the kind that it is popular to use to describe many aspects of galaxy formation and clustering.

3. N-BODY TECHNIQUES

The standard N-body problem involves solving the equations

$$m_i \ddot{\underline{x}}_i = - \sum_{j \neq i} \nabla_i \Phi_{ij}, \quad i = 1, \ldots N, \tag{14}$$

which describe the motions of N point-like particles influencing each other through the potentials Φ_{ij}. In most problems in extragalactic astronomy the number of truly point-like objects is exceedingly large and is far beyond what can be treated computationally. Since this entrains very long collisional relaxation times a natural approach to simulating situations of interest might be via the Vlasov equation, rather than the full N-body equations. Unfortunately the high dimensionality of the phase space required to treat realistic problems precludes a direct solution with reasonable resolution and the only feasible simulation method is usually a Monte Carlo treatment which follows individual particles identified with representative elements of phase fluid. In many ways such a calculation resembles a full solution of the N-body equations; the important practical difference lies in the fact that the potential is calculated using a smooth density of which the actual particle positions are a stochastic representation, rather than the actual particle-particle interaction potential Φ_{ij}. The action of producing this smooth density and potential naturally entrains a coarse graining of the position dependence of the phase-space density and imposes a spatial resolution limit on the calculation. In practice the most efficient way of calculating the smoothed potential required to advance the particles is via a Fast Fourier Transform technique or its analogue, and thus requires a fixed Eulerian coordinate grid within which the particles can move.

Such methods have been hfghly successful for simulations of the dynamics of disk galaxies where most particles remain distributed fairly uniformly in a fixed volume (e.g. Sellwood, this conference) but they are much less well suited to the collision and violent relaxation problems we shall concentrate on below. Miller and Smith (1980) have carried out galaxy collision simulations within a 64^3 coordinate mesh using a FFT approach. Although their grid is the largest on which it has so far been feasible to carry out calculations, their results show it to be too small to follow two well-resolved objects from an initially well separated situation through a collision to either another well separated situation or until the two galaxies have merged and violent relaxation has ceased. Considerably better resolution can be obtained by imposing axial symmetry and so reducing the dimensionality of the problem. Thus van Albada and van Gorkom (1977) were able to study the head-on collision of spherical galaxies with relative ease. The reversibility of their simulations demonstrates that the coarse graining associated with the calculation of the potential has relatively little effect on the results.

A novel technique has been introduced by Villumsen (1982): he represents the potential of a particle at position $\underline{R}$ relative to some chosen origin by its tesseral harmonic expansion,

$$\phi(\underline{r}) = \begin{cases} \dfrac{Gm}{R} \sum\limits_0^{\infty} (r/R)^n P_n(\cos\theta); & r < R, \\[2ex] \dfrac{Gm}{R} \sum\limits_0^{\infty} (R/r)^{n+1} P_n(\cos\theta); & R < r, \end{cases} \qquad (15)$$

where $\cos\theta = \underline{r}.\underline{R}/(rR)$. This expansion is truncated at $n = 4$, say, and the force on each particle due to the first five terms in (15) can then be calculated without requiring a sum over all the other particles. For the purposes of calculating ϕ, each particle is thus represented as a certain surface density distribution on a sphere of radius R, whereas it is moved as a point mass. This method thus retains a number of the advantages of the FFT grid methods while employing an essentially Lagrangian technique that does not suffer from scale limitations. It allows the evolution of systems of several thousand particles to be followed for several dynamical times at fairly modest cost. It should be noted, however, that in some ways it does not behave at all like a normal N-body system.

However, when the expansions are truncated in (15), particle pairs no longer obey Newton's third law (this is easily seen by considering the case where only the first term is retained). As a result momentum will not be conserved. If the expansion centre is allowed to move, energy and angular momentum will not in general be conserved either. This may be quite significant in situations such as tidal encounters where asymmetric particle distributions about each density centre are responsible for important dynamical effects. It is also clear that the scheme will work well only if the origin of the potential expansion is chosen near to the density centre of each significant mass concentration: in practice this means that it is best suited to problems with only a single density centre and becomes completely impracticable for highly inhomogeneous situations.

When a dynamical system has some imposed symmetry the spurious relaxation effects related to the artificially low values of N for which (14) can be solved directly can by reduced by exploiting the symmetry. Thus in a spherically symmetric situation one can follow the evolution of a system of spherical shells thereby removing most of the relaxation due to close encounters (note that small number effects can still introduce significant spurious relaxation near the centre of such systems). This highly idealised approach has been principally applied to star clusters, to galaxies and to the formation of large-scale structures such as galaxy clusters (e.g. Henon 1973a: Silk and Wilson 1979: Hoffman et al. 1980). As we shall see below it is likely that in many cases it may miss the dominant relaxation mechanisms because of the high degree of symmetry imposed. Nonetheless the simplicity of spherical shell models and their close relation to the standard Friedmann cosmological solutions make them a useful tool once their limitations are appreciated.

For the simulation of the dynamical evolution of highly inhomogeneous situations there is still no really competitive alternative to a direct integration of the N-body equations of motion. The Lagrangian nature and the lack of assumed or imposed symmetry conditions make this method very flexible. Present computers and integration techniques are able to follow systems of up to 10^3 particles for several dynamical times at reasonable (though not small!) cost. The N particles can be thought of as a random sampling of the much larger number of stars in the real system. Unfortunately each particle in the simulation contains a much larger fraction of the total system mass than a star in the object being simulated, and as a result 2-body relaxation is much faster in the simulation than it should be (cf. equation 7). This problem can be reduced considerably by replacing the point mass interaction potential by a softened potential such as that given by equation (3). This softening will only have a significant effect provided

$$\varepsilon_{ij} \gg R_{min,ij} = \frac{G(m_i+m_j)}{V^2} = \frac{2(m_i+m_j)}{M} R_g. \qquad (16)$$

Since the introduction of ε introduces a length scale below which results cannot be believed we also require $\varepsilon_{ij} \ll R_g$ to get useful results: a reasonable compromise might be, $\varepsilon^2 = 4R_g^2/N$. A softening of the potential of this kind has several effects which are relevant to the simulation of large N systems. It strongly suppresses large angle deflections in particle encounters and so reduces the evaporative loss of particles from a bound system to a very low rate. It causes a formal increase in the relaxation time in that R_{min} in equation (6) should be set equal to ε rather than to $2Gm/V^2$ (cf. White 1976b, 1978). Further it becomes impossible to form binary particles of sufficient binding energy to affect the overall dynamics, and in fact binary formation becomes entirely negligible.

Some direct tests of the importance of relaxation effects in relatively small N systems (N ~ 100–500) are given by White (1978, 1979b). These show that in spherical or near spherical systems the theoretical formulae appear to be underestimates, if anything, of the timescales on which small particle number gives rise to secular changes in structure. Most such systems remain stable over much longer periods than the dynamical timescale on which violent relaxation takes place.

Simulations like those of Gerhard (1981) show, however, that it is much more difficult to produce a stable disk system out of a small number of particles, even if the disk is embedded in a stabilizing halo and given a velocity dispersion well in excess of that required by the standard local stability criterion. Random fluctuations in the local disk density amplify and coalesce to produce a coherent bar-like structure on a timescale of order half the orbital period near the half-mass radius. This structure redistributes angular momentum and then breaks up to leave a disk with about 20 per cent of its kinetic energy in random motions and a much higher central density than the initial system. For an exponential disk embedded in an equal mass halo of similar extent this instability can be strongly suppressed by increasing the softening parameter ε to $\sim 0.4\ \alpha^{-1}$, where α is the exponential scale factor. The virtues of a stable disk must be balanced against the uncomfortably large softening needed for stability.

The above discussion illustrates an important reservation that must be made about using direct N-body integrations to study the violent relaxation of effectively collisionless systems. The technique, although flexible and powerful, is quite crude and can only give useful results for the grosser aspects of the evolution studied. Direct N-body results will never be very useful for the study of the stability of disks in spiral galaxies: an Eulerian Fourier-type code is much better suited to this kind of work and must therefore be preferred. Direct N-body integration should only be used on problems which require both their large dynamic range and their lack of symmetry restrictions.

A problem of this type which is presently being studied intensively is the tidal interaction between colliding galaxies which, for low velocity encounters, will result in a merger of the two protagonists (cf. equation (13)). Simulations of head-on collisions between spherical galaxies which exploit the axial symmetry of this special case have been published by van Albada and van Gorkom (1977) and by Toomre (1977). The first two authors use an axially symmetric Fourier method and show results for several collisions (cf. also Toomre 1977). The reversability of their calculations, even those in which a single object is formed from two progenitors, demonstrates beautifully the reversability of the violent relaxation process. Evidently the coarse graining associated with the construction of the density from their particle positions is not sufficient to blur out all memory of initial conditions in their calculations. As pointed out by Toomre (1977) a comparison of analytic results with these calculations shows that, if used carefully, the impulsive approximation can yield remarkably accurate results even for low velocity encounters.

Van Albada and van Gorkom (1977) point out that the main interaction mechanism in their head-on collisions is a coherent bounce phenomenon. As two galaxies overlap, each one is impulsively compressed by the gravitational field of the other. One internal dynamical time later the two centres have separated and the rebound within each galaxy is almost unconstrained. In this way considerable orbital energy is fed into the internal energy of the two galaxies and many particles are ejected into weakly bound orbits while the two cores coalesce. Miller and Smith (1980), using a 64^3 grid and a Fast Fourier method, found this effect to be quite important in small impact parameter collisions in the three-dimensional case. Their calculations show quite clearly that the Eulerian Fourier method is not well suited to this problem: they are forced to start their galaxies quite close together in order to have all the particles initially within their active volume, and to terminate most of their calculations before the final merger because of particle losses over the boundary.

At present successful simulations of the fully 3-dimensional collision and merger problem have only been carried out using the tesseral harmonic scheme (Villumsen 1982) and direct N-body methods (White 1978, 1979b: Roos and Norman 1979: Gerhard 1981). The work of White (1978, 1979b) confirmed the coherent bounce phenomenon in head-on collisions, but showed that the main interaction mechanism in glancing collisions is a coupling of the time-dependent tidal potential acting on each galaxy to those of its stars whose orbits are such that they move in phase with the perturbing potential during the encounter. Thus tidal effects act mainly on those stars whose orbital angular momenta within their parent galaxies are approximately aligned with the orbital angular momentum of the pair. After an encounter these particles form broad tails behind their parent galaxies which torque down the orbit. As a result of this coupling, tidal effects on a rotating galaxy depend very strongly on the orientation of its spin vector relative to the orbital angular momentum. This effect is very evident in the restricted 3-body calculations used by Toomre and Toomre (1972) to investigate the formation of bridges and tails in collisions between disk galaxies, and its dynamical consequences in a self consistent system were investigated by White (1979) in a series of simulated collisions between spherical rotating galaxies.

The strong dependence of interaction strength on orientation leads, in general, to substantial asymmetries in encounters between rotating galaxies. This was demonstrated explicitly by Gerhard (1981) in his simulations of parabolic and near parabolic encounters between disk-halo galaxies. Gerhard pointed out that the asymmetry is particularly marked when the extent of mass exchange between systems is considered. A galaxy corotating with the orbit will lose substantial mass to its companion during an encounter while a counter-rotating object will lose very little. Thus in an encounter between corotating and counter-rotating galaxies the latter will gain considerable mass. Notice that this mass transfer means that the stellar make-up of galaxies may change significantly during a collision: it is a rather strange example of the violent relaxation and phase-mixing processes discussed above which go to completion if the system subsequently merges.

A further example of the dependence of tidal interaction mechanisms on the nature of stellar orbits within the individual galaxies is given by Dekel et al. (1980). These authors show that after a tidal encounter mass loss from a galaxy with almost radial orbits occurs mainly in the opposite direction to the position of the perturber at pericentre. If the stellar orbits have an isotropic velocity distribution, however, many of the particles lost are carried away by the perturber and remain bound to it: these are the mass exchange particles discussed above. In general, remarkably few stars become entirely unbound in parabolic or subparabolic encounters between galaxies: these few escapers can, however, carry of a significant fraction of the total angular momentum (White 1979b: Gerhard 1981).

A number of regularities appear in the structure of objects formed by the merger of two or more initially bound or marginally bound systems. The most striking is the fact that the density profile of the remnant is almost always a very close appoximation to the power law $\rho \propto r^{-3}$, regardless of the initial structure of the objects (White 1978, 1979b: Villumsen 1982). Even more surprising is Gerhard's (1981) discovery that in his mergers of disk-halo systems the final distributions of the two components each follow this density law quite closely. So far no theoretical explanation for this result has been found. The only hint of a reason lies in the observation that the power law $\rho \propto r^{-3}$ is the only scale-free mass distribution which diverges both at small and large radii and thus that it allows the mass of the system to be spread over the largest radial range possible while conserving mass and energy. The remnants of mergers are moderately flattened near-oblate structures with some velocity anisotropy. They have flat rotation curves with an amplitude which depends primarily on the angular momentum of the initial orbit.

If elliptical galaxies are all to be identified as merger remnants (Toomre 1977: cf. Ostriker 1980) the problem of explaining their low rotation rate is equivalent to that of explaining why mergers should occur preferentially from nearly head-on orbits (White 1979a). Aarseth and Fall (1980) argue from simulations of galaxy clustering in an expanding universe that just such a situation is to be expected. Though these simulations are dynamically unrealistic in that they put one tenth of the cosmological critical density within the 'optical' boundaries of the 'galaxies' considered, recent work by Findlay (1981, private communication) confirms their results using a more realistic simulation technique.

Tests by White (1980) show that population gradients are only moderately weakened by mixing after a merger between similar spherical systems. In a merger between disk galaxies, a considerable amount of metal rich gas may be expected to fall to the centre of the remnant as a result of shock dissipation thus increasing its central density and partially preserving its metallicity gradient.

4. GALAXY CLUSTERING

Galaxy clustering has been studied extensively using direct N-body codes. Peebles (1970) and White (1976c) considered the formation of individual clusters, while Aarseth, Gott and Turner (1979), Efstathiou, Fall and Hogan (1979) and Efstathiou and Eastwood (1981) considered the growth of the overall clustering distribution in an expanding universe. At present I wish to concentrate only on one aspect of these simulations, and I will ignore many of the interesting results derived from them by their authors.

All these calculations assume implicitly that the spectrum of irregularities in the Universe at some time in the quite recent past ($z \sim 10$–30) may be adequately represented by statistical fluctuations present in a random distribution of point particles. Such fluctuations almost inevitably have largest amplitude variations on small scales (the particles themselves clearly represent highly non-linear local density enhancements). Consequently these simulations should mimic the growth of clustering in the gravitational instability picture which presupposes the existence of isothermal perturbations on all scales at recombination with a power spectrum such that small objects form first and then aggregate into larger systems.

It is considerably more difficult to use N-body calculations to simulate galaxy formation from purely adiabatic fluctuations, because the suppression of small-scale irregularities in such a picture requires that structures larger than individual galaxies are the first things to condense out of the general expansion and smaller objects must be formed by subsequent fragmentation (Zel'dovich 1970). The Russian group studying this theory of galaxy formation has, nevertheless, recently been able to carry out appropriate 2- and 3-dimensional simulations and they find that they are able to reproduce a number of the observed properties of the galaxy distribution (Doroshkevich et al. 1980: Shandarin, this conference). I shall not consider this work further now, but instead will consider relaxation effects in the gravitational instability picture.

White and Rees (1978) used approximate analytic methods of the kind described above to argue that the hierarchical structure, which forms in the gravitational instability picture as successively larger and larger objects separate from the cosmic expansion, is unstable. This instability arises because as a large unit collapses and comes to equilibrium for the first time, the smaller virialised subunits of which it is composed will not in general have a sufficient overdensity to retain their separate identities. Tidal encounters between subunits rapidly inject enough energy to disrupt at least their outer parts and in many cases lead to mergers between them. Subunits which manage to survive the first burst of merging and disruption soon spiral to the centre of the larger object as a result of dynamical friction. The outcome of all these processes is that most objects formed by dissipationless hierarchical clustering smear out their substructure and become smooth centrally concentrated systems within one or two dynamical times of their initial collapse. This picture of relaxation processes in an expanding universe seems to be borne out by the cluster formation simulation of White (1976c) and by an examination of the individual clusters which form in the calculations of Aarseth et al. (1979). It is not consistent with the observed fact that galaxies are seen in clusters with many members and short crossing times, and are clearly able to retain their identity for many cluster dynamical times. White and Rees concluded from this that galaxies cannot have formed in an earlier phase of the dissipationless process which is responsible for galaxy clustering in the gravitational instability picture, but rather that gaseous dissipation effects were very important during galaxy formation and led to a sufficient increase in binding energy for galaxies to survive the epoch of cluster formation.

It is clear, however, that if most of the mass of the Universe is in some dark and dissipationless Population III component which formed at early epochs, then the halos of this material which probably surrounded all galaxies at the time of their formation (and may still surround isolated galaxies) will not long survive cluster formation, but will rather merge together to form cluster 'halos'. Detailed expanding universe simulations by Aarseth and Fall (1980, especially Appendix C) support this conclusion. It remains to be proven that when galaxy halos merge the observed luminous parts are able to miss each other and so to survive as separate members of a cluster. Such an outcome is obviously required if the above picture is to be a viable theory of galaxy and cluster formation.

The relaxation instability of hierarchical clustering described above seems to be at variance with the fact that simple hierarchical models based on the gravitational instability picture can not only produce the observed 2-, 3- and 4-point galaxy correlation functions, but also can reproduce the overall visual impression given by the large-scale galaxy distribution and are consistent with the large empty spaces apparent in recent three-dimensional surveys of the galaxy distribution (Peebles 1980, and this conference).

To test the stability of these hierarchical models Peebles (1978) ran a number of simulations of the evolution of binary hierarchies set up in such a way that virial equilibrium held on each level. He found that the two- and three-point correlation functions changed very little with time and concluded from this that such hierarchies were stable. I have recently begun to run similar simulations of binary hierarchies which I have allowed to evolve ~ 5 times as long as the calculations presented by Peebles. I find that after 2 or 3 dynamical times of the highest level of a hierarchy, the whole system has, in general, relaxed to form a smooth, centrally concentrated structure, while at the same time preserving Peebles' result that the 2-point correlation function (or equivalently the pair count as a function of separation) changes very little on small and intermediate scales. This is a strange result since initially each particle is a member of one and only one close pair, while after considerable evolution the number of close pairs remains the same but only a few particles near the centre of the final structure are members of close pairs. It seems, therefore, that the pair count is an approximate invariant of the relaxation processes which disrupt the clustering hierarchy. Peebles' (1978) results suggest that the 3-point correlation function may also be only weakly affected, at least by the early stages of this disruption.

Other work presented at this conference suggests that if the Universe began in a chaotic state, various dissipative non-gravitational processes may have succeeded in reducing this chaos to the rather orderly situation from which later evolution appears to have taken place. Alternatively if exact symmetry appears to be the most natural initial state of the Universe, there are a number of phase transitions in the state of matter which could have led to the introduction of some structure. Once the requisite small but growing fluctuation component is present, however, gravitational effects gradually make the cosmos more and more lumpy on small-scales until finally structure separates from the general expansion and stars, galaxies and galaxy clusters form. I hope the above discussion has convinced you that non-linear relaxation processes will rapidly restore a smooth structure on all scales where dissipational effects are not dominant.

REFERENCES

Aarseth, S.J. and Fall, S.M. 1980. Ap.J., 236, 43.
Aarseth, S.J., Gott, J.R. and Turner, E.L. 1979. Ap.J., 228, 664.

Alladin, S.M. 1965. Ap.J. 141, 768.
Alladin, S.M., Potdar, A. and Sastry, K.S. 1975. Dynamics of Stellar Systems, ed. A. Hayli, Reidel, Holland, p. 167.
Chandrasekhar, S. 1942. Principles of Stellar Dynamics, Dover, New York.
Dekel, A., Lecar, M. and Shaham, J. 1980. Ap.J., 241, 946.
Doroshkevich, A.G., Kotok, E.V., Novikov, I.D., Polyudov, A.N., Shandarin, S.F. and Sigov, Yu.S. 1980. MNRAS, 192, 321.
Efstathiou, G. and Eastwood, J.W. 1981. MNRAS, 194, 503.
Efstathiou, G. Fall, S.M. and Hogan, C. 1979. MNRAS, 189, 203.
Farouki, R. and Shapiro, S.L. 1981. Ap.J., 243, 32.
Gallagher, J.S. and Ostriker, J.P. 1972. Astr.J., 77, 288.
Gerhard, O. 1981. MNRAS, 197, 179.
Henon, M. 1973a. Astr.Astrophys., 24, 229.
Henon, M. 1973b. Dynamical Structure and Evolution of Stellar Systems, ed. L. Martinet and M. Mayor. Saas-Fee Lectures.
Henon, M. 1975. In Dynamics of Stellar Systems (ed. A. Hayli), D. Reidel Publ. Co., Dordrecht, Holland.
Hoffman, G.L., Olson, D.W. and Salpeter, E.E. 1980. Ap.J., 242, 861.
King, I.R. 1962. Astr.J., 67, 471.
Knobloch, E. 1978. Ap.J., 222, 779.
Lynden-Bell, D. 1967. MNRAS, 136, 101.
Miller, R.H. and Smith, B.F. 1980. Ap.J., 235, 421.
Ostriker, J.P. 1980. Comments on Astrophys. Vol. VIII, No.6, (8, 177).
Ostriker,J.P. and Tremaine, S.D. 1975. Ap.J., 202, L113.
Peebles, P.J.E. 1970. Astr.J., 75, 13.
Peebles, P.J.E. 1978. Astr.Astrophys., 68, 345.
Peebles, P.J.E. 1980. The Large-Scale Structure of the Universe. Princeton University.
Roos, N. and Norman, C.A. 1979. Astr.Astrophys., 76, 75.
Richstone, D.O. 1976. Ap.J., 204, 642.
Silk, J. and Wilson, M.L. 1979. Ap.J., 233, 764.
Spitzer, L. 1958. Ap.J., 127, 17.
Spitzer, L. and Hart, M.H. 1971. Ap.J., 164, 399.
Tremaine, S.D., Ostriker, J.P. and Spitzer, L. 1975. Ap.J., 196, 407.
Tremaine, S.D. 1976. Ap.J., 203, 72.
Toomre, A. 1977. Evolution of Galaxies and Stellar Populations, eds. R.B. Larson and B.M. Tinsley, Yale Univ.Obs., p. 401.
Toomre, A. and Toomre, J. 1972. Ap.J., 178, 623.
van Albada, T.S. and van Gorkom, J.H. 1977. Astr.Astrophys., 54, 121.
Villumsen, J.V. 1982. MNRAS, in press.
White, S.D.M. 1976a. MNRAS, 174, 19.
White, S.D.M. 1976b. MNRAS, 174, 467.
White, S.D.M. 1976c. MNRAS, 177, 717.
White, S.D.M. 1978. MNRAS, 184, 185.

White, S.D.M. 1979a. Ap.J., 229, L9.
White, S.D.M. 1979b. MNRAS, 189, 831.
White, S.D.M. 1980. MNRAS, 191, 1P.
White, S.D.M. and Rees, M.J. 1978. MNRAS, 183, 341.
Zel'dovich, Ya.B. 1970. Astr.Astrophys., 5, 84.

DISC STABILITY AND HALO MASS

J. Sellwood

Institute of Astronomy, University of Cambridge, Cambridge.

1. INTRODUCTION

Studies of the dynamics of disc galaxies tell us something of how the mass must be distributed within these systems to account for their morphology. I will summarise here what we can deduce, from the evident stability of non-barred galaxies, concerning the relative masses of the spheroidal and disc components.

2. ROTATIONALLY SUPPORTED DISCS OF STARS CAN BE UNSTABLE

It has been known since the first N-body models of Hohl and Hockney (1969) and Miller et al. (1970) that discs of stars can be highly unstable to non-axisymmetric perturbations which lead to the formation of a massive slowly rotating bar. The behaviour of a more recent, but similar, computer model is illustrated in Figure 1, reproduced from Sellwood (1981). The bar which develops from the axisymmetric equilibrium state takes no more than five rotation periods to form and seems to survive indefinitely thereafter. (One rotation period near the half-mass radius is 250 in the time units shown in this figure.) I will first describe the main reasons for this instability and will then go on to discuss how it must be suppressed in the majority of disc galaxies.

Bar formation is a form of Jeans instability, i.e. it is a partial gravitational collapse, but where the role of pressure is replaced by rotation. Angular momentum must be lost in order that contraction can occur. Contraction and loss of angular momentum in the central regions must be compensated by expansion and gain in the outer parts since total energy and angular

B. J. T. Jones and J. E. Jones (eds.), The Origin and Evolution of Galaxies, 247–254.

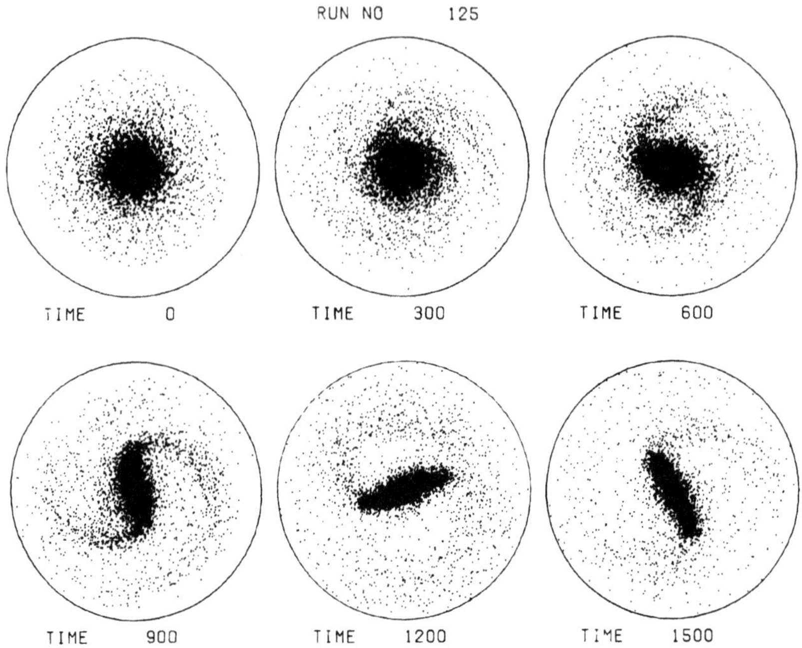

Figure 1. Evolution of model galaxy (Run 125 of Sellwood 1981). The sense of rotation is anticlockwise, and the time unit is Megayears.

momentum have to be conserved in an isolated system. Thus a radial transfer of angular momentum occurs, which requires a torque. In a collisionless stellar system this can best be provided by open spiral arms (Lynden-Bell and Kalnajs 1972). Notice that these develop spontaneously in the model (times 600 and 900 in Figure 1) from infinitesimal perturbations at the start.

3. THE ORIGIN OF SPIRAL ARMS

In order to understand how the instability can be controlled we need first to know why the spiral arms, which enable the bar to form, develop. This vital insight has recently been provided by Toomre and Zang who identify a process known as 'swing amplification'. A full report of their work has yet to appear, although a preliminary description may be found in Toomre (1981). Briefly the process works as follows: density perturbations which are sheared in a differentially rotating disc are strongly amplified as they swing from leading to trailing. Thus any source of low amplitude leading waves will result in more prominent trailing patterns. Toomre shows that a complete feedback cycle can be set up, whereby the amplified trailing waves themselves can become the source of the leading structure, causing open spiral disturbances to grow exponentially. The reader is referred to Toomre's (1981) article for a much fuller explanation.

4. HOW MAY THE INSTABILITY BE SUPPRESSED?

4.1 Resonances

Toomre (1981) makes the point that the feedback cycle will be broken if the spiral pattern possesses an inner Lindblad resonance (ILR). [A rotating density pattern in a stellar disc introduces additional periodic forces perturbing the motion of the stars. Lindblad resonances occur wherever the forcing frequency is equal to the natural frequency, κ, of epicyclic motion for a star.] This happens wherever the angular frequency of the pattern, Ω_p, is equal to $\Omega - \kappa/2$ for a two arm pattern. Typical curves of Ω and $\Omega \pm \kappa/2$ are sketched in Figure 2 together with two possible values of Ω_p (horizontal lines). Two ILR's are present for the lower value of Ω_p while none would be found for the higher value.

If the pattern speed is low enough to allow an ILR the reconversion of trailing waves to leading waves by passage through the galactic centre will be inhibited. The disc will lead a precarious existence, relying on the resonance to break the feed-back cycle.

All rotation curves which slope linearly from zero at the centre have $\Omega - \kappa/2$ rising from zero to some maximum value at a radius near to r_{max}. In these situations, which seem to pertain in all galaxies, it is always possible to choose a pattern speed high enough to avoid any ILRs. A series of computer experiments in which the maximum value of $\Omega - \kappa/2$ was increased from model to model (Sellwood 1981) showed that the angular speed of the bar/spiral pattern which formed in each increased so as to always avoid any ILRs. Thus ILRs will present no obstacles to bar formation in real galaxies.

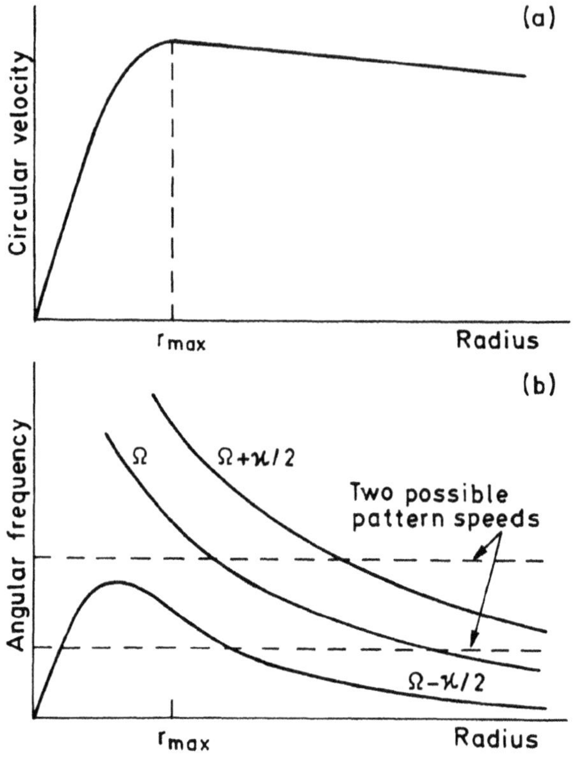

Figure 2. (a) Schematic rotation curve for a model galaxy showing the radius r_{max} at which the circular velocity attains its maximum.
(b) The angular frequencies $\Omega - \kappa/2$, Ω, and $\Omega + \kappa/2$ as a function of radius. The horizontal lines depict possible pattern speeds.

4.2 Velocity dispersion

The gain of the swing amplifier is reduced when the dispersion of velocities amongst the disc stars is increased. Velocity dispersions are usually measured in terms of a parameter Q which is the ratio of the actual velocity dispersion in the radial direction, σ_u, to the minimum required for local stability $\sigma_{u,min}$. i.e. $Q = \sigma_u/\sigma_{u,min}$ where

$$\sigma_{u,min} = \frac{3.36G\Sigma}{\kappa}$$

(Toomre 1964). Here G is the gravitational constant and Σ is the surface density of matter in the disc. Higher velocity dispersion reduces the ability of the disc to respond to any perturbation and Toomre (1981) confirms that swing amplification is no exception. Amplification factors of greater than 100 are possible in one swing if Q is 1.2,i.e. the velocity dispersion is only slightly greater than the minimum for local stability. Even when Q is as large as 2 amplification factors of ~ 10 are still possible. Thus discs would have to be hotter still in order to prevent the growth of bar modes. The velocity dispersions required are much larger than those observed in the solar neighbourhood and those inferred for other galaxies by indirect means, e.g. from the observed thinness of the discs of edge-on systems and the well developed spiral structure in face-on systems both of which argue for low velocity dispersions.

4.3 Massive 'halos'

There remains one final alternative method to suppress the instability: a large quantity of mass in the spheroidal component of the galaxies. Toomre (1981) introduces a new parameter, X, which is the ratio of the wavelength, λ, of the wave of interest, in our case that associated with bar formation, to λ_{crit}, a characteristic wavelength for the disc given by

$$\lambda_{crit} = \frac{4\pi^2 G\Sigma}{\kappa^2}.$$

The value of λ_{crit} will be reduced when unresponsive mass is added to the system thereby increasing κ without increasing Σ. Unresponsive mass could be a component of the disc which has extremely high velocity dispersion but it is more usually thought of as a pressure supported spheroidal component, i.e. bulge or halo. When $X(=\lambda/\lambda_{crit})$ exceeds 3, amplification of bar modes is reduced to almost nothing, and this must be the way in which bar formation is prevented in real disc systems.

5. WHERE ARE DISCS UNSTABLE?

Before studying the implications of the above conclusion, I need again to refer to the computer models this time to show that the bar unstable region of a galaxy is determined by its rotation curve. Sellwood (1981) found that the bar size was controlled by the turnover radius of the rotation curve: the unstable region extends to perhaps ~ 2 r_{max}, and if this central part can be stabilised the rest of the disc should be safe.

Thus, if spherically distributed matter is responsible for stability, it is only the matter contained within ~2 r_{max} of the centre which need be considered. Any spherical matter exterior to this radius cannot affect the dynamics of the centre, since gravitational forces due to a spherical shell of matter are zero everywhere inside the shell.

6. STABILITY CRITERIA

The first criterion to be proposed was that of Ostriker and Peebles (1973) who declared that t, the ratio of rotational kinetic energy to the negative of the total potential energy, should be less than ~ 0.14 for any stellar system to be stable. This criterion is clearly too simple, since extended halos, which are irrelevant to the dynamics of the centre, substantially affect the value of t.

Nevertheless it has been difficult to construct computer models which do not conform to this stability criterion. This is probably because the computer models have been more relevant to the central parts of galaxies: it is difficult for the models to retain adequate spatial resolution if the disc were to extend to radii much larger than 2 or 3 times r_{max}. Thus most N-body experiments (Hockney and Brownrigg 1974: Hohl 1976: Sellwood 1980) indicate that roughly 60-70% of the mass in the models has to be contained in the spheroidal component in order to prevent bar formation in the disc.

Toomre's X parameter predicts a similar mass ratio. If the shape of the rotation curve is not greatly altered by the addition of halo mass, we expect that three times as much mass will be contained in the bulge/halo component as in the disc component within the critical radius (~2 r_{max}). Changing the density profiles of the two components will change this ratio slightly. Of course, the ratio of the total masses of each component integrated out to large radii may be very different.

A new criterion has recently proposed by Efstathiou et al. (1981) which attempts to include all matter within the Holmberg radius. For an exponential disc of total mass M_D and scale length α, having a rotation curve which is made exactly flat outside r_{max} by addition of the right amount of halo material they find

$$V_m \sim 1.1\ (\alpha M_D G)^{1/2}$$

where V_m is the orbital speed in the flat part of the rotation curve. This is the conclusion from a large number of numerical models. For reasonable values of r_{max} the total mass of the halo out to the Holmberg radius is ~ 4 or 5 M_D.

7. DO LUMINOUS BULGES ALONE CONTAIN ENOUGH MASS FOR STABILITY?

One of the principal features of spiral galaxies which changes along the Hubble sequence SO/a-Sm is the prominence of the visible bulge. Early types SO/a-Sb have a substantially larger fraction of their total light emitted by the bulge than do late types. However, the frequency of strong bars is more or less independent of Hubble type. Sandage and Tammann (1981) classify 25% to 30% of disc galaxies in the Revised Shapley Ames Catalog as barred and this fraction is roughly constant for all Hubble types. These morphologists identify fewer bars than most others and we conclude that this is the fraction of strongly barred galaxies.

These two facts seem to argue that the prominence of a visible bulge does not affect the stability of the disc. However, the situation may not be so simple, since there is now indirect evidence (Athanassoula and Martinet 1980) that the bulges of early type galaxies are sufficiently massive to determine r_{max}: the turnover radius of the rotation curve. In late type galaxies this is not the case: frequently the rotation curves of Sc galaxies continue to rise to very large radii (Rubin et al. 1980). In these late systems we expect some other component of the galaxy, probably the dark halo, to be controlling the rotation curve.

It seems therefore, that there is probably sufficient mass within the luminous bulges of unbarred early type galaxies to account for their stability, although we need to appeal to dark matter to account for the stability of later types.

8. WHY ARE ONLY SOME GALAXIES BARRED?

It seems reasonable to assume that strong bars in disc galaxies are the result of a global instability at the time of formation. Thus theories of galaxy formation will have to account for the frequency of galaxies having insufficient spheroidal matter to suppress the instability in the disc.

REFERENCES

Athanassoula, E. and Martinet, L. 1980. Astron.Astrophys. 87, L10.
Efstathiou, G., Lake, G. and Negroponte, J. 1981. MNRAS (to appear).
Hockney, R.W. and Brownrigg, D.R.K. 1974. MNRAS 167, 351.
Hohl, F. 1976. Astr.J. 81, 30.
Hohl, F. and Hockney, R.W. 1969. J.Comp.Phys. 4, 306.
Lynden-Bell, D. and Kalnajs, A.J. 1972. MNRAS 157, 1.

Miller, R.H., Prendergast, K.H. and Quirk, W.J. 1970. Ap.J. 161, 903.
Ostriker, J.P. and Peebles, P.J.E. 1973. Ap.J. 186, 467.
Rubin, V.C., Ford, W.K. and Thonnard, N. 1980. Ap.J. 238, 471.
Sandage, A. and Tammann, G. 1981. Revised Shapley Ames Catalog (Carnegie Inst. Washington).
Sellwood, J.A. 1980. Astr.Astrophys., 89, 296.
Sellwood, J.A. 1981. Astr.Astrophys., 99, 362.
Toomre, A. 1964. Ap.J. 139, 1217.
Toomre, A. 1981. In Normal Galaxies, eds. S.M. Fall and D. Lynden-Bell. Cambridge University Press.

EVOLUTION OF FAINT GALAXIES

Richard Ellis

Physics Department, University of Durham,
Durham DH1 3LE, England

'Errors, like straws, upon the surface flow:
He who would search for pearls must dive below'.

Dryden (1631-1700)

1. INTRODUCTION

It has long been realised that valuable information on the evolution and origin of galaxies lies within reach of our large telescopes, yet after 50 years we have little conclusive evidence of galaxy evolution. Most of the early work was motivated by the search for a value for q_0 - evolution was regarded simply as a further 'correction' in terms of the Hubble diagram. The suggestion that luminosity evolution effects might dominate magnitude - redshift and number magnitude diagrams was made by Beatrice Tinsley in the late 1960's (see Tinsley 1976 for a summary). q_0 does, however, have an important part to play in these diagrams via the linking of cosmological timescales (redshifts and luminosity distances) and galactic timescales (years).

Hubble began observational cosmology by counting galaxies (Hubble 1934, 1936) though the broad range of galaxy luminosities later convinced him to consider magnitude - redshift plots. In 1936 he presented galaxy counts to a limit of 21st magnitude and attempted to derive q_0 by allowing for K-corrections (the effect of redshift on a spectral energy distribution when observing in a fixed passband) and the like. Because of the uncertainties in the photometry and the K-corrections he could only fit the counts in a model having very high curvature: $q_0 \sim 10$. As we shall see,

B. J. T. Jones and J. E. Jones (eds.), The Origin and Evolution of Galaxies, 255–293.

present K-corrections allow the counts to be fitted down to a J magnitude of 21 with an open universe and very little luminosity evolution. (J magnitudes here refer to the Kodak IIIaJ + GG 395 system.)

The frontiers have receded more than 3 magnitudes since 1936. However, it is important to realise that progress in this subject relies equally in studies of nearby objects as well as those of faint ones.

There are two separate methods for extracting evolution from observations. The conventional method is to observe samples of hopefully familiar objects at redshifts $z < 1.0$ where observational selection is under some control. A more ambitious method is to search for galaxies undergoing their first epoch of star-formation. Little is known about the expected appearance of such primeval galaxies - much depends on the redshifts and internal extinctions involved. So far results are disappointing but there are several new ideas to pursue.

Finally although the slant in the lectures will be necessarily observational, useful constraints of current and past star formation rates can be gained from population and abundance analyses of various galaxies. Where appropriate I shall refer to these briefly since they have been reviewed with clarity by Tinsley (1980a), and Pagel and Edmunds (1981).

2. OUTLINE OF EVOLUTIONARY PREDICTIONS

Galaxy evolution can be inferred by matching present day colours with stellar populations that evolve according to tracks on the H-R diagram. Continued star-formation is modelled in a variety of ways, usually by assuming a time-dependent rate that becomes a useful control of the final colours. This approach is cleaner than the synthesis of present day populations (e.g. Faber 1972) since in that method equally acceptable populations could arise from different histories of star formation.

In attempting to summarise the qualitative predictions of the modellers there are a variety of sensitive unknowns but two are critical: namely the form of the initial mass function (IMF) which determines the proportions of low and high mass stars in each new generation (and hence the timescale for evolutionary changes) and the star formation rate (SFR) which together with the formation redshift z_f can be used to control the gas consumption so as to reproduce present day colours. Much of the groundwork is discussed by Tinsley (1968, 1972), Larson and Tinsley (1978) and Tinsley (1980a).

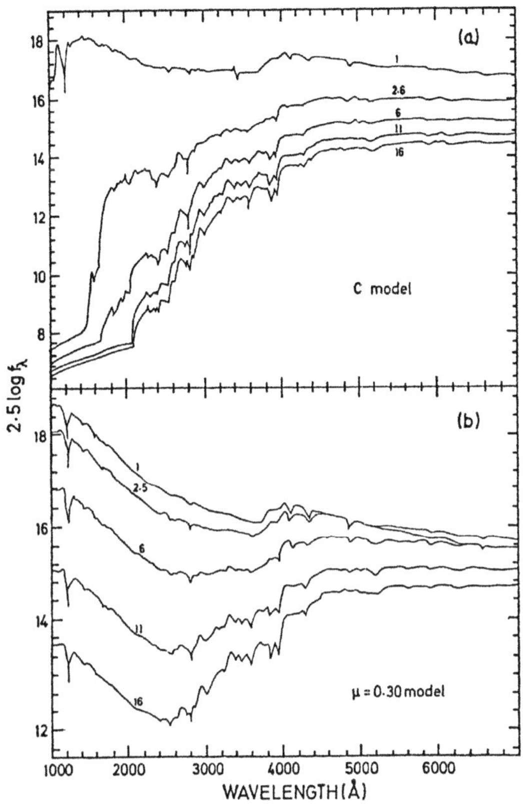

Figure 1. Examples of evolving energy distributions from Bruzual and Kron (1980). Note the different ordinate scales. (a) an early type galaxy modelled by a single burst of star formation, (b) a late type galaxy modelled by a steadily declining SFR. Numbers refer to ages (gyr) since the epoch of formation. On a $H_0 = 60$ $q_0 = 0.0$ cosmological model these refer to redshifts of 0, 0.5, 1.7, 6.2 and 15 respectively.

Briefly the conclusions of the models are that the Hubble sequence can be explained to first order as a variation in the SFR history (Baade 1963). Early types currently showing a near-zero SFR initially consumed gas at a rapid rate (Tinsley and Gunn 1976) whereas the SFR in spirals has remained more or less uniform with time (Tinsley 1978). (The initial burst model for spirals discussed by Tinsley (1977) generates present day spirals that are too red.) Similar models are discussed by Bruzual and Kron (1980).

These simple inferences can be further justified by the age-metallicity relation for Galactic disk stars (Twarog 1980) which confirms the uniform enrichment history. A remarkably elegant analysis of the (B−V) colours of spiral galaxies by Tinsley and Danly (1980) also constrains the past rate of star formation in spirals within certain limits.

The present conclusion therefore is that spirals will have looked much the same over recent epochs whereas early types will have been much hotter and more luminous at epochs when the star formation was vigorous. Figure 1 shows the spectral evolution of an early and late type galaxy as predicted by Bruzual and Kron (1980). The colour evolution over $z < 1$ is not substantial even for early types dominated by old red stars. The evolutionary effect is mainly in luminosity, hence colours are more likely to be useful in classifying objects by morphology and redshift rather than constraining evolution which will be seen best in galaxy counts to various flux limits, or magnitude-redshift diagrams. (Such models are often criticised for their simplistic representaton of the complex factors such as time/spatial variations in the IMF (c.f. Freeman 1977), recycling approximation and the general assumption that all galaxies are the same age.)

Tinsley (1980a) attempted to justify a uniform IMF though optical data cannot provide accurate constraints for masses above $\sim 2\ M_\odot$. For young massive stars the current stellar luminosity function is an accurate representation of the present IMF. IUE spectra of a variety of spiral nuclei have recently shown absorption features indicating that hot main sequence stars account for virtually all of the extreme uV ($\lambda < 2500$Å) flux (Ellis and Gondhalekar 1981a): similar conclusions have been suggested even for early type galaxies though hot old stars cannot be ruled out at this stage (Gunn et al. 1981). With much uncertainty the current IMF can thus be determined for these galaxies using population synthesis techniques. The IMF for a sample of 5 galaxy nuclei observed from 1200Å to 7500Å by Ellis and Gondhalekar (1981a), determined without any 'astrophysical constraints' give some support to the universality of the local IMF in an important mass range.

Two other problems in this simple theory of galaxy evolution: one concerns the nature of SO galaxies which cannot be accounted for by such simple models since they show disk structure without star formation. Quite apart from the issue as to whether such galaxies are spirals which met some dynamical disaster (Larson et al. 1980) or are a bona fide galaxy type intermediate to spirals and ellipticals (Sandage and Visvanathan 1978), S0s are a useful reminder that stellar evolution alone cannot explain the appearance of a disk galaxy.

The second point concerns supposedly young galaxies: these are too blue for standard models e.g. Zwicky compact and Markarian galaxies (Huchra 1977) and are often metal-deficient (Searle and Sargent 1972). Are these really young galaxies or are they simply old ones experiencing a sudden rejuvenation via a burst of star formation? Gordon and Gottesman (1981) recently discovered many blue galaxies have extended halos of neutral hydrogen which may infall and act as a trigger for intermittent star formation. The most sensitive test for an underlying old population lies with infra-red photometry: the problem however is that an infra-red excess might also be related to dust re-radiation associated with the hot stars (Struck-Marcell and Tinsley 1978). Gondhalekar and I recently selected a small sample of blue non-Seyfert Markarian galaxies which, on the basis of the ultra-violet indicator at 2200A (Nandy et al. 1976) showed little or no internal reddening. One might hope to avoid the dust problem in these objects. The data is scarce, but comparison with the models of Struck-Marcell and Tinsley (1978) does not suggest, at this stage, that these galaxies are genuinely young objects.

The formation redshift z_f is a free parameter since as we mentioned the SFR is adjusted to match present colours. Nevertheless it has important effects since for a given cosmological model a large z_f implies a larger redshift and fainter magnitudes before evolutionary brightening becomes apparent. If $z_f < 2$ even the most conservative models would predict a large excess of hot blue ellipticals within easy reach of our telescopes. Furthermore measurements of the extragalactic background light (EBL) (see section 5.2), the integrated extragalactic signal in a random patch of sky (Dube et al. 1977a,b), provide an additional lower limit to z_f, though the precise value is model dependent (Spinrad and Stone 1978). An upper limit to z_f is supplied from considerations of the observed binding energy of galaxies (Binney 1977): observed velocity dispersions are incompatible with dissipational collapse before $z \sim 10$. These constraints are not totally observational and whether the reader agrees will depend on whether or not he/she is a theoretician! Anyway, the general opinion is that $2 < z_f < 10$ is the range to consider.

3. STATISTICAL PROPERTIES OF NORMAL GALAXIES

I have already emphasised the need for precise data on local galaxies. However, the standard galaxy catalogues, the RSA (Sandage and Tammann 1980) and the Second Reference Catalogue (de Vaucouleurs et al. 1976) whilst invaluable generally are not ideal for our particular purpose. These samples are distorted significantly by the Virgo cluster, and in addition the Second Reference Catalogue is a collation of various results and is not statistically uniform or complete except to a very bright limiting magnitude.

Several groups have begun constructing radial velocity surveys of magnitude-limited samples in more distant volumes hopefully unaffected by local inhomogeneities. These are often referred to as 'field surveys' though the adjective really means rich clusters were not specifically chosen to be in the areas surveyed.

Table 1 lists the details for these surveys. The Harvard survey (Davis et al. 1981) will provide a useful data bank for mapping the local Hubble flow and determining the extent of the local supercluster. But as it covers a magnitude region seriously affected in the northern galactic hemisphere by the Virgo cluster (c.f. counts of galaxies discussed by Sandage, Tammann and Hardy 1972: Peterson 1973) the problems discussed above still apply (though the sheer size and homogeneity of the sample will make a tremendous impact on the galaxy luminosity function, particularly the shape of the function in the field (see section 3.2.1)). The Harvard survey also rests at the limit on the Zwicky magnitude system (Zwicky et al. 1961-68) which contains large random and occasional systematic errors (Huchra 1976).

Of the remaining surveys that by Kirschner, Oemler and Schechter (1978) (hereafter KOS) is a magnitude deeper but covers 8 fields with only ~ 20 galaxy redshifts per field. As we discuss below there is evidence for supercluster enhancements in this sample as the authors discuss (Kirschner et al. 1979). The remaining survey carried out on the AAT by a collaboration of Durham and Australian astronomers extends much deeper, the aim being to be doubly sure of avoiding local inhomogeneities. The field areas are similar to those of KOS, consequently each field has many more galaxies. Both the KOS and Durham surveys cover fields in the northern and southern Galactic caps and the N/S differences test the effect of Virgo. At the time of writing the Durham survey has just been completed to J = 16.75 but the provisional analysis presented here was based on an incomplete sample to 16.75 but complete to J = 16.5. The incompleteness was unfortunately wholly in the northern fields so N/S comparisons were not possible for this survey when this material was prepared.

Table 1. Recent field surveys

Survey	Nominal limit	J_{lim}[a]	Area deg^2	Fields	Galaxies	ΔV km/s
Harvard[1]	B=14.5	14.3	7370	1	~2400	~500
KOS[2]	J'=15.0	15.3[c]	123	8	164	~100
Durham[3]	J=16.75	16.75	69	5[b]	342	~40

Notes: references 1. Davis et al. 1981: 2. Kirschner et al. 1978: 3. Bean et al. 1982.

a. J = IIIaJ + GG 395, J' = IIIaJ + Wr4, B = IIaO + GG 385
b. Some Durham fields were extended from J = 16.5 to J = 16.75 at a lower velocity accuracy of ~100 km/s.
c. KOS magnitudes are total, Durhams mags are isophotal.

3.1 Number-magnitude counts and the local supercluster

Figure 2 shows the galaxy counts in the relevant region $13 < J < 18$ (J defined as in Table 1) taken from the KOS and Durham photometry. Also shown are counts from 3 deep Schmidt plates at the south Galactic pole derived via the COSMOS machine (for procedural details see Phillipps et al. 1978). The Durham magnitudes are isophotal and the counts show a smooth trend from J = 13 to 19. The least square fit over $13 < J < 16.75$ is

$$\log N = 0.55 J - 8.73:$$

the slope is flatter than the Eudiclean slope because of the effects of the K-correction. Indeed this slope is a useful check on the mix of local galaxy types and their individual K-corrections (see section 3.3).

The KOS counts refer to total magnitudes in a different J system denoted J' (J_0 by Bruzual and Kron). I have transferred this onto the standard J total system using a transform kindly supplied by D. Koo (this replaces that given by Bruzual and Kron).

$$J = J' - 0.11 + 0.41 (J' - F')$$

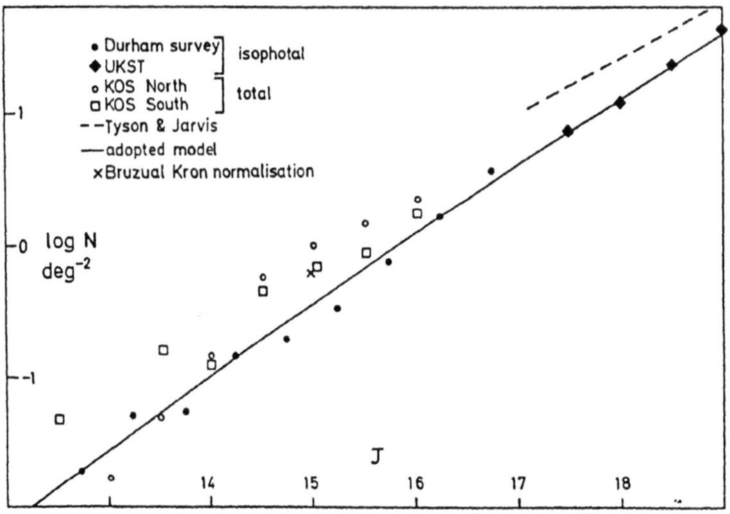

Figure 2. Differential galaxy counts reduced to the J bandpass. The dashed line represents the relation proposed by Tyson and Jarvis (1979) on the basis of their deep counts that supposedly reveal a very large local supercluster. The solid line is the prediction of the adopted galaxy luminosity function using the observed morphological mix in the Durham survey and K-corrections discussed in section 3.2.3 and 3.4.

where (J' - F') ~ (J - F) is tabulated by KOS. The colour term will lead to some incompleteness near J ~ 16 but an inspection of the distribution of KOS colours suggests this effect is not serious.

The overall offset between the 2 surveys is ~ $0^{m}.5$ and is surprisingly large even considering the isophotal/total contribution. The KOS counts are also very irregular (the field to field noise is twice that expected statistically whereas the variation in Durham field counts is as expected statistically) and show an excess in the 14 < J < 16 region. This hump is discussed by KOS (Kirschner et al. 1979) in relation to that seen in the Zwicky catalogue (Zwicky et al. 1961-68) as being due to the local supercluster. In this case Figure 2 can be understood by a photometric shift of ~ $0^m_.25$ (between KOS South and Durham) which is reasonable for the isophotal/total aperture effect. The precise correction will soon be known as we have photometry of a large sample of KOS galaxies in our J system.

3.2 Galaxy luminosity function

To make evolutionary predictions the distribution of true luminosities per unit volume is required. Specifically an accurate absolute magnitude scale (M_J) is required since the apparent magnitude of a distant galaxy can then be placed statistically in luminosity distance or redshift space.

All luminosity functions (LFs) bend over in the log number - M_J plane: this is an expression of the constraint that unit volume of space must yield a finite luminosity (extrapolating a power law LF in either direction would contradict this). Consequently all LFs have a characteristic absolute magnitude M_J^*. A small error in M_J^* can have a large effect on faint galaxy magnitudes since it has a significant effect on the volume and redshift corrections.

The normalisation of the LF is also important and has a useful role elsewhere (e.g. defining the visible luminosity density). For our purposes the normalisation is most usefully expressed via a surface density at a given magnitude (rather than quantities per cubic Mpc). Fortunately the shape of the LF has been shown to have less effect on deep counts (Bruzual and Kron 1980) because the distribution of absolute magnitudes in an apparent magnitude limited sample is dominated by galaxies near the characteristic magnitude.

Because different galaxy types have different energy distributions the galaxy mix is also required to correctly weight the effects of K-correction (and luminosity evolution). Ideally one should determine the LF shape for each morphology (or colour since this subdivision bypasses somewhat the scatter in energy distributions for a given type) but the available data is too scarce: the Harvard survey will provide an excellent opportunity to test this.

Finally it should be noted that most deep surveys have been done in the standard J,F photometric systems (F = IIIaF + RG1/630) yet until now there have been no LFs specifically in these well-used bandpasses.

3.2.1 Shape

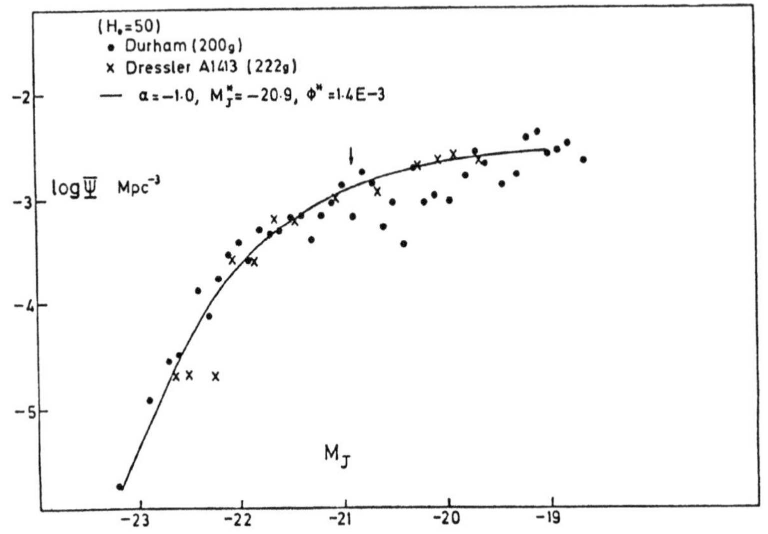

Figure 3. Differential galaxy luminosity function in J passband for the Durham redshift survey as at April 1981. Dressler's LF for the rich cluster A1413 is also shown. A best fit Schechter function with $\alpha = -1.0$ $M_J^* = -20.9$ is shown ($H_o = 50$ km s^{-1} Mpc^{-1} and no Galactic absorption is assumed).

Figure 3 shows the differential LF for a Durham sample of 200 galaxies compared to Dressler (1978)'s function for the rich cluster Abell 1413 (chosen as it contains a similar number of galaxies and was not singled out as anomalous by Dressler in his study of 12 rich clusters). To transfer Dressler's photometry a shift of (J − F) = 1.35 appropriate for E/SO galaxies was adopted. The Hubble constant of 50 km/s^{-1} Mpc^{-1} assumed to plot the LF cancels later when predicting apparent magnitudes.

Were it not for Dressler's (1978) depressing conclusions about the enormous variation in cluster LFs one would be tempted to adopt a single function for field and rich cluster samples (c.f. Schechter 1976). However our field samples are so few compared to those of rich clusters where differences are definitely present that it is premature to make these conclusions (so far we have ~ 1/12th the data!).

Where possible I would prefer to compare LFs directly rather than compare analytical representations particularly Schechter functions (Schechter 1976). The Schechter function in luminosity units is

$$\varphi(L)dL = \varphi^*(L/L^*)^{\alpha}\, e^{-(L/L^*)}dL, \qquad (1)$$

more usefully in absolute magnitudes

$$\varphi(M)dM = 0.92\ \varphi^*\ X^{1+\alpha}\ e^{-X}\ dM$$

$$\text{where } X = 10^{+0.4(M^*-M)} \qquad (2)$$

The shape of the function is given by only two parameters, α and M^*. In our experience the function is particularly inflexible at the bend and we are currently exploring other representations. A least squares fit to Figure 3 yields $\alpha = -1.0$, $M_J^* = -20.9$. Though Dressler chose $\alpha = -1.25$, $M_J^* = -21.03$ there is little practical difference between these representations as inspection of the data confirms.

3.2.2 Normalisation

As Kirschner et al. (1979) found, the normalisation * is best found from bright galaxy counts which generally contain larger samples than those with radial velocities (KOS also adopt an interesting method of fitting the shape which supposedly avoids inhomogeneities in the redshift distributions). In the case of the Durham counts a K-correction must be adopted since a typical redshift at 16.75 is $z \sim 0.1$ and the correction (see section 3.3) can amount to $0^m.5$ for an early type galaxy. The slope and absolute scale of the counts thus checks the first order K-correction and φ^*.

Anticipating the discussion of K-correction and galaxy mix the predicted galaxy count relation for $13 < J < 19$ matches that observed (Figure 2) remarkably accurately - the slopes agree to better than 1% over the range $13 < J < 17$. There is some evidence of a deficiency of galaxies at $J \sim 15$ which may be related to the local supercluster. The normalisation (in terms of the entire galaxy counts and the integrated number per square degree at $J < 16.5$) yields $\varphi^* = 0.0014$ galaxies Mpc^{-3}. In terms of observables we find 3.18 galaxies deg^{-2} to $J = 16.5$: at $J = 15$ the differential number per $0^m.5$ bin is 0.36 (Bruzual and Kron found 0.60 from the KOS total magnitude counts).

Table 2

A summary of field luminosity functions (H_0 = 50)

Ref.	α	M*	M_J^*	Mpc^{-3} φ*	Apole	Source
1.	-1.25	B(0)=-20.6	-21.1	5.0E-3	0.12	Ref Cat
2.	-1.0	Z=-20.8	-21.1	1.4E-3	0.0	Groups
3.	-1.24	B(0)=-20.78	-21.28	2.2E-3	0.25	Review
4.	-1.24	B(0)=-20.67	-21.17	2.9E-3	0.2	Ref Cat and Survey
5.	-1.0	B(0)=-20.5	-21.0	1.8E-3	?	Ref Cat and Survey
6.	-1.10	B_T =-21.0	-21.1	2.3E-3	0.17	Ref Cat and Survey
7.	-1.03	B_T =-20.7	-20.9	-	0.0	RSA
8.	-1.0	-	-20.9	1.4E-3	0.0	Survey

1. Schechter 76
2. Turner Gott 76
3. Felten 77
4. Huchra 78
5. Davis et al. 79
6. Kirschner et al. 79 (KOS)
7. Tammann et al. 80 (TYS)
8. Durham

Means: $\alpha = -1.11 \pm 0.12$: $M_J^* = -21.1. \pm 0.14$

I summarise the various 'field' LF parameters in Table 2. The shape parameters agree very well considering the samples involved - again evidence for universal LFs? - even the normalisation is converging: ours is the lowest as would be expected if local inhomogeneities had plagued the other determinations.

φ* is often used to determine the mean luminosity density which is of importance for constraining the SFR per unit volume. Following Felten's (1977) discussion and adopting B - J ~ 0.23 (B-V) yields $J_\odot$ = 5.34 and a luminosity density

$$\mathcal{L}_J = \Gamma(\alpha+1)\,\varphi^* L_\odot \ \text{dex}\ 0.4\,(M_J^\odot - M_J^*)$$
$$= 4.4 \times 10^7\ L_\odot\ (H_0/50)\ \text{Mpc}^{-3} \qquad (3)$$

In using this a correction of ~ x 1.8 should be made for internal absorption and inclination (c.f. Tinsley and Danly 1980 who obtain $\mathcal{L}_B = 2 \times 10^8\ L_\odot$ for spirals alone). Also since the magnitude system is isophotal this is a luminosity density from sources within the isophotes. The eternal problem of isophotal or total magnitudes will be discussed in section 4.

3.2.3 Morphological variation and galaxy mix

The Durham and KOS samples are insufficient to define an LF shape for each morphological type although various authors have discussed this issue. Christensen (1975) derived individual LFs from a very small sample and more recently Tammann et al. (1980, TYS) fitted Schechter functions to the RSA catalogue divided by morphology. Table 3 summarises their findings.

Table 3

Luminosity functions for different galaxy types

	TYS			Durham	
Class	$M^*_{B_T}$	α	M^*_J	M^*_J	Mix[1]
E/SO	-20.76	-0.90	-20.96	-20.95	28
Sab/Sbc	-20.08	-0.47	-20.28	-20.75	46
Scd/Sdm	-19.00	-0.72	-19.20	-20.30	26
Total	-20.70	-1.03	-20.90	-20.90	-

[1]: the percentage of galaxies of that class brighter than J=16.5.

Whilst the trend in M* is reasonable (though large) the enormous variations in α are quite surprising and bear no relation to the shape defined for the whole sample. For no reason other than caution bearing in mind my earlier remarks about fitting the Schechter function, I have decided to assume a constant α and investigate the M* trend (recall that M* is much more important to nail down than α). One disadvantage with the fainter surveys is that galaxy classification becomes harder: in our case different observers agree to better than 5% at J = 16.5. Eventually the Durham survey will be classified by colour. For the galaxy types E/SO: Sa-Sab-Sbc: Scd-Sdm-Irr in the Durham survey we fitted M^*_J under the assumption $\alpha = -1$ as given by the total LF. A glance at Table 3 shows no large shange in M^*_J such as that found by TYS. Whilst our sample is small this difference in results is significant and cannot be due to our isophotal magnitude system since they find weak bulged galaxies to be fainter than we do.

Table 4

Predicted and observed galaxy mix (percentages)

	E/SO	Sabc	Scdm	Source
Adopted	28	47	25	Model J <16.5
Observed	28	46	26	Durham J <16.5
	23	39	38	Hubble 1926
	23	34	43	van den Bergh 1960
	28	32	40	Sandage 1961
	24	52	24	Dorstal 1977
	45	27	28	KOS 1979

The galaxy mix is simply determined by adopting the individual Schechter functions of Table 3 and scaling the LFs to reproduce the observed proportions at J = 16.5 (including the individual K-corrections to be discussed later). The observed mix is only in moderate agreement with others (Table 4) though comparisons are hampered by alternative classification schemes (e.g. Dostal 1977) and of course different bandpasses. The classical determinations seem to favour later types. The Harvard catalogue will make a valuable contribution in this area provided the effect of the Virgo cluster is controlled. The important implications of the low Scdm contribution for faint magnitude counts will be seen later.

3.3 Ultraviolet spectrophotometry

The problem of determining the visibility of galaxies at high redshift (the K-correction) has been with us for 50 years. The often-quoted solution that one should count galaxies in the red or near infra-red where K-corrections can be inferred from ample visible data (rather than scarce uV data) is fine for Hubble diagrams destined to determine q_0, but as galaxies evolve most at ultraviolet wavelengths (c.f. Figure 1) one has to go to fainter objects with larger photometric/redshift errors to see evolution.

3.3.1 Spiral galaxies

There has been some minimal progress in the ultraviolet (uV) spectrophotometry of normal spiral galaxies. Table 5 summarises the data available.

Table 5

Ultraviolet data on galaxies

Ref.	Satellite	Primary	Aperture	λλ	Resn	Galaxies
1.	OAO-2	0.2m	2/8arcmin	2130 3330	4bb	8 galaxies
2.	TD-1	0.28m	12x17arcmin	1550 2740	4bb	6 spirals
3.	ANS	0.2m	2.5arcmin	1550 3330	5bb	12 galaxies
4.	IUE	0.45m	10x20arcsec	1150 3200	6Å	8 spirals
5.	IUE	0.45m	10x20arcsec	1150 3200	6Å	4 ellipticals

Notes:

1. Code and Welch (1978). Pence (1976) analysed provisional OAO-2 data for K-corrections.
2. Carnochan <u>et al</u>. (1975). Carnochan has catalogued 25 galaxy observations made with TD1 (unpublished).
3. Coleman <u>et al</u>. (1980). Wesselius has catalogued ~ 3000 observations made with ANS (unpublished).
4. Ellis and Gondhalekar (1981a).
5. Bruzual and Spinrad (1980).

An immediately apparent problem is that the large entrance aperture satellites have inadequate spectral resolution for determining K-corrections and examining internal reddening variations, whilst IUE has an ideal wavelength coverage and resolution but an aperture so small that only the nuclei and HII regions of nearby galaxies can be studied (distant galaxies which fall completely within the aperture are too faint for these small telescopes). Accordingly Gondhalekar and I have been surveying several spirals within metric apertures of ~ kpc to determine nuclear K-corrections for use in high resoluton imagery of distant galaxies.

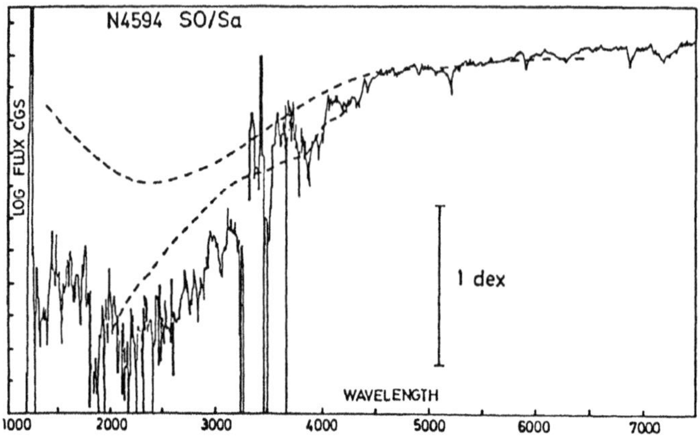
N4594 SO/Sa
LOG FLUX CGS
1 dex
WAVELENGTH
1000
2000
3000
4000
5000
6000
7000

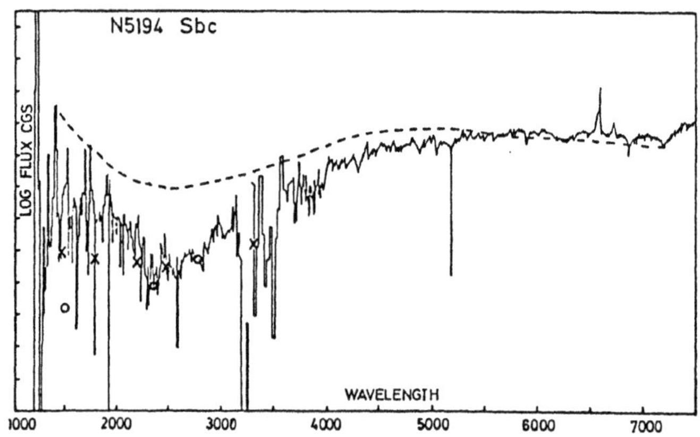
N5194 Sbc
LOG FLUX CGS
WAVELENGTH
1000
2000
3000
4000
5000
6000
7000

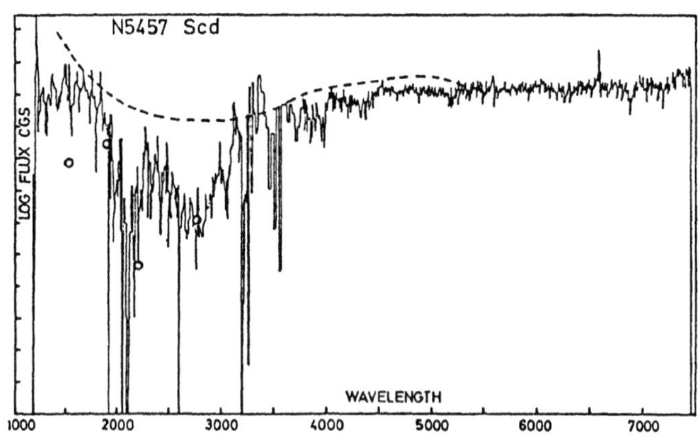
N5457 Scd
LOG FLUX CGS
WAVELENGTH
1000
2000
3000
4000
5000
6000
7000

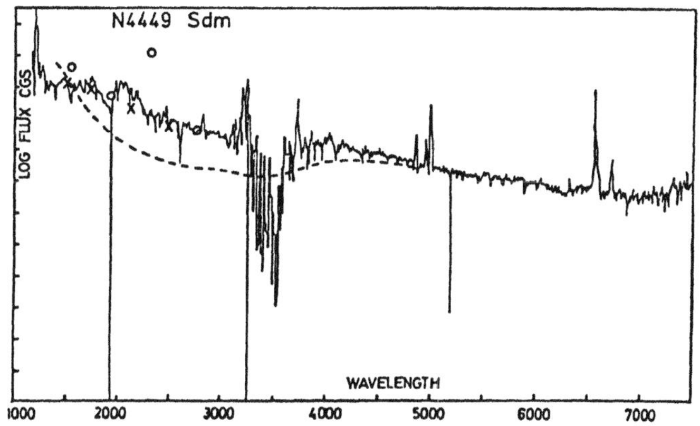

Figure 4. Nuclear and integrated spectrophotometry of 4 galaxies along the spiral sequence. The detailed spectra refer to measurements made with IUE and the Lick scanner through a metric aperture of ~1 kpc. The anomalous fluxes at 3500Å are due to the poor sensitivity of the visible detector in this region. The dotted line represents Pence (1976)'s mean integrated energy distribution for the galaxy type as interpolated from OAO-2 data and visible data given by Wells (1978): N4594 is compared with an Sab and E/S0 distribution. The crosses and circles refer to data for these particular galaxies taken by the ANS and TD1 satellites respectively. Pence's curves are normalised at 5500Å: the satellite data is normalised in the near ultraviolet. NGC 5457 is reddened by E(B-V) ~ 0.4: de-reddening according to a Galactic extinction law improves agreement with Pence's data.

For the latest Hubble types the differences between the nuclear and integrated satellite data are insignificant. In this case the higher resolution IUE data are valuable. It must be remembered that the K-corrections available to date have been estimated by interpolation between the 4 or 5 broad bands often using different apertures and even different galaxies for the uV and visible regions. Another problem is that the widths and short wavelength edges of the bluer passbands UBJ means that the OAO-2 derived K-corrections cannot strictly be extended much beyond z ~ 0.6 without unwarranted assumptions.

The work we have done on IUE, whilst not particularly valuable for K-corrections of the integrated light of galaxies Sab-Sbc has shown the important variations introduced by internal reddening and population fluctuations. Although the calibration uncertainties often attributed to OAO-2 may not be present in the ANS data, Coleman et al.'s sample is too small for general use: a complete analysis of Wesselius' ANS catalogue together with IUE spectra of higher resolution for selected objects would be a valuable interim solution to the K-correction problem.

Figure 4 shows both nuclear and integrated data for 4 spirals along the Hubble sequence compared with Pence's mean energy distributions for these types. The detailed nuclear spectrum of NGC 5457 agrees better with Pence's curve when it is de-reddened assuming a E(B-V) value given by the Balmer spectrum and the strength of the 2200Å feature as calibrated by Nandy et al. (1976). (Strictly speaking one should only de-redden data for the Galactic contribution in determining K since it is the spectrum emerging from the galaxy that is of interest.) To average reddened data for a small sample would be misleading however, for a trivial E(B-V) value produces a significant change in the short wavelength flux. A further complication is the possibility of non-standard extinction laws (Nandy et al. 1978) though some grains (!) of comfort might be drawn from the agreement between E(B-V) estimates using visible and uV features for our sample (Ellis and Gondhalekar 1981a).

The colour gradient across the early type spirals is apparent in Figure 4: the integrated spectral type being much hotter. I have re-determined K(z) for both sets of data using wavelength response curves for the passbands as defined by Kodak (1973) and Schott (1974) sensitivity curves.[1] A standard extinction of 1.0 airmass was also included. (Strictly the atmospheric correction is not needed if using flux-calibrated data reduced to zero airmass: the difference is minimal except for the hottest galaxies when observed in the U band.)

I define the K-correction as:-

$$K(z) = -2.5\log_{10} \frac{\int_0^{\infty} S(\lambda)\, F\left(\frac{\lambda}{1+z}\right) d\lambda}{\int_0^{\infty} S(\lambda)\, F(\lambda) d\lambda} + 2.5\log_{10}(1+z)$$

[1]Tables of the adopted K-terms are available on request.

the latter term being the bandwidth factor. $S(\lambda)$ is the sensitivity of the filter/emulsion combination and $F(\lambda)$ the flux curve.

Considering now the various integrated K-terms available for use: the Code and Welch (1978) values cannot be assessed accurately since their revised OAO-2 data is not published in digital form. As far as I can tell their data is similar to the supposed 'quick look' OAO-2 data used by Pence. I have extended analysis of Pence's data to the newer passbands and been more conservative in the redshift limit. The ANS K-terms are generally larger than Pence's (by $< 0^m.25$ usually) the effect being greatest for late types.

3.3.2 Early type galaxies

The situation here is more favourable because of the abundance of data and the relatively smaller colour gradients. Also nuclei of early type spirals have been shown to represent ellipticals fairly closely (cf Peck 1980). The driving force in making uV measurements of ellipticals has been the need for K-corrections to interpret the Hubble diagram for brightest cluster members (Kristian et al. 1978) though the urgent task for field samples is definitely to provide uV data for spirals.

Figure 5 shows a comparison of uV data for some early type galaxies. Bruzual and Spinrad's (1980) data is the mean of several IUE spectra including that of the bulge of M31 (Johnson 1979). Pence's original data shows a gentler fall-off slope (the normalisation may be uncertain however), whereas the ANS data (the average of the bulges of M81 and M31) is steeper. As M81 is heavily reddened this may explain the steep fall-off. In view of the higher resolution and number of objects studied I favour the Bruzual-Spinrad data though the redshift limit imposed by the cut-off at 2000Å is only $z \sim 0.6$ for the UBJ bands. (There is also evidence from the colour-redshift relation that suggests a steeper fall off in the uV.) Shorter wavelength IUE data is available (c.f. Norgaard-Nielsen et al. 1981) but the sensitivity problems are severe and though an upturn in flux is reported, a larger sample is needed before it can be generally considered in the K-corrections.

3.4 Final Remarks on K-corrections

To summarise therefore, the K-corrections of ellipticals seem adequately defined from IUE data - a larger sample might address luminosity effects (cf. Sandage and Visvanathan 1978) and possible colour gradient effects which are always a worry with small aperture data. The extension to the short wavelength region (1200-2000Å) will also prove valuable though the integration times will be long.

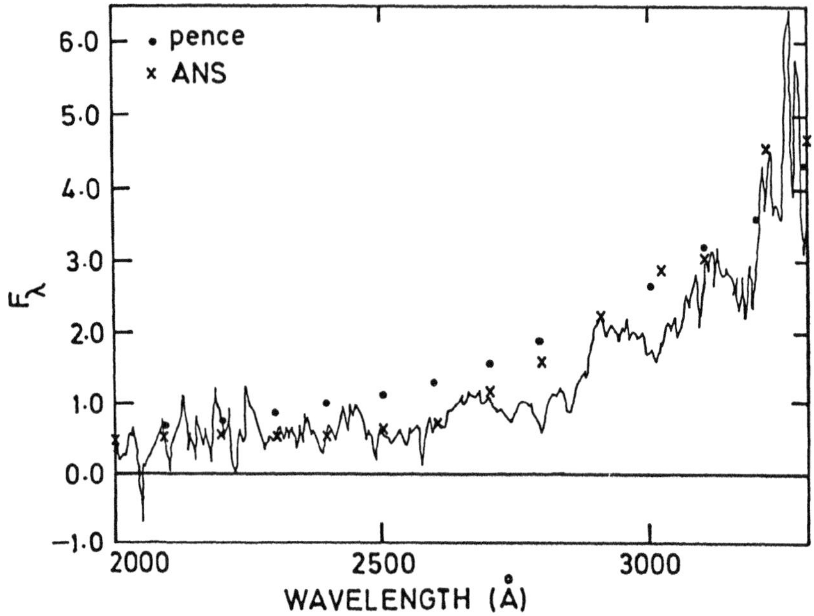

Figure 5. Ultraviolet spectra of early type populations. The solid spectrum is a mean of 4 ellipticals and the bulge of M31 from Bruzual and Spinrad (1980): the crosses represent a mean of M31 and M81 from Coleman et al., and the dots represent the mean E/S0 distribution derived by Pence from OAO-2 data. The fluxes were normalised at 2900 and 3300Å.

In the case of the spirals the main progress has been the large ANS survey catalogued by Wesselius. The IUE data has shown the important effect of reddening which cannot be assessed accurately from the broadband satellite data. For the later types the IUE spectra offer the best K-corrections though comparisons with larger apertures are necessary to check for gradient effects. The combination of IUE and the ANS catalogue offers the only short term hope of clearing up the largest hole in our knowledge of this subject.

As a final check we note that our model, which adopts the

Durham survey LF parameterised by details given in Table 3 and K-corrections described above, reproduces the slope of the observed galaxy counts exactly (Figure 2). The model also reproduces the counts to J ~ 20 to a high degree of accuracy.

4. CONVENTIONAL EVOLUTIONARY TESTS

4.1 Introduction

We are now in the position to explore the faint observations confident that we have some idea of basic galaxy properties and their uncertainties. These is no best way to monitor evolution (except perhaps the H-R diagram in our own Galaxy) since the observational programmes are still guided somewhat by the uncertain predictions. In the case of the Hubble diagram I will inevitably be drawn into the issue of the value of q_0. My aim here is to review the current status of the various methods to see if they can provide a consistent solution. Few would argue against the statement that an equally powerful probe of the cosmic density Ω (=$2q_0$, assuming zero cosmological constant) is the dynamics of nearby galaxies (Peebles 1980). Estimates of Ω from cosmic virial theorems are currently in the range $0.1 < \Omega < 0.5$. An appropriate use of the Hubble diagram is then to assume say $\Omega \sim 0.3$ and infer the evolution. This is far less contrived than the well-established reverse practice of making an evolutionary correction to q_0 (based solely on theoretical models) (Sandage 1961: Gunn and Oke 1975: Kristian et al. 1978).

4.2 Optical photometry of field samples

4.2.1 Techniques

We begin with optical number-magnitude and colour data since these follow directly from section 3. A variety of problems have to be tackled before a list of photometric quantities emerges. Techniques in faint galaxy photometry have been discussed by Jarvis and Tyson (1980), Ellis (1980), Kron (1979) and Godwin (1976). Most work I shall discuss has been done with automatic scanning microdensitometers, some with on-line processors such as COSMOS (Pratt 1975) or APM (Kibblewhite 1980).

The apparent magnitude above an isophote of a galaxy can be modelled taking account of the atmospheric seeing. Alternatively, we could follow Kron (1978, 1980) in attempting to measure all the light from a galaxy by considering the gradient in its light profile. Whichever is chosen, the completeness limit of the observations is model-dependent and that is one of the more miserable aspects of faint photometry.

Kron's (1978, 1980) technique is to sum the light to a point where the profile growth curve derivative reaches a certain value: a small error in radius then produces the minimum error in magnitude. The method operates statistically in that the photometric errors at large radii on a single profile are sizeable. Random errors can be controlled by plate to plate comparisons but systematic errors introduced by sky subtraction must be considered very carefully. CCD magnitudes to the limit of detection for a well-studied field would be a useful check on possible scale errors in photographic photometry (cf. Kron 1979).

The understandable argument against the straightforward isophotal technique is that the fraction of light above the isophote decreases markedly with redshift. Modelling ideally requires information on the light profiles and colour gradients as well as their evolution. The essential decision is whether observation takes precedence over models since a pure experimentalist would argue isophotal quantities are well-defined in terms of their error distributions and theory should apply the necessary corrections (cf. the philosophy outlined in Ellis and Perry 1979). On the other hand Kron would argue the interpretation of data takes precedence and since total quantities are easier to model one should struggle with difficult measurements.

On a differential plot the star count equals the galaxy count at J ~ 21 but the galaxy count slope dlogN/dm is then twice as steep. On a good 4-metre telescope photograph the eye can classify extended and compact objects to J ~ 22.5. The possibility of a large population of nearby compact galaxies at faint magnitudes can only be excluded spectroscopically. A valuable spectroscopic survey by Morton and Tritton (1979) of all objects within 0.3 deg^2 to B = 20 (J = 19.8) revealed 64 galaxies - a surface density in excellent agreement with photographic surveys and well explained by a conventionl LF without a faint extension.

According to the above the star/galaxy classification is most necessary on Schmidt plates reaching to J ~ 21.5. Fainter counts are progressively less seriously affected by stellar contamination: counts fainter than J ~ 23 are only affected to 15% by stellar contamination.

Profile shape alone is not enough to distinguish a galaxy from a star or QSO. Correlations of central surface brightness, profile slope, moments of the light distribution work fairly well (Peterson et al. 1979: Ellis 1980). Multidimensional plots of these are considered by Jarvis and Tyson (1980) though their dimensions are not independent. A probabilistic scheme has also been defined (Sebok 1979).

An alternative scheme is to use colours (Bruzual and Kron

1980) since a stellar population has a flatter energy distribution than that of a single star. Colours are also a valuable extra dimension in classifying galaxies but they do require an accurate zero redshift locus so as to avoid throwing nearby galaxies (e.g. dwarfs) into the star category. (Colour calibrations are also a luxury southern observers can rarely afford!)

4.2.2 Number magnitude counts

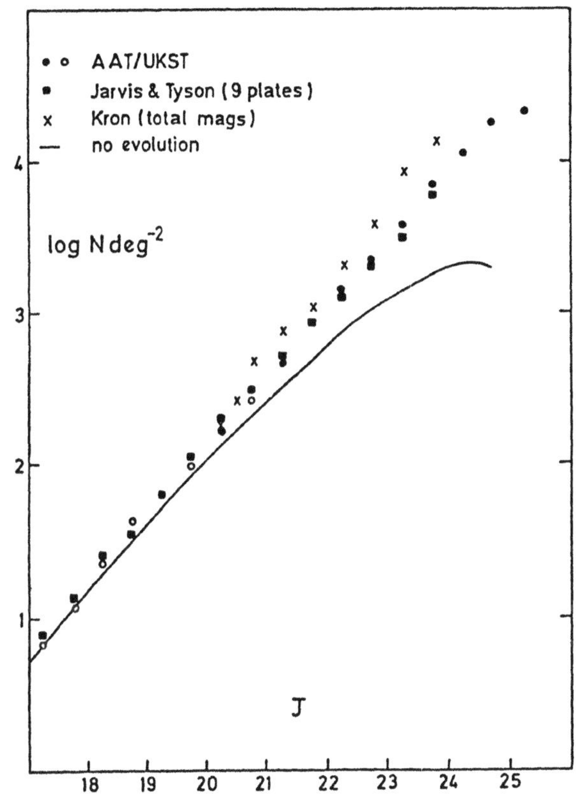

Figure 6. Differential J number magnitude counts in 0.5 mag bins. The Jarvis Tyson counts are based on the mean results from 9 high-latitude fields. The formerly uncertain zeropoint for the Peterson et al AAT/UKST data is now confirmed from photoelectric data. The Kron magnitudes are total: his counts are included for comparison purposes only.

Figure 6 presents galaxy counts in J as published by various observers. I have averaged the 9 highest latitude fields studied by Jarvis and Tyson (1980, JT) and this shows excellent agreement with Peterson et al. (1979) labelled AAT/UKST. The excess of bright galaxies (J ~ 17-18) reported by Tyson and Jarvis (1979, TJ) has now vanished. The AAT data was often neglected as it had an initial zero point calibration accurate only to $\pm$ 0^m.3 (Tinsley 1980b). However recent photoelectric data has established that the initial calibration was accurate to $\pm$ 0.1 mag. (as comparison with JT now clearly shows). The counts of Karachentsev (1980) are also in excellent agreement though the procedures adopted there are not defined in sufficient detail to exclude possible scale errors and incompleteness.

The total magnitude counts of Kron (1978, 1980) cannot be readily compared with the isophotal data since the positive aperture difference (total - isophotal) will be a function of magnitude (cf. JT Table 6) and, more important, the total magnitude has larger random errors which produce an appeciable steep ending of the count slope (discussed by Kron (1978) and Ellis (1980). When Kron reprocesses his data in the isophotal manner the agreement with the published isophotal data is good (Kron 1979).

Our standard model (section 3) reproduces the counts to J ~ 20. A slightly higher φ^* would improve the fit for J ~ 21 but would cause problems at the bright end (φ^* is in this case better set by our 5 survey fields at J = 16.75 where patchy Galactic obscuration might average out to that expected for all high latitude fields). For J > 21.5 the model falls below the data, the deficiency is a factor of 5 at the completeness limit of the AAT (J = 24.5). The count slope for the no evolution model at J = 23.0 $\pm$ 1.5 is α ~ 0.3, substantially flatter than previous models. The Peterson et al. no evolution model gave α = 0.40 but was later revised to α = 0.36 on the publication of the KOS mix. This drop in the predicted counts stems partly from the move away from galaxy mixes dominated by late type galaxies (e.g. those adopted by Tinsley 1977, 1978) but is also due to an improved characteristic magnitude scale now a LF is available in the J passband (cf. M* in Table 3 with Table 1 of Phillips et al. 1980). Figure 7a shows the effect of uncertainties in our model: there is no denying the need for substantial evolution over recent epochs (cf. more pessimistic conclusions in Ellis 1979a,b, 1980). Tinsley's slow-star formation (SSF, Tinsley 1978) models are consistent with the low extragalactic background light measurements (Dube et al. 1977a,b) and were discussed earlier (section 2). The quiescent versions of the SSF models where the initial burst of star formation is obscured (e.g. by dust) fall short of the data even when the redshift of galaxy formation is quite small (Figure 7b). The visible burst models however produce large excess at magnitudes within reach. For $z_f = 2$ such models produce large humps at J ~ 19–21 and can be clearly ruled out. For $z_f = 5$ the data seems to rise more gradually than the models but this is probably due to the abrupt representation of the end of star formation in Tinsley's models.

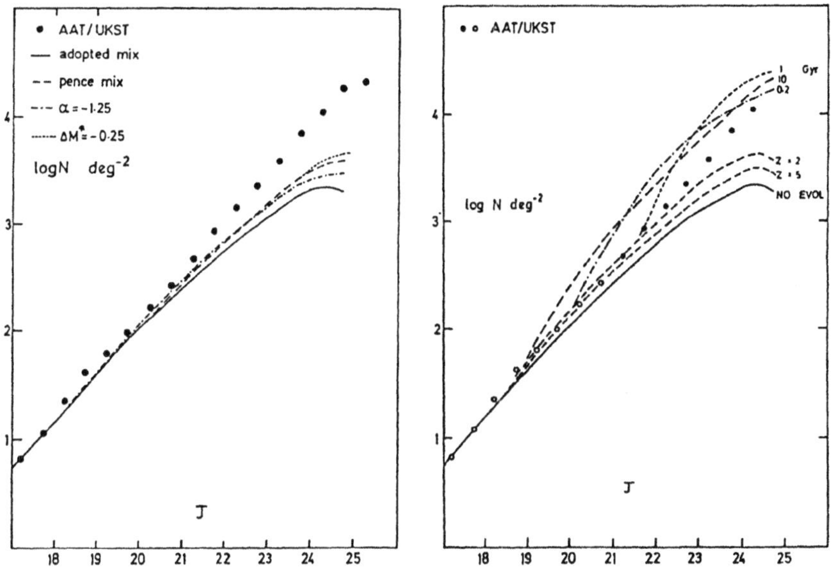

Figure 7. (a) Uncertainties in the standard model: the curve labelled Pence mix changes the adopted mix to that compatible with the Second Reference Catalogue: the curve labelled $\alpha = -1.5$ assumes a steeper Schechter function slope. The curve labelled $\Delta^* = 0.2$ brightens each M^* by $0\overset{m}{.}2$.

(b) Comparison of number magnitude counts with Tinsley's (1978) slow-star formation evolutionary models. The curves labelled $z_f = 2.5$ refer to models where E/SO galaxies are obscured during their initial burst of star formation at redshift z. The other models have $z_f = 5$ with visible bursts of varying duration in Gyr. The evolutionary timescale assumes $q_o = 0.05$, $H_o = 50$.

4.2.3 Colour information

As previously mentioned, colour is a valuable additional classifier though its evolution is likely to be slight. Indeed, if colour changes reflect only morphology and redshift, the real benefit of colours will be in creating subsamples for further analysis (Kron and Koo have already attempted to isolate galaxies on redshift slices). However, since the errors on colours are larger than those for a single galaxy magnitude, model comparisons on colour-colour scatter diagrams are not straightforward.

Work done by Bruzual and Kron (1980), Phillips et al. (1981) and Karachentsev (1980) show that the mean galaxy colour moves blueward with fainter magnitudes as expected but the evolutionary inferences with regard to the distribution of colours are not significantly clear as of yet. Certainly the present conclusions rest mainly on the inferences which could have been derived from number counts in the individual passbands.

In discussing the frequently quoted 'puzzle' of faint blue galaxies attributed to Kron it should be noted that most reasonable models including ones with no evolution can explain the colour data both in terms of the mean and the distribution (cf. figure in Bruzual and Kron 1980): that is, the (J - F) K-corrections are sufficient to explain the overall trends. Again the interpretation rests heavily on the distribution of intrinsic colours, relations between K-corrections and both intrinsic colour and luminosity.

Turner (1979) examined 49 colour selected galaxies in Kron's sample with $19 < J < 21$. This redshift distribution is anomalous in that there is a $z > 0.3$ tail implying strong evolution in the bluest galaxies. The fact that this can be shown to be atypical of the field via number counts simply underlines the fact that a very large or complete radial velocity survey at faint magnitudes would be needed to draw useful evolutionary conclusions.

4.3 Spectroscopic data

4.3.1 Spectrophotometric evolution

The continuous energy distribution of a distant object would be the clearest evolutionary test since it avoids any assumptions about K-corrections or absolute magnitudes. Apart from the overwhelming integration times the catch is the definition and selection of a standard object.

Brightest cluster members are a favourite choice but the order of rank in a cluster at $z \sim 0.5$ would be seriously affected by the presence of spirals whose smaller K-corrections (~ 1.5 mag less in B) would make them apparently brighter than many of their elliptical counterparts. Furthermore it is actually difficult to find very distant clusters (Spinrad 1980) except by random searches on small field 4-metre telescope plates or via radio data. The more elaborate the method for selecting the objects the more likely unknown selection effects are involved.

Over the range $0 < z < 0.5$ changes in rank should not be too serious a problem. Figure 8 shows Spinrad's (1977) comparison of the mean restframe spectral energy distribution (s.e.d.) for three $z \sim 0.5$ ellipticals with the average s.e.d. of NGC 4486 (M87) and NGC 4472. I have added the IUE data of Bruzual and Spinrad (1980) and Norgaard-Nielsen et al. (1981). The comparison, normalised at restframe 4500Å shows a consistently bluer spectrum by ~ $0^m.15$ in (B-V). For comparison, the SSF model $z_f = 5$, $\tau = 0.2$ Gyr which was too strong for the blue field counts (Figure 7b) shows evolution in (B-V) over and above the K-term of ~ $0^m.05$ so these observational estimates are not consistent if most of the field evolution occurs in ellipticals.

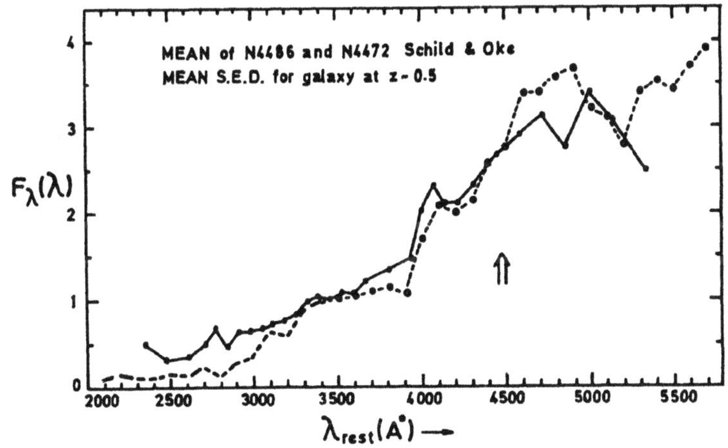

Figure 8. Spinrad's comparison of spectral energy distributions of ellipticals at $z \sim 0.5$ and $z \sim 0$: the local data has been extended into the ultraviolet using recent IUE observations.

Incidentally, one remark here about redshift and morphological uncertainties in faint spectra: the uV features dicussed by Spinrad and colleagues as being useful redshift indicators for $z \sim 1$ early type galaxies are also apparent in the IUE spectra of nuclei of Sbc and Scd galaxies (Ellis and Gondhalekar 1981a). The presence of such features may give reliable redshifts if the 4000Å break can be excluded but should not be taken as proof that these galaxies are ellipticals.

Comparisons such as those of Figure 8 would be invaluable if many galaxies at various z could be studied: the spectra could be compared in detail, for example, with Bruzual and Kron's predictions. The telescope time required, however, is currently beyond the scope of most individuals though multi-object spectroscopy (Hill et al. 1980) and CCD photometry through narrow-band filters are practical alternatives.

4.3.2 Hubble diagrams

This is a large subject and my aim here is to simplify it addressing the question as to whether evolutionary implications from magnitude/colour-redshift plots are consistent with those determined from field counts assuming a cosmological model ($q_o = 0.15$, $\Omega = 0.3$ see section 4.1). Any differences might indicate an incorrect cosmological model, some stellar/dynamical evolution peculiar to clusters or (most likely) unaccounted selection effects. Already Figure 8 indicates evolution in clusters is stronger than expected: other examples of colour evolution will be discussed in section 4.4.

We have already introduced some of the problems of the Hubble diagram: how can one be sure a distant standard candle is the same kind of object as its nearby comparison? Any systematic error is more likely to be correlated with redshift than anything else. The rank changes at $z \sim 1$ may be overwhelming even if ellipticals evolve significantly. $K_B(E/SO) - K_B(Scd)$ is about 4 magnitudes at $z \sim 1$ whereas E/SO galaxies may only have brightened by ~ 1.5 mags if $z_f = 5$ and $\tau = 0.2$ Gyr. Other selection effects include the cluster richness-luminosity bias (Scott effect) and photometric procedure problems such as the widespread use of metric magnitudes which require a cosmology before they can be defined (since the angular diameter distance as a function of z is a strong function of q_o) (see Peach 1970).

The mounting evidence for cannibalism in clusters as proposed by Ostriker and Hausman (1978) is supported by observations of rising stellar velocity dispersions across cD galaxies (Dressler 1979: Carter et al. 1981) and the failure of the Faber-Jackson luminosity-velocity dispersion relation for giant ellipticals (Kirschner and Malumuth 1981). These results further indicate there may be enormous systematic errors involved in traditionally straightforward interpretations of the Hubble diagram.

When one throws in the sheer amount of telescope time needed to construct the Hubble diagrams the advantages of the methods discussed previously for field galaxies seem overwhelming. However, as much data exists in this form we shall now examine it for evolutionary trends.

Most of the data is contained in the articles by Hoessel (1980), Kristian et al. (1978) and Gunn and Oke (1975) though some of the samples are radio galaxies with possible non-thermal components (and hence non-standard K-corrections). The fact that the 3 groups of authors respectively derive $q_0 = -0.55 \pm 0.45$ ($z < 0.1$), 1.6 ± 0.4 ($z < 0.4$) and 0.31 ± 0.68 ($z < 0.5$) when evolution is neglected indicates the sensitivity of the various assumptions made.

Gunn and Oke (GO)'s data is spectrophotometric and thus avoids the K-correction uncertainty (though today this is one of the smallest problems). Since early types are brighter at large z we always overestimate q_0 if we neglect evolution. If q_0(true) = 0.15 this immediately rules out sensible discussion of Hoessel's data though the redshifts involved in that data are modest and a reliable estimate of q_0 would require extremely accurate photometry and corrections.

In the case of GO's data their apparent q_0 is so close to our true value that an evolutionary brightening in V of only a few 1/10ths of a magnitude at z = 1 is implied, in quite marked contrast to the implications of the field counts ($\Delta M_V(z = 1) < -1.4$ mag). Of course the Hubble diagram authors quote large error bars and another way of viewing the problem is to calculate from the evolutionary model that approximates to the field counts the apparent q_0 given that q_0 (true) = 0.15. In this case for the data sets with $z < 0.5$ the q_0 expected would exceed 2.0 and again only Kristian et al.'s sample seems consistent with the conclusions derived from field counts.

Kristian et al. also present colour-redshift diagrams which minimise some of the uncertain selection effects but introduce other difficulties such as the insensitivity to evolution considering the photometric errors. Kristian et al. concluded from (B-V) measures that there was either significant evolution beyond $z \sim 0.3$ (at variance with Bruzual and Kron's 1980 field survey) or that a typical cluster member had ultraviolet spectrophotometry more similar to that of M87 than NGC 4472. Bruzual and Kron and Kristian et al. are able to reproduce the colour changes from $0 < z < 0.3$ with little difficulty, though I have some difficulty reproducing such a large change in (B-V) from available K-corrections (Figure 9).

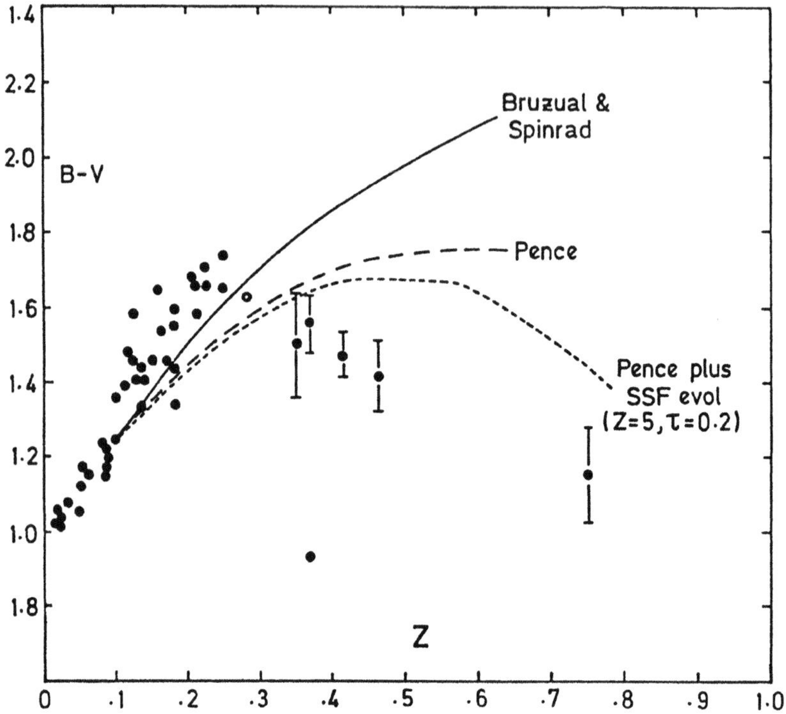

Figure 9. (B–V) colours for cluster galaxies plotted against redshift. Data is taken from Kristian et al. (1978). The bluer colours for z > 0.3 cannot be reproduced with reasonable models allowing for K-corrections and luminosity evolution.

Beyond z ~ 0.3 the colours are significantly bluer but cannot be easily reproduced with reasonable models (though it must be stressed that such fine changes in colour are a 'strain' on the accuracy of the K-corrections particularly beyond z ~ 0.5: the difference, for example, between the Bruzual-Spinrad and Pence curves is an indication of the remaining uncertainty). The conclusion that there is a significant colour evolution for giant galaxies in clusters inconsistent with field studies (such as might be expected for a variety of reasons) will require more data (e.g. for non-radio galaxies) and thereby a better understanding of the other effects involved.

A final remark on our choice for q_0: Lebofsky (1981) and others have examined the infra-red Hubble diagram which is more immune to evolution, uncertain K-terms and aperture/extinction corrections. It might yield a believable q_0. In the absence of any evolutionary correction her data for 33 clusters with $0 < z < 1.18$ yields $q_0 \sim 0.7$. With a simple evolutionary correction derived from Tinsley's SSF models q_0 reduces to -0.05 ± 0.30, a range just covering our preferred value of 0.15.

4.4 Colour evolution in clusters

The bluer colours claimed by Kristian et al. for giant E's differ of course from the well-known claim by Butcher and Oemler (1978, 1980) that a significant proportion of galaxies in clusters with $0.2 < z < 0.5$ have colours bluer than expected from the effects of redshift alone. The B-O effect relates to the distribution of colours and the presence of a blue tail when compared, for example, to that for the Coma cluster (Spinrad 1977) or structurally similar clusters. The implication that at surprisingly recent epochs blue spirals undergo a gas removal process followed by a remarkably rapid quenching of star formation (Larson et al. 1980) has been questioned both in terms of the difficulty in its understanding and its observational validity. Now is, sadly, not a ripe time for a major review of these important issues since several groups are in the process of analysing similar but higher quality data for many clusters (see below).

In their initial study B-O obtained ISIT cluster photometry in V and R (λ = 7050Å) for 2 clusters but derived field counts from a combination of sources. An obvious concern is the degree of field contamination: the spiral-rich field could easily produce a significant blue tail in the cluster distribution if the background were slightly underestimated. As Mathieu and Spinrad (1981) point out, the counts used by B-O to estimate the field contribution are anomalously low when compared to the red field counts of Kron (1978). By examining the cluster 3C295 with (J,F) prime focus plates they were able to derive the transform F-R = 0.2 and obtain field counts consistent with those of Kron on the same plates. They conclude there is no evidence for a blue tail and that the surface overdensity of this cluster had been overestimated by a large factor by Butcher and Oemler. For example doubling the field density reduces the overdensity of galaxies bluer than V-R ~ 1.33 from 4:1 to 0.1:1. The cluster 3C295 was the least rich of B-O's first set of clusters but even in the case of 0024 + 1654 the overdensity is seriously affected by the revised background.

On the other hand Butcher and Oemler (1980) have continued their analysis for 7 clusters confirming the effect and claiming that background corrections are negligible since the cluster overdensities are ~ 5–10:1 (though I am not sure whether this figure refers to the blue galaxies or the entire sample). Furthermore Newell and Couch (private communication) have homogeneous data on 14 clusters with $0.2 < z < 0.4$ including 0024 + 1654 which, they state, has a large overdensity. They find a surface density of field galaxies lower than Kron (using his total magnitude technique) and tend to support the work of Butcher and colleagues. Finally, Carter and I recently obtained CCD photometry of 3 richness class 3 clusters with $0.22 < z < 0.31$ through 6 narrow band filters within $4200 < \lambda < 8700$Å. This long baseline will allow fairly accurate estimates of morphology in and out of the clusters. There are, therefore, good reasons for believing that the observational status of the B-O effect will be finalised very soon.

Finally an ingenious explanation that goes most of the way to explaining the B-O effect even if their background corrections were shown to be correct has been discussed by De Gioia-Eastwood and Grasdalen (1980). They show that the luminosity-colour relation for early type galaxies (Sandage and Visvanathan 1978) introduces K-corrections sufficiently dependent upon the average luminosity sampled to explain all but the bluest cluster galaxies ($V-R < 1.2$) at $z \sim 0.4$, which can be readily dismissed in terms of background uncertainties.

4.5 Conclusions on evolutionary tests

Most of our attempts to reconcile the evolution inferred from the various methods look pretty dismal! The Hubble diagram and cluster data is so varied there are as yet no two sets of observations that agree. On the other hand the observations of the field are in excellent agreement and the uncertainties in their interpretation, whilst still uncomfortably large are far less depressing than they were 2 years ago (and no worse than in the case of the Hubble diagrams). As our knowledge of galaxy properties has improved the implied luminosity evolution has increased and now seems consistent with Tinsley's $z_f = 5$ models where young galaxies should be visible (though I would hardly call this an unique conclusion). I believe the right direction for the future continues to be studies of field galaxies (multi-object spectroscopy and infrared colours will yield valuable redshift distributions if the statistical errors can be beaten down with large enough samples). After several decades and countless months of telescope time what have Hubble diagrams taught us? After only a few years of effort and a small number of prime focus plates we are already able to exclude the no-evolution case with a high degree of certainty.

5. SEARCHES FOR PRIMEVAL GALAXIES

5.1 Introduction

The subject of primeval galaxy (PG) searches was recently reviewed by Davis (1980). PG detection is limited by two things: the sensitivity of the available detectors and our knowledge of the likely properties of a PG essential if we are to recognise one. Both these factors have changed fairly substantially recently so a brief refresher course is in place.

5.2 Extragalactic background light (EBL) measurements

A direct measure of the visibility of unresolved luminous PGs comes from experiments measuring the EBL. The difficulties in these experiments lie not in the signal to noise but in the detailed corrections necessary for atmospheric and Galactic light contributions. The two popular methods are discussed by Mattila (1976) and by Dube *et al.* (1977,1979). Table 6 summarises the most recent results.

Table 6

EBL Estimates

Source	EBL (S_{10})*	Å Wavelength	Method
Spinrad and Stone 78	<5	4000	Dark Cloud
Dube *et al*. 77,79	<3.4	5115	Mask
Mattila 76	12±4	4000	Dark cloud
Lillie 68	<1	4100	Satellite
Roach and Smith 68	<5	5300	Direct Photometry
Peterson *et al*. 79	0.17±0.03	4400	Galaxy counts extrapolated
Tinsley 78	<1	5500	SSF models

* S_{10} = 10 mag AOV star per sq deg = 7.55 x 10^6 cgs at 5100Å

Both the dark cloud and focal plane mask methods make assumptions about the magnitude distribution of faint images not excluded but these enter at only a few 1/10th S_{10} and can be modelled to a few 1/100ths.

The large positive result from Mattila is generally suspect since the cloud he hoped to be totally dark was not and complex corrections were necessary to allow for scattered light from the cloud. The table also shows the flux from galaxy counts integrated to J = 25. Clearly the experimental EBL upper limits allow much freedom for fainter galaxy counts (for example if the ~ 0.4 slope of log N − m continued indefinitely the contribution to EBL would still be a mere 0.3 S_{10} − a factor of 10 below most upper limits). Only at this level of a few 1/10ths of a S_{10} unit would EBL data usefully constrain the visibility of PGS since Tinsley's SSF models predict values well below 1 S_{10} even for z_f = 2. This accuracy may be possible in the infrared where the redshifted galaxy flux is higher compared to the other unwanted contributions, though as yet comparatively little is known about normal galaxies in the infra-red.

5.3 Properties of a primeval galaxy

The pioneering models of Partridge and Peebles (1967) predicted PGs to be large (~ 10 arcsec) red low surface brightness features with spectra revealing intense Lyman series emission. The objects were expected to be large because the fastest star formation was thought to occur before the collapse had completed.

Early searches for extended objects by Davis and Wilkinson (1974), Partridge (1974) failed largely because the flux limits reached were not very low (a few percent of sky). As Davis shows, a modern ground-based CCD system would offer a gain of ~ 10 and the Space Telescope in a much darker background sky would offer a gain of ~ 100.

Meier (1967a,b) also considered the expected appearance of PGs and concluded many (but not all) QSOs (e.g. OH 471) might be PGs. Such objects would have stellar appearance (since the Larson collapse models suggested active star formation within only 10 kpc) but would be distinguished from other QSOs via thermal continua, broader than usual stellar absorption lines (CIV, SiIV at ~ 500 km/s i.e. 15Å at z ~ 5) and neither polarisation nor variability.

Sunyaev et al. (1978) reviewed the observable tests in detail. A major issue they raised was the possible presence of enormous quantities of dust which could mask the early periods of active star formation. Models predicted a steep thermal continuum followed by a sharp Lyman discontinuity: reddening would modify this considerably and although a re-radiation excess might emerge at 100 (1+z) μm (Kaufman 1976) how would one select an object for study? Individual PGs at ~ 300-600 μm are predicted to be ~ 100 times fainter than our current sensitivity limit. EBL measurements would be the only way of detecting this signal.

Recently Meier and Terlevich (1981) examined the uV spectra of active star forming regions (extragalactic HII regions) with sufficiently high redshifts for the Lyα line to be separated from its geocoronal equivalent. They confirm hot star absorption features of CIV and SiIV similar to those seen in many late-type spirals (Ellis and Gondhalekar 1981a) but Lyα is much weaker than expected, particularly in metal-rich systems. The conclusion that such regions are dusty may explain the null results of slitless spectroscopy surveys for high z Lyα (Koo and Kron 1980) and cast doubt on proposed long slit (Hogan and Rees 1979) or narrow-band searches for line emission.

The whole question of whether a PG appears stellar is clearly important to any observers planning a long-term search. The implication of our discussion on galaxy counts is that a substantial fraction of 23-24th magnitude galaxies are young ellipticals at high redshift. Koo and Kron (1981) have just completed a search for faint star-like objects with anomalous UJFN colours and do not find any significant extragalactic component which could be identified with PGs. The next step will be to repeat the procedure with multi-colour galaxy data (Koo, thesis). Studies of blank fields may show blue objects with breaks corresponding to the Lyman limit. Narrow band photometry with CCDs will be an important technique and will reach Davis' calculated limits. The problem is how can we prove such objects, if found, are PGs? Spectroscopy is out of the question. Perhaps the sheer number and large-scale isotropy of such objects (~ 20 per typical 4-metre telescope CCD frame to J ~ 25.5) will be enough evidence: it may have to be!

6. ACKNOWLEDGEMENTS

In compiling this analysis and review I have benefitted from discussions with Dave Koo. I also thank Prab Gondhalekar and my colleagues at Durham for discussions and allowing me to use (in provisional form) results of the galaxy surveys obtained respectively with IUE and the AAT.

REFERENCES

Baade, W. 1963. Evolution of Stars and Galaxies. (Harvard University Press).

Balzano, V.A. and Weedman, D.W. 1981. Ap.J., 243, 756.

Bean, J., Ellis, R.S., Shanks, T., Efstathiou, G., Peterson, B.A. and Zou, Z.L. 1982. In preparation.

Binney, J. 1977. MNRAS, 181, 737.

Brown, G. 1974. Ph.D. Thesis, University of Texas.

Bruzual, G. and Kron, R. 1980. Ap.J., 241, 25.

Bruzual, G. and Spinrad, H. 1980. The Universe in Ultraviolet Wavelengths. Reidel: Dordrecht. (BS)

Butcher, H. and Oemler, A. 1978. Ap.J., 219, 18.

Butcher, H. and Oemler, A. 1980. IAU Symposium 92, Objects of High Redshift. Reidel: Dordrecht. p. 49.

Carnochan, D., Navach, T. and Wilson, R. 1975. MNRAS, 172, 27P.

Carter, D., Efstathiou, G., Ellis, R.S., Inglis, I. and Godwin, J. 1981. MNRAS, 195, 15P.

Christensen, C.G. 1975. Astron.J., 80, 282.

Code, A.S. and Welch, G.A. 1978. Ap.J., 228, 95.

Coleman, G.D., Wu, C.C., and Weedman, D.W. 1980. Ap.J.Suppl. 43, 393.

Davis, M. 1980. IAU Symposium 92, Objects of High Redshift. Reidel: Dordrecht.

Davis, M. and Wilkinson, D.T. 1974. Ap.J., 192, 251.

Davis, M., Geller, M.J. and Huchra, J.P. 1979. Ap.J., 221, 1.

Davis, M., Latham, D.L. and Tonry, J.L. 1981. In preparation.

De Gioia-Eastwood, K. and Grasdalen, G.L. 1980. Ap.J., 239, L1.

Dostal, V.A. 1979. Sov.A.J., 23, 135.

Dressler, A. 1978. Ap.J , 223, 765.

Dressler, A. 1979. Ap.J., 231, 659.

Dube, R.R., Wilkinson, D.T. and Wickes, W.C. 1977. Ap.J., 215, L51.

Dube, R.R., Wilkinson, D.T. and Wickes, W.C. 1979. Ap.J., 232, 333.

Ellis, G.F.R. and Perry, J.J. 1979. MNRAS, 187, 357.

Ellis, R.S. 1979a. Phil.Trans.R.Soc. London.A., 296, 355.

Ellis, R.S. 1979b. IAU Symposium 92. Objects of High Redshift. Reidel: Dordrecht. p. 23.

Ellis, R.S. 1980. Two-Dimensional Photometry. Ed. Crane and Woltjer. (ESO Publications), p. 339.

Ellis, F.S., Fong, R. and Phillipps, S. 1977. MNRAS, 181, 163.

Ellis, R.S. and Gondhalekar, P. 1981a. MNRAS submitted.

Ellis, R.S. and Gondhalekar, P. 1981b. In preparation.

Faber, S.M. 1972. Astron.Astrophys., 20, 361.

Felten, J.E. 1977. Astron.J., 82, 861.

Freeman, K.C. 1977. Evolution of Galaxies and Stellar Populations. eds. R.B. Larson and B.M. Tinsley. (Yale). p.33.
Godwin, J. 1976. D.Phil. thesis, Oxford University.
Gordon, D. and Gottesman, S.D. 1981. Astron.J., 86, 161.
Gunn, J.E. and Oke, J.B. 1975. Ap.J., 195, 255. (GO)
Gunn, J.E., Stryker, L.L. and Tinsley, B.M. 1981. Ap.J. in press.
Hill, J.M., Angel, J.R.P., Scott, J.S., Lindley, D. and Hintzen, P. 1980. Ap.J., 242, L69.
Hoessel, J.C. 1980. Ap.J., 241, 493.
Hogan, C. and Rees, M.J. 1979. MNRAS, 188, 791.
Hubble, E.P. 1926. Ap.J., 64, 321.
Hubble, E.P. 1934. Ap.J., 79, 8.
Hubble, E.P. 1936. Ap.J., 84, 517.
Huchra, J.P. 1976. Astron.J., 81, 952.
Huchra, J.P. 1977. Ap.J.Suppl. 35, 171.
Huchra, J.P. 1978. IAU Symposium 79. Large Scale Structure of Universe. eds. M. Longair and J. Einasto. Reidel:Dordrecht, p. 271.
Jarvis, J.F. and Tyson, J.A. 1980. Astron.J., 86, 476. (JT)
Johnson, H.M. 1979. Ap.J., 230, L137.
Karachentsev, I.D. 1980. Sov.A.J.Lett., 6, 1.
Kaufman, M. 1976. Astrophys.Sp.Sci., 40, 336.
Kibblewhite, E.J. 1980. APM Manual. (Institute of Astronomy, Cambridge).
Kirschner, R.P. and Malumuth, J. 1981. Preprint.
Kirschner, R.P., Oemler, A. and Schechter, P.L. 1978. Astron.J. 83, 1549. (KOS)
Kirschner, R.P., Oemler, A. and Schechter, P.L. 1979. Astron.J. 84, 951.
Kodak. 1973. Technical Information for Scientific Films.
Koo, D. and Kron, R. 1980. PASP 92, 537.
Koo, D. and Kron, R. 1981. Preprint.
Kristian, J., Sandage, A. and Westphal, J.A. 1978. Ap.J., 221 383.
Kron, R. 1978. Ph.D. Thesis, University of California, Berkeley.
Kron, R. 1979. Physica Scripta, 21, 652.
Kron, R. 1980. Ap.J.Suppl. 43, 305.
Larson, R. and Tinsley, B.M. 1978. Ap.J., 219, 46.
Larson, R., Tinsley, B.M. and Caldwell, C.N. 1980. Ap.J., 237, 692.
Lebofsky, M.J. 1981. Ap.J., 245, L59.
Lillie, C.S. 1968. Ph.D. Thesis, University of Wisconsin.
Mathieu, R.D. and Spinrad, H. 1981. Ap.J. In press.
Mattila, K. 1976. Astron.Astrophys., 47, 77.
Meier, D. 1976a. Ap.J., 207, 343.
Meier, D. 1976b. Ap.J. 203, L103.
Meier, D. and Terlevich, R. 1981. Preprint.
Morton, D.C. and Tritton, K.P. 1979. AAT Symposium (Edinburgh).

Nandy, K., Thompson, G.I., Jamar, C., Monfils, A. and Wilson, R.R. 1976. Astron.Astrophys., 51, 63.
Nandy, K., Thompson, G.I., Carnochan, D. and Wilson, R. 1978. MNRAS, 184, 733.
Norgaard-Nielsen, H.V. and Kjaergaard, P. 1981. Astron. Astrophys., 93, 290.
Ostriker, J.P. and Hausman, M.A. 1978. Ap.J., 217, L125.
Pagel, B.E.J. and Edmunds, M. 1981. Ann.Rev.Astron.Astrophys. in press.
Partridge, R.B. 1974. Ap.J., 192, 241.
Partridge, R.B. and Peebles, P.J.E. 1967. Ap.J., 147, 868.
Peach, J.V. 1970. Ap.J., 159, 753.
Peck, M.L. 1980. Ap.J., 238, 79.
Peebles, P.J.E. 1980. Texas Symposium, Baltimore, December 1980.
Pence, W.D. 1976. Ap.J., 203, 39.
Persson, S.E., Frogel, J.A. and Aaronson, M. 1979. Ap.J.Suppl. 39, 61.
Peterson, B.A. 1973. IAU Symposium 58, Formation and Dynamics of Galaxies. Reidel: Dordrecht. p. 75.
Peterson, B.A., Ellis, R.S., Kibblewhite, E.J., Bridgeland, M., Hooley, T. and Horne, D. 1979. Ap.J., 233, L109.
Phillipps, S., Fong, R., Ellis, R.S., Fall, S.M. and MacGillivray, H.T. 1978. MNRAS, 182, 673.
Phillipps, S., Fong, R. and Shanks, T. 1981. MNRAS, 194, 49.
Pratt, N.M. 1975. Image Processing Techniques in Astronomy Reidel: Dordrecht. p. 217.
Roach, F.E. and Smith, L.L. 1968. Geophys. JRAS, 15, 227.
Sandage, A. 1961. Ap.J., 133, 355.
Sandage, A. and Tammann, G.A. 1980. Revised Shapley-Ames Catalog. Carnegie: Washington.
Sandage, A., Tammann, G.A. and Hardy, E. 1972. Ap.J., 172, 253.
Sandage, A. and Visvanathan, N. 1978. Ap.J., 225, 742.
Schechter, P.L. 1976. Ap.J., 203, 297.
Schott. 1974. Technical Manual of Scientific Filters.
Searle, L. and Sargent, W.L.W. 1972. Ap.J., 173, 25.
Sebok, W.L. 1979. Astron.J., 84, 1526.
Shanks, T., Fong, R., Ellis, R.S., and MacGillivray, H.T. 1980. MNRAS, 192, 209.
Spinrad, H. 1977. Evolution of Galaxies and Stellar Populations. Yale, p. 301.
Spinrad, H. 1980. IAU Symposium 92. Objects of High Redshift. Reidel: Dordrecht. p. 39.
Spinrad, H. and Stone, R.P.S. 1978. Ap.J., 226, 609.
Spinrad, H., Stauffer, J. and Butcher, H. 1981. Ap.J. in press.
Struck-Marcell, C. and Tinsley, B.M. 1978. Ap.J., 221, 562.
Sunyaev, R., Tinsley, B.M. and Meier, D. 1978. Comm.Astrophys., 7, 183.

Tammann, G.A., Yahil, A. and Sandage, A. 1980. Ap.J., 234, 775. (TYS)
Tinsley, B.M. 1968. Ap.J., 151, 547.
Tinsley, B.M. 1972. Ap.J., 178, 319.
Tinsley, B.M. 1976. IAU Colloquium 37. Decalages vers le Rouge et Expansion de l'Univers. (Paris), p. 223.
Tinsley, B.M. 1977. Ap.J., 211, 621. erratum Ap.J., 216, 349.
Tinsley, B.M. 1978. Ap.J., 222, 14.
Tinsley, B.M. 1980a. Fund.Cosmic Phys., 5, 287.
Tinsley, B.M. 1980b. Ap.J., 241, 41.
Tinsley, B.M. and Danly, L. 1980. Ap.J., 242, 435.
Tinsley, B.M. and Gunn, J.E. 1976. Ap.J., 203, 52.
Turner, E.L. 1979. IAU Symposium 92. Objects of High Redshift. Reidel: Dordrecht, p. 71.
Turner, E.L. and Gott, J.R. 1976. Ap.J., 209, 6.
Tyson, J.A. and Jarvis, J.F. 1979. Ap.J., 230, L153.
Twarog, B.A. 1980. Ph.D. Thesis, Yale University.
van den Bergh, S. 1960. Publ. DDO 2, 159.
de Vaucouleurs, G. 1963. Ap.J.Suppl., 8, 31.
de Vaucouleurs, G., de Vaucouleurs, A. and Corwin, H. 1976. Second Reference Catalog of Bright Galaxies. (University of Texas Press).
Wells, D. 1978. Ph.D. Thesis, University of Texas.
Zwicky, F., Herzog, E., Wild, P., Karpowicz, M. and Kowal, G. 1961-68. Catalog of Galaxies and Clusters of Galaxies. (Caltech). Vols. 1-6.

COSMOLOGICAL EVOLUTION OF QSOs AND RADIO GALAXIES

J.V. Wall

Royal Greenwich Observatory, Herstmonceux Castle, Hailsham, Sussex, U.K.

1. INTRODUCTION

These lectures represent a personal account of present understanding of the spatial distribution of QSOs and radio galaxies - the luminosity functions and their dependence on cosmic epoch. I suggest five interrelated reasons why such studies may be of interest.

(1) Classical cosmology? Early identifications of high-galactic-latitude sources led to the conclusion that the majority lay at cosmologically significant distances. Could they thus be used to map the framework of the Universe, to determine H_0, q_0, Λ ? Many broad luminosity functions provide something of a deterrent, but another factor provides the ultimate deterrent: since the pioneering investigations of Ryle, Davidson and coworkers in the late 1950s and early 1960s, it has become clear that gross evolution of the radio-source populations far overwhelm the geometrical/relativistic effects. We do not understand this cosmological behaviour in physical terms, but we can at least map it as a step towards producing the physics. It is this mapping process with which my lectures are principally concerned.

(2) Do numerical models work? Schemes to map the spatial distribution of QSOs and radio galaxies were developed during the mid to late 1960s (Longair 1966, Schmidt 1968). At the same time, radio surveys at different frequencies together with continued programmes of optical identification and redshift measurement demonstrated that more than one population of

B. J. T. Jones and J. E. Jones (eds.), The Origin and Evolution of Galaxies, 295–328.

extragalactic radio sources existed. Numerical models - models which simply purport to describe the data - must produce results consistent with all relevant observations. If the numerical schemes do not work, then the basic assumptions are wrong. For example, there may be an unrecognized population of sources lurking in the cosmic depths which results in counts of faint radio sources disagreeing with the numerical predictions made from our knowledge of the statistics of bright radio sources. A more interesting possibility is that the modelling might show up an error in a fundamental assumption such as that of the cosmological nature of the redshifts.

(3) Do physical models work? We do not know how extragalactic radio sources work. We do not know the nature of the central powerhouse, although a massive spinning black hole is a good bet (Rees 1979). We do not know the fuel supply, the mechanism(s) of energy transport to the familiar double-lobe configuration, or how the plasma is confined in these lobes. These three factors may be strongly dependent on epoch for a variety of reasons (cf section 5). But no model of QSOs or radio galaxies which purports to describe the birth-life-death cycle can truly do so unless it produces a map of the bulk cosmological evolution which matches those derived from the data. Conversely, these evolution maps may point the way towards physical models for these objects.

(4) What observations? Most of the sky has now been covered by radio surveys for extragalactic sources at a wide range of frequencies. Large-scale searches at optical wavelengths have also been carried out. Nevertheless it remains easy to find survey/identification/redshift programmes to pursue, by surveying any number of different regions at different frequencies to ever fainter radio and optical limits. To what purpose? The data synthesis which I discuss here at least serves to define our areas of ignorance: it serves to demonstrate in an unambiguous way which parts of the luminosity functions are poorly defined, and which are the critical observations.

(5) Impact on other problems? Consider, for instance, the isotropic X-ray background, for which it is necessary to know the luminosity functions for both radio and optically-selected QSOs in order to estimate the QSO contribution. Present indications are that QSOs account for a significant fraction (Kembhavi and Fabian 1981, Cavaliere et al. 1981), but cannot account for it all. An accurate calculation of the QSO contribution is therefore vital to determine the intensity and spectrum of the unexplained remainder. A second example is the epoch of galaxy formation , studies of QSOs and radio galaxies provide our best chance of seeing a redshift cutoff, corresponding to an epoch of creation for these objects. This switch-on time may not correspond precisely to the epoch of formation of galaxies, but it does at

least define an epoch at which nuclear activity commenced, an epoch at which there were aggregates capable of defining gravitational potential wells.

2. METHODS AND RESULTS

There are three basic methods for investigating the spatial distribution of QSOs and radio galaxies: (i) counts of sky surface density as a function of apparent intensity, (ii) the V/V_{max} or 'luminosity-volume' test, and (iii) direct evaluation of the luminosity function at different redshifts.

2.1 Surface Density Counts

The counts of sky surface density as a function of intensity are the fundamental and simplest data from radio and optical surveys. In integral form, the counts are the 'log N – log S' curve at radio wavelengths, and the 'log N – log m' curve at optical wavelengths.

What might we expect to observe? For simplicity, consider initially a static Euclidean Universe in which the volume density of objects as a function of luminosity P – the luminosity function – is $\rho(P)$. Then the flux S observed from an object at distance R is

$$S = P/R^2, \tag{1}$$

and the radius out to which an object of luminosity P has a flux > S is

$$R(S) = (P/S)^{1/2} \tag{2}$$

The number of objects of power P in the shell R to R+dR is

$$N(S) = \rho(P)dV(R) = \rho(P).4\pi R^2 dR \tag{3}$$

Hence the number of objects with luminosity P and with flux > S is

$$\begin{aligned} N(P,>S) &= \rho(P)\int_0^{R(S)} 4\pi R^2 dR \\ &= \frac{4\pi}{3}\rho(P)P^{3/2} . S^{-3/2}, \end{aligned} \tag{4}$$

and when the contribution of all objects is considered,

$$\begin{aligned} N(>S) &= \frac{4\pi}{3}\int_0^{\infty} \rho(P)P^{3/2}\, dP.S^{-3/2} \\ &= \text{const. } S^{-3/2}, \text{ the well-known } -3/2 \text{ power law.} \end{aligned} \tag{5}$$

The calculation for an FRW Universe ($\Lambda = 0$) follows identical steps. In (1), S and P respectively represent monochromatic flux and luminosity as emitted at the frequency of observation, while R^2 is replaced by

$$R^2 = D^2(1 + Z)^{(1+\alpha)} \tag{6}$$

where α is the spectral index ($S \propto \nu^{-\alpha}$), $(1 + Z)^{(1+\alpha)}$ is the so-called K-correction, and D is the 'effective distance'

$$D = \frac{2c}{H_o\Omega^2(1+Z)} [\Omega Z - (\Omega-2)\{(\Omega Z+1)^{1/2} - 1\}], \tag{7}$$

with (8)

$$\Omega = 2q_o = \frac{\rho}{\rho_{crit}} = \frac{8\pi G\rho}{3H_o^2}$$

In (2), instead of solving for R(S) we solve for Z(S), the redshift out to which an object of luminosity P has a flux density > S. In (3), dV(R) becomes

$$dV(Z) = 4\pi\, D^2\, dr, \tag{9}$$

where dr is the element of co-moving coordinate distance

$$dr = \frac{cdZ}{H_o(1+Z)(\Omega Z+1)^{1/2}} \tag{10}$$

Finally, then,

$$N(>S) = \int_0^\infty dP \int_0^{Z(S)} \rho(P)\; 4\pi D^2\; dr, \tag{11}$$

and for $Z \ll 1$ this reduces to $N(>S) = \text{const}.S^{-3/2}$ while for larger Z, $N(>S) = \text{const}.\ S^{-\beta}$ where $\beta < 3/2$ (see Longair 1971, Scheuer 1975).

In source-conserving Universes, then, we expect the integral counts of QSOs and radio galaxies at optical and radio wavelengths to have a slope of -3/2 at the highest observable fluxes, and to flatten progressively to the lower intensities as the geometrical effects of K-dimming and and co-moving volume elements come into play. In fact the observations look nothing like this. Look for example at Figure 1, which shows the radio source count at 408 MHz. This count is in 'relative differential' form, as described in the caption: the integral form is dangerous from a statistical point of view (Jauncey 1967) because the $\sqrt{N}$ error bars are not independent. The initial portion of the count at 408 MHz is significantly steeper than the -3/2 law. This steep portion is followed by a plateau, a region in which the integral

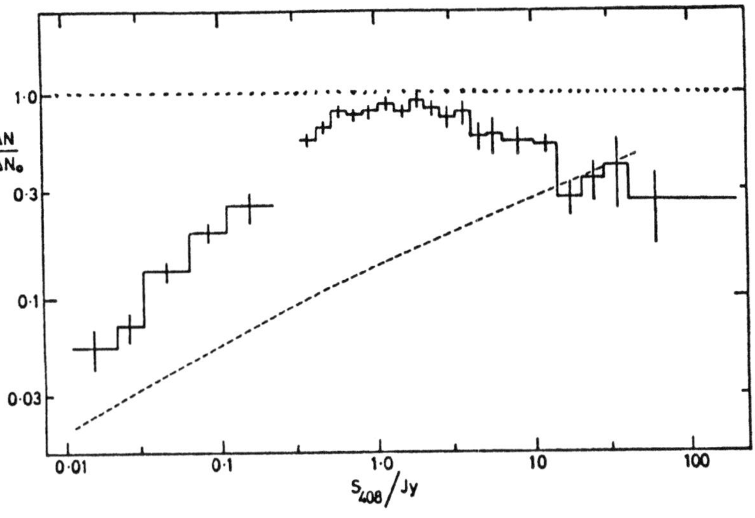

Figure 1. The surface density counts of radio sources at 408 MHz, plotted in 'relative differential' form, i.e. the observed counts divided by counts obeying the -3/2 power law of a uniform Euclidean model universe. Such a model thus yields the horizontal dotted line: for an FRW Universe in which sources are uniformly distributed, equation 11 yields the dashed curve.

slope is close to -3/2, which is in turn followed by a tail-off to the faint intensities at a rate faster than the -3/2 law. The count computed as described above, and with a luminosity function giving the correct proportion of luminosities amongst the brighter sources catalogued at 408 MHz is shown as the dotted line. It is a hopeless description of the data, and it is important to note that it is the plateau in the observed count which is not reproduced, in addition to the initial steep rise which caused the early source-count controversies. It is also important to note that non-cosmological redshifts do not alleviate the difficulty. If luminosities lower than those corresponding to the 'cosmological distances' are adopted, curves between the dotted and the dashed curves result, all of which decrease monotonically and none of which look like the observed count. The general conclusion from Figure 1 - largely independent of redshift interpretation - is that extragalactic radio sources found in surveys at 408 MHz cannot be uniformly distributed in co-moving volume. In order to produce a count like that of

Figure 1, the relevant objects must have been vastly more plentiful or more luminous at earlier epochs of the Universe.

At different survey frequencies, different populations of radio sources dominate the source counts. Figure 2 shows counts at five different radio frequencies, from 151 to 5000 MHz. Some change in form with frequency is evident in the sense that the Euclidean region (the plateau) increases in extent as survey frequency is raised. Details of the change in populations with frequency, together with quantitative explanation of the change in form of the radio counts, will be given in section 3 and section 4. Here it is sufficient to note that the change in form is not enough to avoid the general conclusion of evolution for the bulk of radio sources detected at each frequency.

At optical wavelengths, surface-density counts become known as log N - log m, or log N(m) relations, and because $m_2 - m_1 = 2.5\log(S_1/S_2)$, the -3/2 power law translates to $\log N(>m) = 0.6m + \text{const}$. Log N(m) relations have been constructed for galaxies, and for optically-selected QSOs. Galaxy counts are the subject of Ellis's lectures and are not considered here. (It is important to note that optical counts cannot be constructed for radio galaxies or radio QSOs because plate limits impose cutoffs which generally dictate the forms of the counts. The counts must be at the frequency of the finding survey.) The counts for optically-selected QSOs show evolutionary effects which are even more extreme than those apparent at radio wavelengths. Source-conserving counts calculated in FRW Universes show coefficients of ~ 0.4 for brighter magnitudes, rather than the Euclidean 0.6: in fact values of over 0.9 have been reported (Green and Schmidt 1978). The evolution of optically-selected QSOs is discussed in section 5.

2.2 The V/V_{max}, or 'Luminosity-Volume' test

The test was developed independently by Schmidt (1968) and Rowan-Robinson (1968) to explore the cosmology of a particular sample of objects: the QSO identifications in the 3CR catalogue of radio sources. It is simple to apply:

(1): Obtain a complete sample of objects, a sample in a given region of sky of all objects with fluxes at the finding survey frequency which exceed a given intensity limit S_o.

(2): In addition to flux at the finding frequency, measure redshift Z and spectral index α for each object.

(3): 'Push' each object outward by 'increasing' its redshift, to Z_{max}, the limit at which its 'observed' flux S drops to S_o, as calculated via equations (1), (6), and (7).

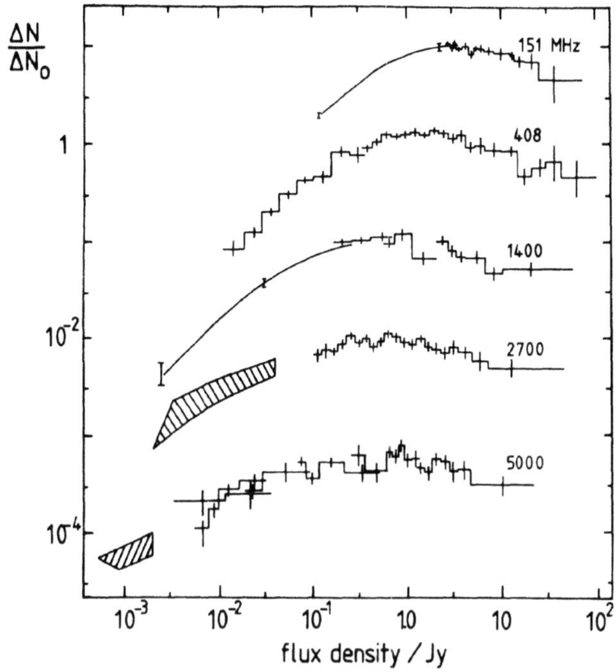

Figure 2. Source counts from surveys and P(D) analyses (section 4.2.2) at 151 to 5000 MHz. References to the observations appear in Wall (1980).

(4): From (7), (9) and (10), calculate the co-moving volumes V(Z) and $V(Z_{max})$ for each object, and hence form the ratio V/V_{max}. Thus obtain $\langle V/V_{max}\rangle$ for the sample.

The test is easily generalized when more than one survey limit is involved, as for its first application in which a sample of radio sources was detected above a (radio) survey limit, and the optical counterparts were subsequently identified above some optical limit. The appropriate V_{max} corresponds to the <u>first</u> limit encountered in the pushing process.

Again let us ask what we might expect to observe if the objects are uniformly distributed in space. Firstly it is clear that under this circumstance, no object has a predilection for any region of its own V_{max}, the quantity V/V_{max} should be

TABLE 1

RESULTS $\langle V/V_{max} \rangle$ TEST

Reference	Sample	No.	$\langle V/V_{max} \rangle$
Schmidt 1968	3CR QSOs	33	.70+.05
Lynds and Wills 1972	4C QSOs	31	.67+.05
Fanti et al. 1973	3CR	33	.70+.05
	4C	30	.67+.50
	optically-selected QSOs	23	.63+.06
Schmidt 1976	NRAO 5GHz QSOs:		
	α < .05	17	.50+.06
	.05 < α < .45	17	.60+.07
	α > .45	17	.71+.05
Masson and Wall 1977	PKS 2.7 GHz QSOs:		
	α < 0.0	16	.50+.07
	0.0 < α < .50	26	.57+.06
	α > .50	15	.68+.07
Schmidt 1977	optically-selected QSOs		
	Sandage and Luyten	34	.66+.05
	Braccesi and Formiggini	20	.70+.06
Wills and Lynds 1978	3CR QSOs	34	.69+.04
	4C QSOs	76	.70+.02
	PKS 2.7 GHz QSOs		
	α < 0.0	18	.62+.06
	0.0 < α < 0.50	27	.60+.05
	α > 0.50	15	.76+.05
Laing et al. 1978	3CR QSOs	23	.71+.06
	3CR Radio Galaxies:		
	$P_{178} > 10^{26}$ W Hz^{-1}sr^{-1}	23	.61+.06
	$< 10^{26}$	10	.52+.06
Savage 1978	optically-selected QSOs	148	.60+.02
Blake 1978	Flat-spectrum QSOs, BDFL	50	.72+.05
Peacock et al. 1981	Flat-spectrum QSOs, PWI	50	.68+.04

uniformly distributed between 0 and 1, and $\langle V/V_{max} \rangle$ should be 0.5 $\pm$ $(12N)^{-1/2}$, where N is the number of objects in the sample. Secondly it is easy enough to show this arithmetically for the Euclidean case. Because $S \propto R^{-2}$ an object at $S > S_o$ can be observed $(S/S_o)^{1/2}$ times further away. Thus

$$V/V_{max} = \left(\frac{S_o}{S}\right)^{3/2}, \tag{12}$$

and from (5), the -3/2 power law, we find that $\bar{S}$, the mean value of flux density for a sample complete to S_o is

$$\bar{S} = 0.5^{-2/3}\ S. \tag{13}$$

Substituting $\bar{S}$ for S in (12) yields $\langle V/V_{max} \rangle = 0.5$. Thirdly, it is possible (though messier arithmetically) to show that the same result obtains if FRW geometries are used, the relevant equations being

$$V/V_{max} = \frac{\int_o^Z D^2\, dr}{\int_o^{Z_{max}} D^2\, dr} \tag{14}$$

and

$$\left\langle \frac{V}{V_{max}} \right\rangle = \frac{\int_o^Z (V/V_{max})\ D^2\, dr}{\int_o^{Z_{max}} D^2\, dr} \tag{15}$$

V/V_{max} analyses have been performed on many samples of extragalactic objects, and a number of results are collected in Table 1. Without exception, the $\langle V/V_{max} \rangle$ exceed 0.5. In not all cases do they exceed 0.5 by more than 1σ, and for some samples, the excess is significantly less than for others. Again, the details are addressed in the following three sections, here it being sufficient to note that the high values of $\langle V/V_{max} \rangle$ indicate extremely non-uniform radial distributions, a preference of the objects for the more distant regions of their accessible volumes. Typically $\langle V/V_{max} \rangle = 0.70$ implies an increase over present-day numbers at $1 < Z < 3$ of ~ 1000, as Longair (1978) showed with a simple analysis.

2.3 Direct Investigations of Luminosity Functions

A third method for investigating the luminosity functions of QSOs, radio galaxies, and their epoch dependence, is to do so directly. If a sample is available complete to some given flux limit, then each sample member contributes a volume density to the luminosity function of

$$\rho_i = \frac{1}{V_{max}} \tag{16}$$

where V_{max} is computed as described in section 2.2. (Again if there are two limits, as for a complete sample of radio sources identified with galaxies above a magnitude limit, then the V_{max} is the smaller of the two limiting volumes.) If the sample members are then divided into redshift ranges, the volume contributions can be used to construct luminosity functions for different epochs. It should be clear that the procedure is an extension or variation of the V/V_{max} test.

The most notable achievements for the technique have been in investigation of luminosity functions for radio galaxies at both radio and optical wavelengths (e.g. Colla et al. 1975, Auriemma et al. 1977, Meier et al. 1979). These investigations have resulted in construction of the so-called 'bivariate luminosity function', a two dimensional luminosity function ρ (P_{radio}, $P_{optical}$) which demonstrated directly the proclivity of the intrinsically brighter galaxies to form the more powerful radio sources.

More recently, Katgert et al. (1979) have used the technique to investigate the epoch dependence of the radio luminosity function for powerful radio galaxies. By taking deep plates at the positions of faint radio sources, and estimating redshifts from magnitudes of faint galaxy identifications, they were able to show directly the large enhancement in space density of radio galaxies at epochs corresponding to Z = 0.4 to 0.8.

2.4 What Happened to QSOs and Radio Galaxies?

Surface-density counts and V/V_{max} results, then, attest to an earlier epoch of violent nuclear activity: QSOs and radio galaxies even more luminous or more numerous at such times. To describe the evolution in general terms, we can write the luminosity function for each class of object as

$$\rho(P,Z) = F(P,Z) \,.\, \rho(P,Z = 0) \tag{17}$$

where F is the 'evolution function' giving the density enhancement factors in the P-Z 'evolution' plane, and $\rho(P,Z = 0) = \rho_o$ is the 'local' luminosity function. Now the counts for a particular class of object are given by

$$N(>S) = \int_o^\infty dP \int_o^{Z_{lim}} F(P,Z) \;.\; \rho_o = 4\pi\, D^2\, dr \tag{18}$$

while $\langle V/V_{max} \rangle$ is

$$\left\langle \frac{V}{V_{max}} \right\rangle = \frac{\int_o^{Z_{lim}} F(P,Z) \;.\; (V/V_{max}) \;.\; D^2\, dr}{\int_o^{Z_{lim}} F(P,Z) \;.\; D^2\, dr} \tag{19}$$

The essential role of surface-density counts, V/V_{max} investigations, and direct investigation of luminosity functions, becomes the determination of the evolution function F for each population of objects. In this each method has strengths and weaknesses, and its own distinct contribution to make. The strength of the direct investigation process is its directness, but for radio galaxies it can examine the luminosity function out to small redshifts only. This is because redshifts are required for all members of the sample out to the redshift limit, not just for the intrinsically luminous members. For the same reason, the V/V_{max} test is hard work, but the problem is somewhat alleviated because the sample can be restricted to essentially luminous objects only. This restriction is a weakness - only a small portion of the luminosity function is investigated - but nevertheless there is no doubt that the V/V_{max} method provides the most powerful way of directly investigating the radial distribution of a particular class of objects. The surface-density counts contain information on more than one class of object, and the great range in flux provides a wider sampling of the luminosity functions. The counts are easy to observe, but complementary information, redshifts of complete samples in particular, is crucial for interpretation. The mixture of populations and the chronic incompleteness of the complementary information, mean that interpretation in terms of evolution functions is less direct, and that proper statistical procedures are required. However the beauty of the counts is that the basic result - evolution - is essentially model-independent (section 2.1).

3. COSMOLOGY OF EXTENDED RADIO SOURCES

The earliest searches for extragalactic radio sources - indeed the searches for discrete sources at times when the extragalactic nature of high-latitude radio sources, was not established - consisted of sky surveys at frequencies at or below 408 MHz. Most of the sources were found to have steep radio spectra, roughly power-law in form ($S \propto \nu^{-\alpha}$) with spectral indices $\alpha \sim 0.8$ (Shimmins 1968: Scheuer and Williams 1969). The radio morphologies were shown to be double-lobed with linear separations of 10's to 100's of kpc (Moffet 1968).

The most (radio) luminous sources show the classical Cyg A structure (Hargrave and Ryle 1974) with bright spots at the extremities of the lobes, and a weak, compact central component coincident with the nucleus of the optical counterpart. The less radio-luminous sources with steep spectra, less powerful than ~ 10^{25}W Hz^{-1} sr show more diffuse structures (Fanaroff and Riley 1974) which generally do not have the regions of highest suface brightness at the extremities. The 'bent-doubles', 'radio trails', and 'giant' radio sources (Simon 1978, Willis and Miley 1979) are in the latter category. The lower-power radio sources are invariably identified with galaxies, while the high-power sources of Cyg A type may have either giant ellipticals or QSOs as optical counterparts. It is widely accepted that the double lobes are powered by a nuclear engine which beams energy continuously to the lobes (Scheuer 1974, Blandford and Rees 1974). The nuclear engine needs to be both an efficient energy convertor and a massive rotator to provide observed gyroscopic and collimating effects: massive black holes accreting gas or whole stars have been suggested (Rees 1979).

Strenuous and prolonged efforts have been made to identify complete samples of steep spectrum sources, and to measure redshifts for the optical counterparts. The extensive data base which therefore exists allows the cosmic history of extended sources of both high and low luminosity to be studied effectively - provided an efficient and statistically satisfactory scheme is available to synthesize the data into a description of the luminosity function and its epoch dependence.

3.1 A Systematic Scheme for Investigation

The usual procedure in determining the evolution function F for a given class of objects is as follows:

(1): Adopt a geometry, H_o, q_o, Λ .

(2): Guess a local luminosity function ρ_o

(3): Guess an evolution function F(P,Z)

(4): Compute surface density counts or V/V_{max} via (18) or (19)

(5): Compute luminosity distributions at flux levels where redshift data are available for complete samples. For the Euclidean, source-conserving case, the luminosity distrbution is given by (4): for the relativistic, evolutionary case, N is (cf. (18))

$$N(P,>S)dP = dP\int_0^{Z(S)} F(P,Z).\ \rho_o(P).dV \qquad (20)$$

Note that for comparison purposes, it is not necessary to know the redshifts of all objects in the sample. Useful constraints on F can result if the proportion of objects with redshifts/luminosities less than some value is known.

(6): Adjust ρ_o and/or F (or even geometry): try again.

The standard difficulty with the foregoing sequence is that the guesses are not independent. There are not enough high-luminosity objects at non-cosmological distances ($Z < 0.1$) to define a local luminosity function, and if we use objects at larger distances, F must be known to determine ρ_o.

Wall, Pearson and Longair (1980: hereafter WPL) suggested a way of rationalizing the sequence. In (20), ρ_o does not depend on Z, so that

$$\rho_o(P)\, dP = \frac{N(P,>S)\, dP}{\int_o^{Z(S)} F.dV} \qquad (21)$$

Thus the local luminosity is <u>defined</u> if a luminosity distribution and the evolution function are known. The sequence advocated by WPL consequently

(1) Adopt a geometry,

(2) Define (observationally) a luminosity distribution,

(3) Guess F

(4) Compute ρ_o (21), and hence N(S) (18), V/V_{max} (19), and luminosity distributions at flux densities other than that used in step (2).

(5) Compare these with observations: adjust F and repeat (2) to (5).

The WPL technique was first employed at 408 MHz to determine F for the extended radio-source population. Because almost all sources catalogued in surveys at frequencies of 408 MHz and lower are of the steep-spectrum, extended, type, the source counts at this frequency may be taken to represent the counts for extended sources. Moreover, because of the extensive efforts to obtain identifications and redshifts for bright, steep-spectrum sources in the northern hemisphere - in particular those in the 3CR catalogue - a luminosity distribution could be constructed (at S_{408} = 10 Jy and over 6 sterad of sky) which was large enough in size to guarantee statistical accuracy. The results of this first application of the WPL technique are summarized as follows:

(1) As found in the early source-count investigations by Ryle, Davidson, and co-workers, strong cosmological evolution is evident, the numbers of the most luminous, extended radio sources at epochs corresponding to $1 < Z < 3$ exceeding present-day numbers by factors of 1000 or more.

(2) As found in Longair's (1966) pioneering investigation, 'differential' evolution is essential: only the most luminous of the extended sources can partake in the evolution. Figure 1 shows why. The range in <u>luminosity</u> for extended radio sources exceeds 10^5, and yet there are <u>bends</u> in the counts of apparent intensity, over a range $< 10^2$. The region of radio luminosity function which evolves to shape the humps in the counts must therefore be narrower than a factor of 10^2 in power, and both the shape of the counts and the identification content of the bright sources at 408 MHz decree that this evolving region of the luminosity function is at the high-power end. Thus the evolution function depends on power as well as epoch, F = F(P,Z), and there is a transition between the low-power radio galaxies which show little or no evolution, and the high-power radio galaxies and

QSOs which show very strong evolution. The transition occurs at powers ~ 10^{25} or 10^{26} WHz^{-1} sr^{-1}, and it is interesting to note that such values are close to the transition power dividing 'classical doubles' (Cyg A type morphologies) from the more diffuse morphologies of the lower power sources (Fanaroff and Riley 1974).

All V/V_{max} investigations of radio QSOs with steep spectra and extended structures (see Table 1: Schmidt 1968, Lynds and Wills 1972, Fanti et al. 1973, Wills and Lynds 1978, Laing et al. 1978) yield $\langle V/V_{max} \rangle$ ~ 0.70. These objects have high radio luminosities, and the high values of V/V_{max} indicate precisely the degree of powerful evolution deduced from source count analysis. Recently it has become possible to demonstrate differential evolution with V/V_{max} data. Using CCD data from the 200-in reflector of Hale Observatories, Laing et al. carried out a V/V_{max} investigation for steep-quantum radio galaxies, and although the samples are small (Table 1), the $\langle V/V_{max} \rangle$ for the lower-power group is less than that for the high-power group, and is consistent with a uniform distribution in co-moving volume for low-luminosity radio galaxies.

(3) Although the general evolution features deduced from earlier analyses were supported by the WPL procedure, no models from the literature provided statistically adequate fits to the data. Power law evolution models of the form $F(Z) = (1+Z)^{\beta}$ were found to be particularly unsatisfactory.

(4) For illustrative purposes only, WPL put forward two quite different evolution models which provided adequate statistical description of the data. The models are more complex than previous models: the present data base makes it far from easy to dream up satisfactory analytic representations of the P-Z plane. Nevertheless, the models provide vastly different populations of the P-Z plane. At high flux-density levels, the models yield identical luminosity distributions (or redshift distributions), as demanded by the WPL technique. At low flux densities, the models predict widely divergent redshift distributions (Figure 3). It is therefore clear that very powerful constraints on the evolution can result if identifications and redshifts are available for a complete sample of steep-spectrum sources at faint flux densities. This conclusion was the genesis of the 5C12 survey.

3.2 5C12: Preliminary Results

The 5C (Fifth Cambridge) surveys represent an ideal way of providing samples of steep-spectrum sources complete to faint flux densities (Pooley, Kenderdine and Ryle 1965). They are carried out with the 3-element One-Mile Telescope (OMT) using the

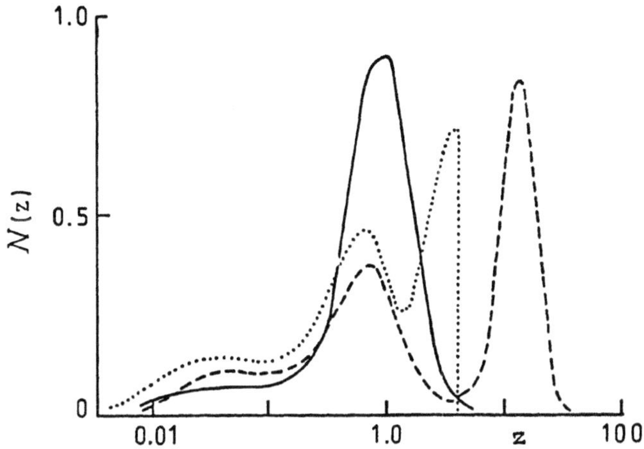

Figure 3. The distribution of redshifts for samples of sources with S_{408} = 10 mJy, as predicted by the evolution models of WPL.

earth-rotation synthesis technique: 64 days of observation at the frequency of 408 MHz yields 128 spacings and a map of a region 4 degrees in diameter containing some 250 sources complete to an apparent intensity of ~ 10 mJy. In principle any of the first 11 5C surveys could be applicable to the present investigation. In practice, each suffered some fatal flaw - low galactic latitudes, systematic position/flux density errors, or field centres which are certainly not 'cosmologically representative', e.g. M31, Coma Cluster, Hercules Supercluster.

5C12 is centred at 12^h58^m, +35°, the Sandage-Veron-Braccesi field. It is close to the North Galactic Pole, to minimize obscuration at optical wavelengths and chance coincidences with foreground stars. It coincides with the first Westerbork deep survey (Katgert et al. 1973). The 5C12 survey (Benn et al. 1981) has produced a list of 299 sources with S_{408} > 10 mJy. Super-cleaning techniques and calibration of individual phases have resulted in positions accurate to 3 arcsec and a noise level of 1.8 mJy rms, parameters considerably better than those achieved in previous 5C surveys. Comparison with data from the first Westerbork survey showed that no systematic position errors are present, so that optical identifications may be attempted with some degree of confidence.

With regard to optical indentifications, it is an attractive feature of 5C surveys that their 4-degree-diameter area precisely fits the unvignetted area of a 6.5 x 6.5 degree plate from the

Palomar 48-in Schmidt telescope. The complementary optical survey for 5C12 has consisted of deep and repeated plates from this telescope in each of 3 broadband colours. Magnitude calibration has been achieved by using a Racine wedge (Δm = 5.0 mag), and by taking deep plates of SA57, plates which overlap the 5C12 plates by a strip ~ 1.5 degrees wide. On each plate in each colour, a 1.0 x 1.0 arcmin field centred on each radio position has been scanned and digitized with a microdensitometer. The fields in each colour have been added ('stacked') to gain ~ 1 mag over the magnitude limit of a single plate.

Preliminary results from 5C12 are as follows.

(1) An identification rate of 33 per cent has been obtained from preliminary searches at the positions of all sources on a single V plate. This is higher than the rate found by Perryman (1979a,b) for the 5C6 and 5C7 surveys and the difference is probably due to some underestimation of position errors for sources found in these surveys. The 5C12 identification rate agrees with luminosity distributions at 10 mJy predicted by WPL models of type 4. Models of type 5 ascribe considerable evolution to sources of lower powers than do models of type 4: models of type 5 accordingly predict many more identifications at 10 mJy levels, far more than appear in 5C12. Such models are therefore rejected.

Figure 4

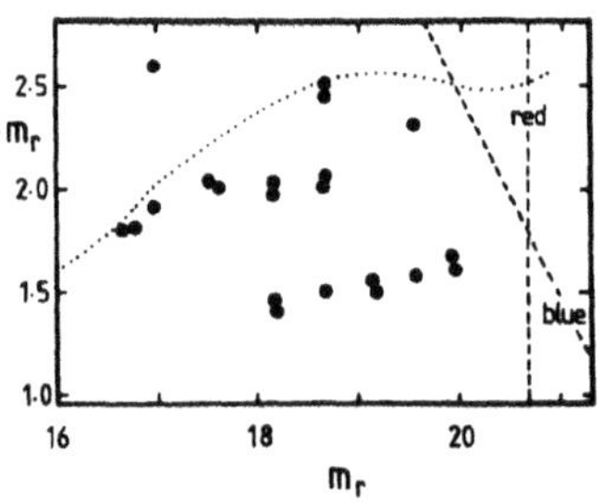

(2) An initial investigation of the colours of 5C12 radio galaxy identifications is shown in Figure 4. The fainter galaxies, those whose magnitudes correspond to Z > 0.4, show a blue excess, but the effect does not exceed $\Delta(m_b-m_r)$ = 0.7 mag. This is in disagreement with the suggestion by Katgert *et al*. (1979), that at Z > 0.4 radio galaxies become drastically bluer by some 2 magnitudes in m_b-m_r. The effect which they noted may be due to lack of magnitude calibration for their deep plates.

A great deal more will emerge from 5C12 when analyses are completed of the optical data, and of the additional radio data obtained with the VLA and MERLIN.

4. COSMOLOGY OF COMPACT RADIO SOURCES

In the mid 1960's, it was recognized that an extensive population of radio sources existed with flat or inverted radio spectra. This form of spectrum is due to a compact component or components of high surface brightness in which synchrotron self-absorption is occurring. The picture is confirmed both by monitoring the cm-wavelength flux densities of such sources and by VLBI observations. The former indicate variations on time-scales as short as weeks, while the latter indicate angular sizes on milliarcsec scales, overall structures of parsecs to 10s of parsecs in linear extent. The great majority of such sources have QSOs as their optical counterparts. Kellermann and Pauliny-Toth (1981) give a recent review.

Compact sources are grossly under-represented in metre-wavelength (low frequency) surveys of the 3C and 4C type, and in the late 1960s, large-scale radio surveys at cm-wavelengths were initiated to detect and catalogue them. Of these, the most extensive are the Parkes 2.7 GHz (11 cm) survey and the NRAO/MPIfR 5 GHz (6 cm) survey (summarized respectively by Wall (1977) and Pauliny-Toth (1977)). At high flux densities, about 50 per cent of sources catalogued at these frequencies have spectra which demonstrate that the structures are dominated by compact components.

Given the drastic difference in physical structure between the compact and the extended radio sources, there is no reason to believe that the spatial distribution should be similar. Might not the very active compact sources be expected to show even greater cosmological activity or evolution than that of the extended sources? But the preliminary indications from V/V_{max} data for compact sources are shown in Table 1 (Schmidt 1976, Masson and Wall 1977). These results - lower values of $\langle V/V_{max} \rangle$ for the compact QSOs than for the extended - imply a more uniform spatial distribution, and it is this apparent difference in evolution which has motivated the analyses and observations described in section 4.

4.1 An Initial Investigation

The WPL procedure was used by Wall, Pearson and Longair (1981) in an initial determination of the epoch dependence of the radio luminosity function for compact sources. The basic results are shown in Figure 5: it appeared that only moderate evolution is demanded, and that compact radio sources are therefore more uniformly distributed in space than the most luminous extended sources, as suggested by early V/V_{max} tests (Schmidt 1976, Masson and Wall 1977). However, the analysis was carried out with data sets suffering severe statistical limitations, and the

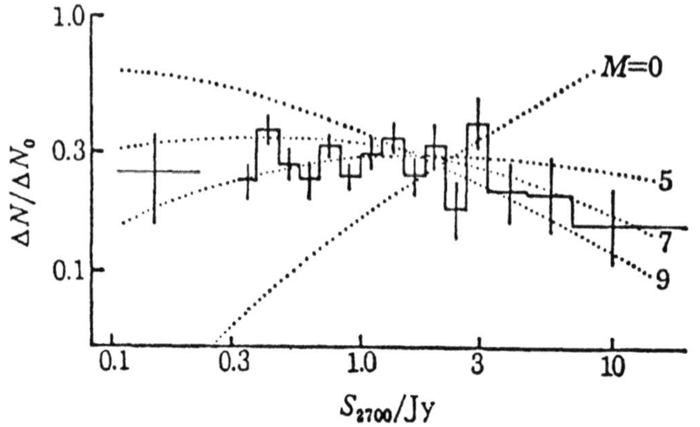

Figure 5. The 2700-MHz source count for compact radio sources, together with results of model calculations (Wall, Pearson and Longair 1981) in which it is assumed that the entire luminosity function for compact sources evolves according to exp[M.(look-back-time)]. A value of M = 0 corresponds to uniform distribution (no evolution): Wall, Pearson and Longair (1980) showed that M > 9 is required for the most luminous extended radio sources.

results are re-considered in section 4.3. The most important aspect of the investigation was to reveal which observations were crucial to improve knowledge of spatial distribution of this class of object. There were three areas:-

(1) An improved luminosity distribution for compact sources, a luminosity distribution which parallels the original (408 MHz) WPL luminosity distribution (for extended sources) in that it is compiled over several steradians of sky and contains 50-100 objects rather than the 20 objects of Wall, Pearson and Longair (1981):

(2) Improved definition of the total counts of sources at faint flux densities for both 2.7 and 5.0 GHz: and

(3) Improved definition of the proportions of compact and extended sources constituting these total counts at cm wavelengths, so that counts of compact sources may be constructed to much lower flux densities than in Fig. 5.

4.2 Improved Data

4.2.1 Large-area samples at cm wavelengths

With an improved luminosity distribution for compact sources as a prime objective, Peacock and Wall (1981) compiled a 'Northern Hemisphere' sample of the brightest sources at 2.7 GHz. The sample is complete to the limits $S_{2.7} > 1.5$ Jy, $\delta > +10°$ and $|b| > 10°$: it is intended as the high-frequency counterpart of the 3CR '166 sample' (Jenkins, Pooley and Riley 1977). The 2.7 GHz sample contains 168 sources, of which 156 (93%) are identified, and 108 (64%) have measured redshifts. More than half the sources are also in the 3CR sample, and the very high identification rate is due to a considerable extent to prolonged efforts to identify 3CR sources. In the unidentified sources not in 3CR and lacking previous optical identifications, a programme of direct CCD exposure with the Hale 200-in telescope (Peacock et al. 1981) has yielded 7 further optical counterparts in 12 attempts. All 168 sources in the sample have been mapped with the Cambridge 5km telescope at 2.7 or 5.0 GHz.

There are 51 flat-spectrum ($\alpha < 0.5$) sources in the sample: 49 of these have optical identifications and 41 have measured redshifts. Figure 6 demonstrates that these data provide a luminosity distribution which indeed is a considerable improvement over that used by Wall, Pearson and Longair (1981).

Several further points of interest have emerged from studies of this sample.

(1) It was recognized that the composition in relation to that of low frequency samples (e.g. 3CR) would be affected by two factors: (a) the number of flat-spectrum, compact sources would increase at the expense of steep-spectrum sources, and (b) the P-α correlation (see e.g. Kellermann, Pauliny-Toth and Williams 1969) would increase the proportion of low-luminosity steep-spectrum sources, and so broaden the luminosity distribution of steep-spectrum sources defined at a high frequency. Both these effects were observed, but a third effect was not anticipated - 33 sources had steep spectra, were not resolved by the 2 arcsec beam of the 5km telescope, and are not in the 3CR sample. These sources generally show some flattening or turnover of the spectrum towards the lower frequencies, enough to keep them out of the 3CR sample. Their nature is unknown. Are the very powerful, very distant double sources of extreme Cyg A type? Or are they 'semi-compact' sources, some kind of missing link between extended and compact sources? High-resolution observations will be of interest.

(2) The 50 flat-spectrum QSOs in the sample have a $\langle V/V_{max} \rangle$

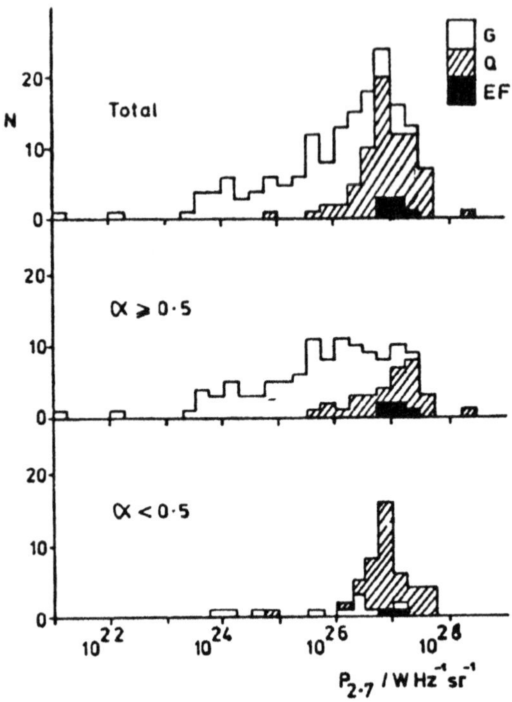

Figure 6. The 2.7 GHz luminosity distributions compiled by Peacock and Wall (1981), and divided according to spectra index ($S \propto \nu^{-\alpha}$: steep-spectrum (extended) sources have $\alpha >$ 0.5, and flat-spectrum (compact) sources $\alpha < 0.5$). G and Q stand for galaxy and quasar identifications respectively, while EF implies empty field, no identification.

of 0.68 + 0.04 (Table 1): Peacock et al. 1981). This is significantly higher than the early values for such objects (Table 1: Schmidt 1976, Masson and Wall 1977): contrary to these and to results from the preliminary source count analysis described above, it suggests strong cosmological evolution for this class of object. Again the matter is deferred to section 4.3.

4.2.2 The faint ends of cm-wavelength source counts

Radio luminosity functions are at least 7 orders of magnitude in width. To define these functions and their epoch dependence, it is clearly desirable to have surface-density counts over ranges of apparent intensity which exceed this width. In fact we fall well short of this goal. The finite surface area of the sky stops definition of the bright ends of radio source counts at ~ 20 Jy (Figure 2), while the most sensitive surveys for faint sources reach mJy levels, some 4 orders of magnitude fainter. Previous sections have shown how cosmological information is nevertheless derived: the analyses show that the 'faint ends', although not so faint as we would like, are crucial in constraining cosmic evolution.

At frequencies of 408 and 1400 MHz, the source counts have been extended to mJy levels by synthesis of small areas of sky with the Cambridge and Westerbork interferometers (Figure 2). At higher frequencies, sensitivities have been inadequate to carry out such syntheses, and we have had to do as best we can with the big filled-aperture telescopes (Parkes 64m, NRAO 91m, MPIfR 100m). Recently the latter two telescopes have been used to carry out 5 GHz surveys which reach 15 mJy (Ledden et al. 1980, Pauliny-Toth, Steppe and Witzel 1980). This is close to the limit to which these telescopes can push the source counts via direct surveys. The limit is set by confusion, the bumpy background which results as the beam of the telescope integrates the multitude of faint radio sources into continuum.

In fact, this bumpy background provides a way of estimating the source count to levels far below 15 mJy. Such estimates are made with the P(D) technique, first described by Scheuer (1957). P(D) stands for Probability of Deflection, and the deflections are the bumps or holes in the confusion continuum. The frequency distribution of these deflections contains information on the source count below the level at which individual sources remain detectable. Scheuer's original (1957) analysis applied to a phase-switched interferometer: much later Scheuer (1974) and Condon (1974) independently extended the analysis to pencil-beam telescopes. This analysis was applied by Wall and Cooke (1975) to confusion-limit data obtained at 2.7 GHz with the Parkes 64m telescope, and the results extended the 2.7 GHz count to ~ 1 mJy as shown by the hatched area in Fig. 2.

At 5 GHz, Wall (1978) carried out a second P(D) analysis, again using data from the Parkes telescope, to estimate the source count at a level of ~ 1 mJy. The results, unlike the 2.7 GHz data, suggested an extreme extension of the source count plateau which proved particularly difficult to understand in terms of the mix of radio-source populations constituting the bright sources at 5 GHz. It seemed necessary to postulate a 'new' population of low-luminosity sources with inverted spectra. Subsequently, though, deep 5 GHz surveys with the MPIfR 100m reflector (Pauliny-Toth et al. 1978, Pauliny-Toth, Steppe and Witzel 1980), and with the Westerbork Synthesis Telescope (Willis and Miley 1979), suggested that Wall (1978) had overestimated the source surface density at mJy levels. However, the counts from these surveys also disagreed with each other (see Figure 2): and it is fair to say that in 1979 and 1980, the faint end of the 5 GHz count was plagued by extreme uncertainty.

Two recent experiments have rectified the situation. They were carried out with the NRAO 91cm reflector (Ledden et al. 1980) and the MPIfR 100m reflector (Wall et al. 1982): they are thus completely independent, and the results are in excellent agreement. The result from the MPIfR experiment is shown in Figure 2, again as a hatched area. These experiments, together with further deep surveys at both MPIfR and NRAO, show conclusively that:

(i) the differences between counts from the original deep surveys at MPIfR and Westerbork were statistical only,

(ii) the Wall (1978) estimate is high by ~ 2.5σ, due to an instrumental effect discussed in Wall et al. (1982), and

(iii) the 5 GHz counts of extragalactic radio sources show a smooth, monotonic decrease below the Euclidean law as flux density is decreased below 0.1 Jy. No 'mystery' population need be invoked, therefore, to explain its form, but the dominant population at mJy levels is still an interesting matter, as discussed in the following section.

4.2.3 Deep surveys at cm-wavelengths: 5C12 again

Although P(D) analyses can define total counts to very faint levels, they are of course no substitute for deep surveys yielding direct counts and individual sources which can be identified, and whose radio spectra and structures can be measured. Such observations are requisite to determine which population comes to dominate the counts at mJy levels, or (equivalently) to allow construction of source counts for the different populations over the widest possible range in intensity. The deep Westerbork and Cambridge synthesis surveys at 408 and 1400 MHz have shown that the dominant population at mJy levels for the longer wavelengths is basically the same steep spectrum, extended sources which comprise the overwhelming majority at Jy levels. The cm-wavelength surveys select a mix of compact and extended sources: at 5 GHz some 60 per cent at Jy levels are of the compact, flat-spectrum type. Does this same mix persist down to mJy levels?

The question has now been answered. Several surveys at the ~ 20 mJy level, 5 GHz, have been carried out recently, together with programmes of optical identification and radio spectral measurement:

(i) Ledden et al. (1980), Condon and Ledden (1981) - NRAO 91cm reflector, 300 sources, complete to 15 mJy.

(2) Maslowski et al. (1981) - MPIfR 100m reflector, 104 sources, complete to 20 mJy (5C6 area).

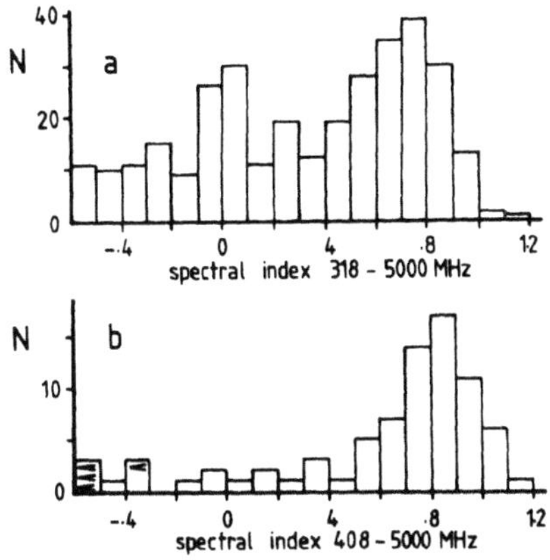

Figure 7. Spectral index ($S \propto \nu^{-\alpha}$) distributions for samples selected at 5 GHz: (a) $S_5 > 0.6$ Jy (Condon and Jauncey 1974), and (b) $S_5 > 18$ mJy (5C12 area).

(3) Wall et al. (1981), Savage, Bolton and Wall (1981) - Parkes 64m reflector, 75 sources, complete to 32 mJy.
(4) C.R. Benn, G. Grueff, M. Vigotti and Wall (in preparation) - MPIfR 100m reflector, 99 sources, complete to 18 mJy (5C12 area).

The last, a 5 GHz survey of the 5C12 area, yields a particularly rich store of information of the 5 GHz faint-source population. The 5C12 (408 MHz) survey of the area (section 3.2) provides (automatically) comprehensive radio spectra data over a wide frequency range. The 5C12 Schmidt plates permit very faint optical identifications to be made for the sources in the deep 5 GHz survey as well as for the 5C12 sources. Preliminary results are summarized in Figure 7: and these are in basic agreement with results from the other deep 5 GHz surveys listed above.

Figure 7 demonstrates that as flux density at 5 GHz is decreased, an astonishing change in the population content takes place. The spectral index distribution for bright sources at 5 GHz (Figure 7a) shows the well-known double-peak structure, with similar contributions from flat- and steep-spectrum sources. But at low flux densities, only a single peak remains: the flat-spectrum sources have almost disappeared, the remnants of this population spilling out in a long tail to negative spectral indices. Paralleling this, the proportion of QSO identifications has fallen drastically - most of the remaining identifications are with faint galaxies. It is clear, then, that the source count for the flat-spectrum compact sources drops off below 0.1 Jy at a rate very much faster than that for the steep-spectrum, extended sources, so that at mJy levels only the latter are left.

We are left, then, with the following picture for the 5 GHz source count. Despite earlier indications, it shows a smooth drop-off below Euclidean values of source surface density. The total count is composed essentially of <u>two</u> source counts, one for the compact sources which has its plateau centred on ~ 1 Jy, and which drops rapidly below 0.1 Jy, and the second for the steep-spectrum sources which has its plateau centred on ~ 0.2 Jy and whose drop-off to low intensities is essentially mapped by that of the total count, Figure 2. The latter basically represents the 408 MHz count shifted to lower flux densities by virtue of the steep mean spectrum of its constituent sources. The increasing width of plateau of the total source counts which is so apparent as survey frequency is raised (Figure 2) is thus due to the increasing prominence of the count of compact sources, whose plateau centre is shifted to higher flux densities than that for the count of steep-spectrum sources.

4.3 Multifrequency/Dual-Population Models

The final episode in this present tale of the cosmology of compact radio sources concerns development of an improved technique for synthesizing luminosity functions and their epoch dependence. The WPL technique has statistical virtues as described in section 3.1, but has a serious drawback in requiring analytic functions to describe the evolution. Robertson (1980) generalised the WPL procedure to some extent, but the type of evolution which he adopted for the extended sources was not 'free-form' as he suggested.

Peacock and Gull (1981) have developed a truly free-form approach in which entire <u>luminosity functions</u> (not just the <u>evolution functions</u>) for both extended and compact sources are written as series expansions across the (P,Z) plane:

$$\log \rho = \sum_{i=0}^{n} \sum_{j=0}^{n-i} A_{ij} \, [\log P]^i [f(z)]^j \tag{22}$$

Although it is fair to describe the analysis as free-form, there remains a choice of z- or epoch-dependent coordinate: Peacock and Gull tried f(z) = log (1+z), and f(z) = look-back-time. The χ^2 statistic was used to test goodness-of-fit, and an algorithm was developed to optimize expansion coefficients to yield best fit and maximum smoothness in the luminosity functions. As data, Peacock and Gull used (i) observations of bright galaxies (to establish the local RLF and low radio powers), (ii) source-counts at 408, 1400, 2700 and 5000 MHz, with sub-counts constructed from the latter two total counts using radio spectral data to divide extended and compact sources (Kuhr 1980, Wall 1980), (iii) luminosity of redshift distributions from complete samples at several different frequencies and flux-density levels, (iv) V/V_{max} results from complete samples again at several different frequencies and intensity levels, many of which are listed in Table 1, and (v) (for a recent discussion) to relate the steep-spectrum populations at low and high frequencies.

In its first application, this beautiful technique has provided several results of major importance.

(1) The technique provides quantitative measures of how well data sets fit the synthesized evolution models. Basically, all data sets - including the disparate $\langle V/V_{max}\rangle$ results of Table 1 - are in statistical agreement. The few disagreements which are larger than expected have ready physical explanations, such as differences in flux-density scales.

(2) The different formulations of the series expansion resulted in different populations of the (P,Z) plane, and the disagreements were used by Peacock and Gull to map our areas of ignorance (Figure 8). The qualitative differences between the (P,Z) plane populations produced by WPL's models demonstrated that these areas were large: the quantitative differences found by Peacock and Gull allowed them to be delineated. It should be noted in Figure 8 how particularly difficult it is to determine if a redshift cutoff (an epoch of creation?) exists at $Z > 3$. These are the 'Dark Ages' (Wall 1980).

(3) Most crucial to the present discussion is the question of evolution for the compact sources. Here the results of Peacock and Gull are unequivocal: despite the earlier indications (cf. section 4.1), strong cosmological evolution is undoubtedly required for the compact radio sources of high luminosities.

How did we come to get this wrong? The short answer is that we did not, in the sense that all the results are reconcilable - or a picture of differential evolution for the compact radio sources. The values of $\langle V/V_{max}\rangle$ obtained by Schmidt (1976) and Masson and Wall (1977) indicated milder cosmological evolution for radio-compact QSOs than that exhibited by the most luminous, extended radio sources. The values of $\langle V/V_{max}\rangle$ obtained by Blake (1978) and Peacock et al. (1981) disagreed: they were higher, indicating evolution for radio-compact QSOs every bit as drastic as that for the luminous, extended radio sources. But the latter two samples contain higher-luminosity objects than the former two, because the latter are defined at higher flux densities over large areas of sky. There was thus a direct indication of differential evolution for compact radio sources as early as 1978, and our mistake was to label the differences as controversy rather than science.

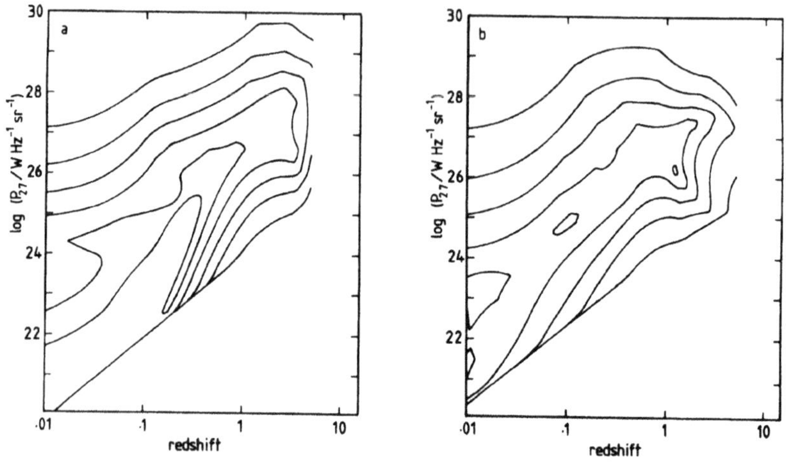

Figure 8. 'Uncertainty' maps of the (P,Z) plane (Peacock and Gull 1981) for (a) steep-spectrum sources, (b) flat-spectrum sources. The contours represent factors to which the space density is defined with the outermost at 10,000, and the remainder at 100, 10, 2.5 and 1.25.

In fact, Peacock et al. showed that the V/V_{max} data for their sample of flat-spectrum sources suggested luminosity-dependent evolution. They pointed out that the results further confirmed the strong link between source-counts slope and $\langle V/V_{max} \rangle$ (Longair and Scheuer 1970): the Masson-Wall sample, complete to $S_{2.7}$ = 0.35 Jy, is from a ~ Euclidean portion of flat-spectrum source-count (Figure 5), while the Peacock et al. sample, complete to $S_{2.7}$ = 1.5 Jy, is from the steep (Peacock and Wall 1981) initial portion of this count.

With regard to interpretation of the flat-spectrum source count, the initial investigation (section 4.1) indicated moderate evolution in agreement with the low $\langle V/V_{max} \rangle$ values of Schmidt (1976) and Masson and Wall (1977). But the counts computed by ascribing moderate evolution to the entire luminosity function do not provide good representations of the observed count (Figure 5). The initial slope is reproduced by M = 7, but this model predicts too many faint sources: M = 5 yields the right surface density for faint sources, but fails to follow the high-flux-density count. Differential evolution - ascribing strong evolution to the

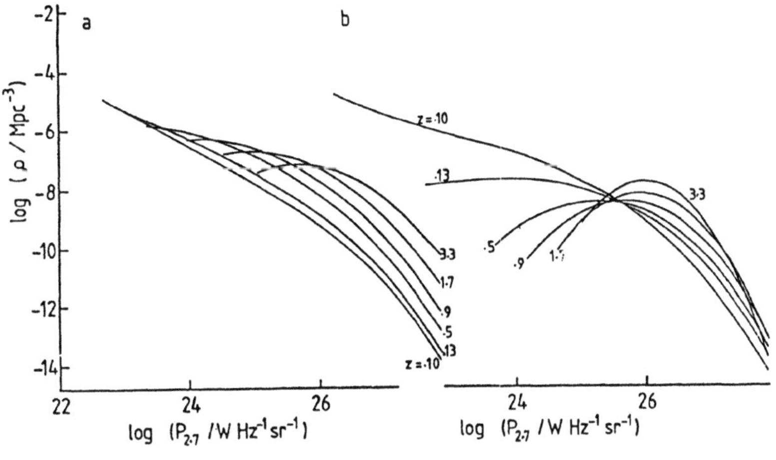

Figure 9. Luminosity functions and their epoch dependence as given by Peacock and Gull (1981) (a) for steep-spectrum sources, (b) for flat-spectrum sources.

high-power end of the luminosity function alone - alleviates the problem of fitting the hump, just as it did for the steep-spectrum sources.

Figure 9 shows this differential evolution: the luminosity functions determined by Peacock and Gull are drawn as a function of redshift. The qualitative similarity in cosmological behaviour for flat and steep-spectrum sources is evident. Some quantitative differences may exist, and it is not easy to establish the reality of these in the face of the uncertainties shown in Figure 8. The question is clearly of high priority, because it bears on a bigger question. If the flat- and steep-spectrum populations co-existed with the same density ratio throughout cosmic time, do they constitute a single population in which the members appear in different guises?

5. SOME PHYSICAL CONSIDERATIONS

These lectures have been concerned with the processes of observation and data-synthesis to produce models of spatial distribution describing the cosmology of extragalactic radio sources. It is important to emphasize two things about this type of work. Firstly, the models are nothing to do with proper physical models. The models of WPL and Peacock and Gull (1981) represent arithmetical descriptions of cosmological behaviour which fit the data: nothing more. As such, they really are data-syntheses, not models. Secondly, and on the positive side, I have tried to show how the sequence

data synthesis (e.g. WPL) - better data (Section 3.2, 4.2) - improved data synthesis (e.g. Peacock and Gull)

results in steady and systematic advances in our knowledge of the extragalactic radio-source populations and their cosmology. Of course the ultimate aim is physical understanding of both radio sources and their cosmology. We are far from this: but nevertheless, another set of lectures would be necessary to describe the physics, or at least physical speculation, which has resulted from the mapping of the radial distribution of radio sources. Here I can only state a few considerations, and leave the references to be picked up by those who are interested.

(1) Optically-selected QSOs. Section 4 closed by asking if compact and extended radio sources were members of a single population: the role of the radio-quiet QSO is crucial in any such picture. In surface density, these outnumber the radioactive QSOs by factors of 10 to 100 and it is clearly essential to explore their cosmology in the same systematic way, as the studies of radio sources. However, there is a fundamental difficulty: it is still impossible to construct samples for 'source' counts, luminosity distributions, V/V_{max} tests which are complete in any satisfactory sense. No matter what selection process is used, there is severe discrimination against some known spectral type of QSO. Objective-prism surveys (Savage and Bolton 1979, Osmer and Smith 1980) discriminate against QSOs with weak emission lines, and of certain redshifts for which strong emission lines fall outside the visible spectral region. Searches for objects with ultraviolet excess (e.g. Braccesi et al. 1980) discriminate against the redder QSOs while contaminating the sample with white dwarfs. As a result there has been continued controversy about statistics (see e.g. Setti and Woltjer 1973). A recent compilation of source counts (Cheney and Rowan-Robinson 1981) demonstrates the disagreements - the disparity in surface densities at a given magnitude obtained by different observers is vast.

Nonetheless, there is unequivocal evidence from the samples for strong cosmological evolution. For instance the initial slope for the m-logN relation obtained by Braccesi et al. (1980) is 0.86, while that found by Breen and Schmidt (1979) is 0.93. The Euclidean value (section 2.1) is 0.6, and on a cosmological interpretation of redshifts, the conclusion that strong evolution has taken place is inescapable. This conclusion is backed by the $\langle V/V_{max} \rangle$ results for ~ complete samples (Table 1): values in the range 0.6 to 0.7 have been obtained. Recent attempts to analyse the spatial distribution have been carried out by Bracessi et al. (1980), Turner (1979) and Cheney and Rowan-Robinson (1981). There are difficulties with either pure density evolution or pure luminosity evolution. Most authors conclude that luminosity-dependent density evolution - 'differential' evolution of a type very similar to that required by both flat and steep-spectrum radio-source populations (section 4.3), provides the best description of the data.

Philosophically this may seem satisfactory, the parallel space densities suggesting that all QSOs belong to a single population in which the radio phase is a short-lived (1-10%) era in the lifetime. But physically, it requires a connection between the optical and radio powers. This is easier to envisage for the flat-spectrum, compact radio QSOs in which both radio and optical emissions are essentially from the 'core', whatever this may be. But for extended sources there is a difficulty: and we are forced to accept that the optical and radio emissions, widely separated spatially, are correlated manifestations of the efficiency and fuel supply of the nuclear engine. Schmidt (1970) demonstrated the dependence of the radio luminosity function on optical luminosity for steep-spectrum QSOs. The nature of the connection needs much more consideration.

(2) Relativistic beaming. The sample of compact (flat-spectrum) radio sources which has been mapped by VLBI techniques is not large, but it is large enough to show that the majority of them show (i) asymmetric 'core-jet' structures, and (ii) apparent transverse velocities of blobs in the jets in excess of c. Scheuer and Readhead (1979) and Blandford and Konigl (1979) put forward a model for such sources in which pairs of radio blobs are ejected quasi-continuously in opposite directions at speeds ~ c along an axis close to the observer's line-of-sight. Such a model accounts for (i) the observed core-jet structure, because of the Doppler enhancement of the approaching jet relative to the retreating one, and (ii) for the apparent transverse velocities greater than c. But the supreme attraction of the model is that it provides a natural explanation of the observed numbers of radio-quiet versus radioactive QSOs. If the radio axis is misaligned - say, transverse to our line-of-sight - then the object is radio-quiet: when the axis aims at us, the Doppler enhancement makes it radio-loud. The distinction becomes a simple geometric matter of aiming the radio axis properly. Scheuer and Readhead show that the ratio of surface densities is 10-100:1 in favour of radio-quiet QSOs, is as predicted with an ejection speed of $\gamma \sim 5$, the speed required to produce the observed apparent transverse velocities of 3-10c.

There are difficulties with this attractive model. One of these is the role of the steep spectrum radio QSO, and indeed I.W.A. Browne (personal communication) has suggested that the Scheuer-Readhead model unites the QSOs of compact and extended radio structures by a range in jet speed, rather than the radio-quiet and the compact radioactive QSOs. A second difficulty is that the optical luminosities of radio QSOs appear to be higher than those of radio-quiet QSOs (Smith and Wright 1980). One major difficulty has been that similar spatial distributions would be expected for radio-quiet and flat-spectrum radio QSOs: initial indications were that the former showed strong evolution, while the latter showed a much more uniform distribution. But the recent analyses (section 4.3) now indicate strong evolution for compact QSOs. A major hurdle for the theory has thus been cleared: and it is now worth while to carry out some more detailed calculations including luminosity functions and ranges in γ to test the hypothesis against data in a more quantitative way.

As for the evolution itself, it could be caused by the ambient medium (points (3) and (4) below), by fuel supply (5), or it could even be a trick-of-the-light (6).

(3) Ram-pressure confinement. Is it possible that the evolution of steep-spectrum, extended radio sources is due to better ram-pressure confinement of the diffuse structures at earlier epochs? The density of the IGM is $\propto (1+Z)^3$ and denser epochs might favour higher formation rates and/or longer lifetimes. In fact no strong dependence of space density with epoch has been demonstrated theoretically. Rees and Setti (1968) calculated a density dependence of spherical structures of $(1+Z)^{1.5}$. There are indeed observational indications that the effect is not of primary importance. Firstly, the luminosity functions for cluster and non-cluster radio galaxies are similar (Jaffe and Perola 1976): the enhanced IGM within clusters clearly has little effect on radio sources. Secondly, the strong and similar evolutions for optically-selected QSOs and compact radio sources are clearly not due to efficiency of confinement. And thirdly, there is the dependence of radio luminosity function on optical luminosity for extended radio QSOs (Schmidt 1970) which suggests so strongly that it is activity at the nucleus - epoch dependence of some property of the central engine - which governs the evolution.

(4) Compton snuffing. With regard to the extended sources again, there is a mechanism which has been studied as a candidate to provide differential evolution. Inverse Compton losses must be suffered by relativistic electrons in the lobes of radio sources, as these electrons collide with the photons of the 3K MWB. The

energy density of the MWB is $\propto(1+Z)^4$, and differential evolution would arise because the less luminous, more diffuse sources would have the shorter lifetimes at earlier epochs. Several authors have considered the effect, and find (Rees and Setti 1968, Rowan-Robinson 1976, Scheuer 1977) that it does not produce enough differential evolution. (In fact it produces some overall evolution in the wrong sense, and in this regard Rees and Setti noted that a redshift cutoff for extended sources could arise at Z > 4 by complete Compton snuffing. Hence if such a cutoff were observed, it might not herald the epoch of galaxy formation.) The fact that radio-quiet QSOs and compact radio QSOs both show differential evolution in modern analyses again suggests that Compton snuffing, a mechanism which can affect only extended sources, is not a dominant factor in shaping evolution of radio-source populations.

(5) Fuel supply. The most obvious property of the central engine which could give rise to evolution is its fuel supply. In this regard, Gisler (1976) considered the fate of gas shed by stars cooling in elliptical galaxies, and showed that the rate of production is $\propto 1/t$ where t is the age of the galaxy, and that the accumulation must exceed observed limits unless there is a removal mechanism. If nuclear activity is fuelled by continuous gas supply, then the decline of gas production at 1/t is too gentle to produce the observed density evolution. But if the necessary removal mechanism is more efficient at later times, then the correspondingly steep drop in fuelling rate could produce the evolution. Sweeping the gas from ellipticals by ram-pressure interaction with the IGM is just such an epoch-dependent mechanism. Gisler found that the density of galaxies accumulating enough gas to drive a nuclear engine was very different for high and low-mass ellipticals, and that the density law closely follows the evolution derived for steep-spectrum radio sources. These are predominantly identified with ellipticals, so that Gisler's scheme describes both the evolution and its differential nature, as radio sources of lower powers are associated with ellipticals of lower luminosities.

(6) Gravitational lensing. Is the apparent evolution just a trick-of-the-light: is it All Done With Mirrors? The general acceptance of the double QSO 0957+561 (Walsh, Carswell and Weyman 1979) as the product of a gravitational lens provides a second effect of relativistic geometry which may bear on the cosmology of QSOs and radio galaxies. Turner (1980) has suggested that the apparent evolution of these objects may be due to lensing: the steep slope of the luminosity function coupled with the increased space density of galaxy lenses at early epochs could give rise to apparent differential evolution similar in nature to that observed. Detailed calculations by J.A. Peacock (personal communication) indicate that the effect is negligible on the

statistics of radio galaxies and radio QSOs. For radio-quiet QSOs, however, the situation is unclear, because as Peacock points out, objects as small as 10^{-3} M_{0} can act as lenses for them. The very steep source counts for optically-selected QSOs suggest that they have apparent evolution somewhat in excess of the radio QSOs. Could it be that the 'excess' evolution is due to lensing? Is it special pleading to suggest that the radio QSOs are thus the true tracers of the cosmic history of the most energetic constituents of the Universe?

Speculation aside, I trust to have shown that although much remains to be done, we have tools, both observational and analytical, to make steady progress towards an understanding of radio-source cosmology.

References

Auriemma, C., Perola, G.C., Ekers, R., Fanti, R., Lari, C., Jaffe, W.J. and Ulrich, M.H. 1977. Astron.Astrophys., 57, 41.

Benn, C.R., Grueff, G., Vigotti, M. and Wall, J.V. 1982. MNRAS, (in press).

Blake, G.M. 1978. MNRAS, 183, 21P.

Blandford, R.D. and Konigl, A. 1979. Astr.J., 232, 34.

Blandford, R.D. and Rees, M.J. 1974. MNRAS, 169, 395.

Braccesi, A., Zitelli, V., Bonoli, F. and Formiggini, L. 1980. Astr.Astrophys., 85, 80.

Cavaliere, A., Danese, L., De Zotti, G. and Francheschini, A. 1981. Contribution at 15th ESLAB Symposium 'X-ray Symposium'.

Cheney, J.E. and Rowan-Robinson, M. 1981. MNRAS, 195, 497.

Colla, G., Fanti, C., Fanti, R., Gioia, I., Lari, C., Lequeux, J. Lucas, R. and Ulrich, M.H. 1975. Astr.Astrophys., 38, 209.

Condon, J.J. 1974. Astrophys.J., 188, 279.

Condon, J.J. and Jauncey, D.L. 1974. Astron.J., 79, 1220.

Condon, J.J. and Ledden, J.E. 1981. Astron.J., 86, 64.

Elsmore, B., Kenderdine, S. and Ryle, M. 1966. MNRAS, 134, 87.

Fanti, R., Formiggini, L., Fanti, C., Padrielli, L., Katgert-Merkelijn, J.K. and Katgert, P. 1973. Astr.Astrophys., 23, 161.

Fanaroff, B.L. and Riley, J.M. 1974. MNRAS, 167, 318.

Gisler, G.R. 1976. Astr.Astrophys., 51, 137.

Green, R.F. and Schmidt, M. 1978. Astrophys.J., 220, L1.

Hargrave, P.J. and Ryle, M. 1974. MNRAS, 166, 305.

Jaffe, W.J. and Perola, G.C. 1976 Astr.Astrophys., 46, 275.

Jauncey, D.L. 1967. Nature, 216, 877.

Jenkins, C.J., Pooley, G.G. and Riley, J.M. 1977. Mem.R.astr. Soc., 84, 61.

Katgert, P., Katgert-Merkelijn, J.K., Le Poole, R.S. and van der Laan, H. 1973. Astr.Astrophys., 23, 171.

Katgert, P., de Ruiter, H.R. and van der Laan, H. 1979. Nature, 280, 20.
Kellermann, K.I. and Pauliny-Toth, I.I.K. 1981. Ann.Rev. Astron.Astrophys., 19, 373.
Kellermann, K.I., Pauliny-Toth, I.I.K. and Williams, P.J.S. 1969. Astrophys.J., 157, 1.
Kembhavi, A.K. and Fabian, A.C. 1982. MNRAS, 198, 921.
Kuhr, H. 1980. Ph.D. Thesis, Bonn University.
Laing, R.A., Longair, M.S., Riley, J.M., Kibblewhite, E. and Gunn, J.E. 1978. MNRAS, 184, 149.
Laing, R.L. and Peacock, J.A. 1980. MNRAS. 190, 903.
Ledden, J.E., Broderick, J.J., Condon, J.J. and Brown, R.L. 1980. Astr.J., 85, 780.
Longair, M.S. 1966. MNRAS, 133, 421.
Longair, M.S. 1971. Rep.Prog.Phys., 34, 1125.
Longair, M.S. and Scheuer, P.A.G. 1970. MNRAS, 151, 45.
Lynds, R. and Wills, D. 1972. Astrophys.J., 172, 531.
Maltby, P. and Moffet, A.T. 1962. Astrophy.J.Suppl.Ser., 7, 141.
Maslowski, J., Pauliny-Toth, I.I.K., Witzel, A. and Kuhr, H. 1981. Astron.Astrophys., 95, 285.
Masson, C.R. and Wall, J.V. 1977. MNRAS, 180, 193.
Meier, D.L., Ulrich, M.H., Fanti, R., Gioia, I. and Lari, C. 1979. Astrophys.J., 229, 25.
Osmer, P.S. and Smith, M.G. 1980. Astrophys.J.Suppl.Ser., 42, 333.
Pauliny-Toth, I.I.K. 1977. In Radio Astronomy and Cosmology, IAU Symp. No. 74 (ed. D.L. Jauncey), p. 63. Dordrecht:Reidel.
Pauliny-Toth, I.I.K., Steppe, H. and Witzel, A. 1980. Astr. Astrophys., 85, 329.
Pauliny-Toth, I.I.K., Witzel, A., Preuss, E., Baldwin, J.E. and Hills, R.E. 1978. Astron.Astrophys.Suppl. 34, 253.
Peacock, J.A. and Gull, S.F. 1981. MNRAS, 196, 611.
Peacock, J.A. and Wall, J.V. 1981. MNRAS, 194, 331.
Peacock, J.A., Perryman, M.A.C., Longair, M.S., Gunn, J.E. and Westphal, J.A. 1981. MNRAS, 194, 601.
Perryman, M.A.C. 1979a. MNRAS, 187, 223.
Perryman, M.A.C. 1979b. MNRAS, 187, 683.
Rees, M. 1979. Ann.N.Y.Acad.Sci., 302, 613.
Rees, M.J. and Setti, G. 1968. Nature, 219, 127.
Robertson, J.G. 1980. MNRAS, 190, 143.
Rowan-Robinson, M.M. 1968. MNRAS, 138, 445.
Rowan-Robinson, M. 1970. MNRAS, 150, 389.
Savage, A. 1978. D.Phil.Thesis, University of Sussex.
Savage, A. and Bolton, J.G. 1979. MNRAS. 188, 599.
Savage, A., Bolton, J.G. and Wall, J.V. 1982. MNRAS (in press).
Scheuer, P.A.G. 1957. Proc.Camb.Phil.Soc., 53, 76.
Scheuer, P.A.G. 1974. MNRAS, 166, 329.
Scheuer, P.A.G. 1975. In Galaxies and the Universe, vol.9 (Stars and Stellar Systems), ed. A. Sandage, M. Sandage and J. Kristian, p. 725. University of Chicago.

Scheuer, P.A.G. 1977. In Radio Astronomy and Cosmology, IAU Symp. no. 74, ed. D.L.Jauncey, p. 343. Dordrecht:Reidel.
Scheuer, P.A.G. and Readhead, A.C.S. 1979. Nature, 277, 182.
Scheuer, P.A.G. and Williams, P.J.S. 1968. Ann.Rev.Astr. Astrophys., 6, 321.
Schmidt, M. 1968. Astrophys.J., 151, 393.
Schmidt, M. 1970. Astrophys.J., 162, 371.
Schmidt, M. 1976. Astrophys.J., 209, L55.
Schmidt, M. 1977. In Radio Astronomy and Cosmology, IAU Symp. no. 74, ed. D.L. Jauncey, p. 259. Dordrecht:Reidel.
Setti, G. and Woltjer, L. 1973. Ann.N.Y.Acad.Sci., 224, 8.
Shimmins, A.J. 1968. Astrophys.Lett., 2, 157.
Simon, A.J.B. 1978. MNRAS, 184, 537.
Smith, M.G. and Wright, A.E. 1980. MNRAS. 191, 871.
Turner, E.L. 1979. Astrophys.J., 231, 645.
Turner, E.L. 1980. Astrophys.J., 242, L135.
Wall, J.V. 1977. In Radio Astronomy and Cosmology, IAU Symp. no. 74., ed. D.L. Jauncey, p. 55. Dordrecht:Reidel.
Wall, J.V. 1978. MNRAS, 182, 381.
Wall, J.V. 1980. Phil.Trans.R.Soc.Lond., A296, 36.
Wall, J.V. and Cooke, D.J. 1975. MNRAS, 171, 9.
Wall, J.V., Pearson, T.J. and Longair, M.S. 1980. MNRAS, 193, 683.
Wall, J.V., Pearson, T.J. and Longair, M.S. 1981. MNRAS, 196, 597.
Wall, J.V., Savage, A., Wright, A.E. and Bolton, J.G. 1982. MNRAS (in press).
Wall, J.V., Scheuer, P.A.G., Pauliny-Toth, I.I.K. and Witzel, A. 1982. MNRAS, 198, 221.
Walsh, D., Carswell, R.F. and Weymann, R.J. 1979. Nature, 279, 381.
Willis, A.G. and Miley, G.K. 1979. Astr.Astrophys., 76, 65.
Willis, A.G., Strom, R.G. and Wilson, A.S. 1974. Nature, 150, 625.
Willis, D. and Lynds, C.R. 1978. Astrophys.J.Suppl.Ser., 36, 317.

CLUSTERS OF GALAXIES

Piotr Flin

Jagiellonian University Observatory
ul. Orla 171, 30-244 Krakow, Poland

1. INTRODUCTION

The subject of this school, 'The Origin and Evolution of Galaxies', should include 'Clusters of Galaxies' in the title! Not only is galaxy clustering an essential aspect of the galaxy formation process, but I shall show that there is considerable evidence that galaxy morphology is not independent of is cluster environment. To understand galaxy evolution we must recognise the environmental effects, and before we can do that we have to recognise the galaxy cluster themselves. In these lectures I shall address this aspect of the problem, and I shall restrict the discussion to the observational data based on optical studies of rich clusters of galaxies. Moreover, I shall leave the theoretical explanations of the data to the imagination of the reader: relatively little theoretical work has been done on the environmental aspects of galaxy formation and I hope that this summary may provide some encouragement for theorists.

It is quite clear that my presentation expresses the point of view of the traditional, optical astronomer. I shall present only selected topics. This choice is subjective, and there are many interesting papers which will be not even mentioned. I apologise for such unfair treatment of many works, but many gaps in my presentation are covered by two excellent reviews dealing with clusters of galaxies by Bahcall (1977a) and van den Bergh (1977). (Throughout this paper I shall use H_0 = 50 km sec^{-1} Mpc^{-1}.)

There are two ways of investigating the distribution of galaxies. The first one is to select regions of sky where

B. J. T. Jones and J. E. Jones (eds.), The Origin and Evolution of Galaxies, 329–346.

condensations of galaxies are observed and then to study various properties of these condensations. The second manner is a statistical one: this approach was described by Peebles during the School. The analysis of the distribution of galaxies leads to the conclusion that the distribution of galaxies is not random. The two basic models for the large scale distribution of galaxies were described by de Vaucouleurs (1970, 1971).

The first of them is a continuous clustering model. The clustering occurs on all scales and there is no typical, characteristic length of the structure size. The second is an hierarchical model: the galaxies are in pairs and multiple systems, as well as groups which are themselves grouped into clusters, and so on.

In the modern astronomical terminology the notions of continuous clustering and of hierarchical clustering are used synonymously. This causes confusion, because the 'continuous clustering' describes an aspect of the distribution of galaxies, whereas the second notion suggests the existence of individual structures at different levels of the hierarchy, having preferred sizes. This is not a merely linguistic question as can be seen from the work of Rudnicki (1978) and the debate between Shanks (1979) and Peebles (1979). Bearing in mind the distinction between these two notions, continuous and hierarchical clustering, we can ask what contemporary results tell us. From the works of Peebles (1980, and references therein) and his collaborators we know that the distribution of galaxies on scales between a few kpc and 20 Mpc is continuous without any preferred scale. Thus structures with different sizes contribute to the observed shape of the covariance function.

So, if we know the shape of the covariance function why should we try to use another approach? The answer to this important question has been stressed by Peebles (1979):

'One must bear in mind that the two statistics, $w(\theta)$ and $z(\theta_1, \theta_2, \theta_3)$ do not fix the pattern of galaxy clustering, they only test the candidate models. The clustering hierarchy model seems to be the simplest that fits the observations. Decisions among models with the same w and z of course still is possible through the use of other measures ... or the detailed study of specific objects.'

Thus the detailed studies of separate aggregates of galaxies is of considerable importance.

2. THE DEFINITIONS OF CLUSTERS OF GALAXIES

A cluster of galaxies is seen as an enhancement of galaxies in one region of the sky in comparison to a neighbouring region. The question arises as to precisely what kind of enhancement would be called a 'cluster', so we need a definition of what is meant by a cluster of galaxies. There are two different definitions of clusters in common use, and these are related to the surveys of Abell (1958) and Zwicky (Zwicky et al. 1961-1968). Both Catalogues are based on the Palomar Observatory Sky Survey.

2.1 Abell's definition

A rich cluster of galaxies is an aggregation of galaxies having at least 50 members in a two magnitude interval (starting from the third brightest galaxy in cluster) lying within a circle of radius r from the center of the cluster. The value of r on Palomar Sky Survey plates corresponds to 1.5 Mpc. Two additional, selection criteria are imposed: the clusters ought to have the velocities of recession between 6000 and 60000 km sec^{-1} (z = 0.02-0.2) and the cluster ought to be in the investigated area of the Palomar Sky Survey. The distances to clusters and the number of galaxies within the circle, (corrected for background), were coded as 'distance' and 'richness' group. These are defined in Table 1.

Table 1

Richness group	0	1	2	3	4	5
Counts	30-49	50-79	80-129	130-199	200-299	over 300

Distance group	1	2	3	4	5	6
Magnitude of tenth brightest galaxy	13.1-14.0	14.1-14.8	14.9-15.6	15.7-16.4	16.4-17.2	17.3-18.0

In the whole investigated area of 30206 square degrees Abell found 2712 clusters, among which 1682 are the so-called 'rich clusters' (richness group not less than 1) which form a homogeneous sample.

Table 2

Distribution of Abell clusters

	Richness					Distance					
Group	1	2	3	4	5	1	2	3	4	5	6
No. of clusters	1224	383	58	6	1	9	2	33	60	657	921

There are selection effects that have to be considered. For example, the correlation between richness and distance can be attributed to selection effects. Furthermore, Dressler (1978) showed that richness as quoted in the Catalogue is affected by the presence or lack of bright galaxies.

Abell (1965) divided objects selected as clusters into two classes, regular and irregular clusters. The regular clusters are highly centrally condensed and they are spherically symmetric, having a great number of galaxies (about 10^3). The brightest members are ellipticals and lenticulars. The giant spirals are lacking or are very rare. Examples: Coma Cluster (A1656), and CrB Cluster (A2065). The irregular clusters do not show central condensation and they are not spherically symmetric. They contain a few hundreds of galaxies or less, sometimes a few dozen. Examples: Virgo Cluster, Hercules Cluster (A2151), Local Group (!).

2.2 Zwicky's definition

A cluster of galaxies is a region of the sky where the surface density of galaxies in a given magnitude range is twice the density of neighbouring regions. The magnitude range is m_{max}, $m_{max} + 3^m$, where m_{max} denotes the magnitude of the brightest cluster galaxy and within this range of magnitudes the number of galaxies inside the cluster contour, after correction for background, should be greater than 50.

Zwicky divides all clusters into five groups according to their distances from the observer (velocity of recession V) (see Table 3) and three morphological (structural) types.

The morphological type of a cluster reflects the distribution of galaxies within the cluster contour: compact, medium compact and open. In compact clusters only one central condensation of bright galaxies is observed. In this region ten or more of the

Table 3

Zwicky distance groups

Near	N	V < 15000 km/sec
Medium Distant	MD	15000 km/sec < V < 30000 km/sec
Distant	D	30000 km/sec < V < 45000 km/sec
Very Distant	VD	45000 km/sec < V < 60000 km/sec
Extremely Distant	ED	60000 km/sec < V

brightest galaxies seem to be in contact. These clusters show high degree of spherical symmetry.

Medium compact clusters are characterized by the occurrence of single condensation, where the brightest members are not in contact, or by several condensations. Some of them can be highly compact. Open clusters have no central condensation, but the number of galaxies per square degree inside the cluster is five times greater than in the neighbouring area, which is not considered as a cluster.

Table 4

Distribution of Zwicky Clusters

	N	MD	D	VD	ED
Open	292	348	495	447	336
Medium compact	219	517	850	1415	1790
Compact	26	54	143	664	1923

Zwicky's Catalogue of Galaxies and Clusters of Galaxies (Zwicky et al. 1961-68) contains objects with $\delta > -3°$.

Strong selection effects are observed in the distribution of clusters, as was pointed out by Zwicky himself. For example, he suspected that some of ED clusters may be low surface brightness nearby dwarf galaxies. This was confirmed by Rudnicki et al. (1978). Burns and Owen (1977) pointed out that the Catalogue is not complete over the full range of distances and morphological types. Abadi and Edmunds (1978) claim that the reduction in the number of non-compact clusters observed behind nearby clusters is considerable.

2.3 The definition of cluster (formal approach)

Possible formal definitions of clusters of galaxies, based on the two-dimensional distribution of galaxies were discussed by Rudnicki (1964). He considered five parameters, which are used in defining a cluster:

I) The considered range of magnitude of galaxies, 1) without restriction, and 2) with restriction.

II) The morphological types of galaxies. 1) Definition disregarding the morphological type of galaxies. 2) Definition taking into account the morphological type.

III) Features on which the definition is based. 1) The number of galaxies. 2) Number and area considered together. 3) Some other structural features.

IV) Correction for background. 1) All galaxies considered (no correction). 2) Excess of galaxies beyond the statistical number of background galaxies. 3) Number of galaxies in proportion to background.

V) Additional restriction of number of galaxies and/or considered area. 1) No restriction. 2) There are restrictions.

All definitions and classifications of clusters taking into account parameters III.3, IV.1 and IV.3 are distance dependent. Thus Abell and Zwicky both invoke I.2: Abell limits his sample to two magnitudes faintward of the third brightest and Zwicky limits his to 3 magnitudes fainter than the brightest. They both disregard morphological types of galaxy (II.1) and both recognise clusters via surface density enhancement (III.2). A major difference is in dealing with the background: Abell uses method IV.2 whereas Zwicky uses the distance dependent IV.3. What difference does this make to their respective catalogues of clusters?

Figure 1 shows the same area of sky with the clusters found by Zwicky (1959) and Zwicky (1961). The principal difference lies in the choice of parameters I and V: in the older map there is no restriction on the number of galaxies and nor is there any minimum membership (some clusters have fewer than 50 members). The main difference between the Abell and Zwicky catalogues lies in the correction for background. Thus it is often the case that Zwicky found one medium compact cluster where Abell selected two. It is slightly surprising then that the mean density of Zwicky compact clusters (0.14 clusters per sq. deg) is higher than the density of Abell clusters (0.09 clusters per sq.deg).

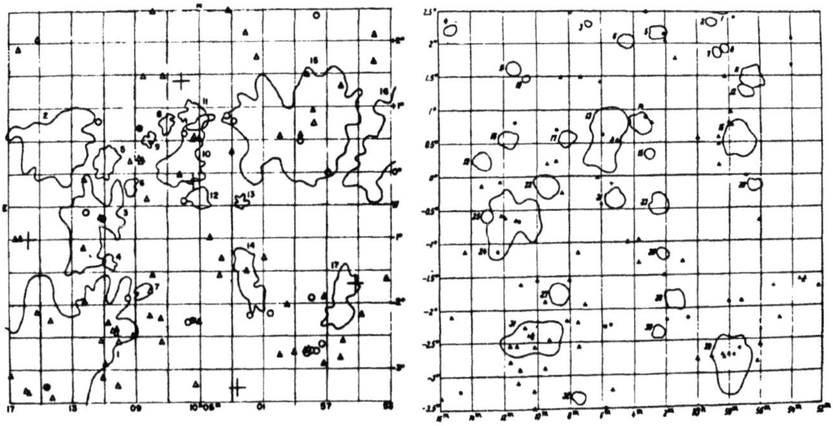

Figure 1. The same area surveyed by Zwicky (1961) (left) and Zwicky (1959) (right). The reasons for the difference are discussed in text.

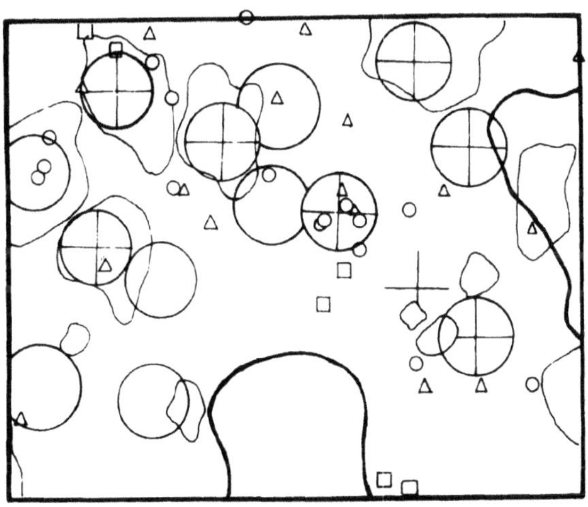

Figure 2. Clusters selected by Abell and by Zwicky on the same area of sky. Of Abell's 13 clusters only 7 were selected for the catalogue.

If we look at the particular region of sky shown in Figure 2, we see that Abell marked 13 clusters, but selected only 7 for the catalogue. Considering that Abell and Zwicky used the same material, the differences are remarkable, and this is a reflection on the great variety of structures that exist on the sky.

3. THE CLASSIFICATION OF CLUSTERS

The two most popular classifications of clusters were proposed by Bautz and Morgan (1970) and Rood and Sastry (1971). The Bautz-Morgan classification is based on the relative contrast between the brightest cluster galaxy and the other members.

Type I - cluster containing a centrally located cD galaxy (example: A 2199).

Type II - cluster where the brightest galaxy (or galaxies) is intermediate between a cD galaxy and giant ellipticals (example: Coma).

Type III - cluster without a dominant galaxy (example: Virgo).

Two intermediate types: I-II and II-III were also introduced into this classification scheme. A new Abell catalogue containing B-M classification of 1889 clusters was prepared by Leir and van den Bergh (1977). They pointed out some difficulties connected with this classification: the wrong assignation of the brightest galaxy changes the type from I to III (or from III to I), and clusters having very bright galaxies with multiple nuclei do not fall naturally within this scheme.

The decrease of the number of B-M type I, I-II and II with distance is caused by observational selection effects.

The Rood-Sastry (1971) classification is based on the observed distribution of the ten brightest galaxies in the cluster.

cD - the cluster contains cD galaxy (example: A401).

B - the cluster contains a bright binary system in the centre (example: A1656).

L - three or more brightest members are arranged in a line (example: A426).

C - four or more brightest galaxies are located in the centre forming a core, numerous fainter galaxies are distributed around them (example: A2065).

Table 5

The distribution of clusters according to B-M types

B-M type	I	I-II	II	II-III	III
% of clusters	3	4	14	22	57

Table 6

The distribution of B-M types with distance groups

B-M type	I	I-II	II	II-III	III	No. of clusters
Distance						
1-4	0.016	0.086	0.211	0.192	0.404	104
5-6	0.030	0.035	0.133	0.224	0.578	1579

F - several bright galaxies and a majority of fainter members form a flat configuration (example: A2151).

I - galaxies are distributed irregularly without a well defined centre (example: A400).

They classified 111 clusters in the distance groups 0-3. The distribution of cluster type is: cD - 23 clusters, B 10 clusters, L - clusters, C - 15 clusters, F - 20 clusters, I - 32 clusters. Their classification can be represented by a tuning fork diagram (Figure 3). The brightness and diameter of the brightest galaxy decreases along the tuning fork, remaining similar at corresponding places of the tuning fork arms.

Oemler (1974) studied 15 clusters of galaxies. For 10 of them besides the photometric data of member galaxies he had their morphological types. He suggested three morphological types of clusters.

cD cluster - has one (or two) outstanding galaxy. The ratio E:SO:S 3:4:2 (examples: A1413, A1656).

spiral-rich cluster. The ratio E:SO:S 1:2:3 (examples: A1226, A2151).

spiral-poor cluster. The ratio E:SO:S 1:2:1 (example: A194). The relation between the shape of the cluster and spiral content was that cD clusters are spherically symmetric while the spiral rich clusters are amorphous.

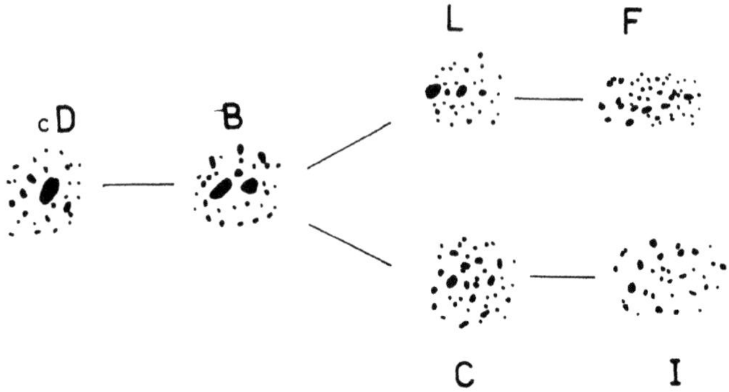

Figure 3. The Rood-Sastry (1971) classification of galaxy clusters represented as a 'tuning-fork' diagram.

The sample of clusters studied by Dressler (1980) containing photometric data and morphological types of galaxies in 55 clusters showed that Oemler's division was somewhat crude, owing to the smallness of his sample. The galaxian content changes continuously from cluster to cluster, without distinct separation into three groups.

4. GALAXIAN CONTENT OF CLUSTERS

Oemler (1977) showed that the ratio of elliptical to spiral galaxies depends on the cluster density: this ratio is greatest in the densest regions. He pointed out also (Oemler 1974) that the frequency of spirals is a decreasing function of cluster density, while the frequency of elliptical and lenticular galaxies increases. Melnick and Sargent (1977) noticed that the decrease of spirals toward the cluster centre is accompanied by the increase of lenticulars. White (1978) showed that when spirals occur among brightest members they are generally 'early' types, whereas in non-compact clusters 'late' type spirals are found among the brighter members.

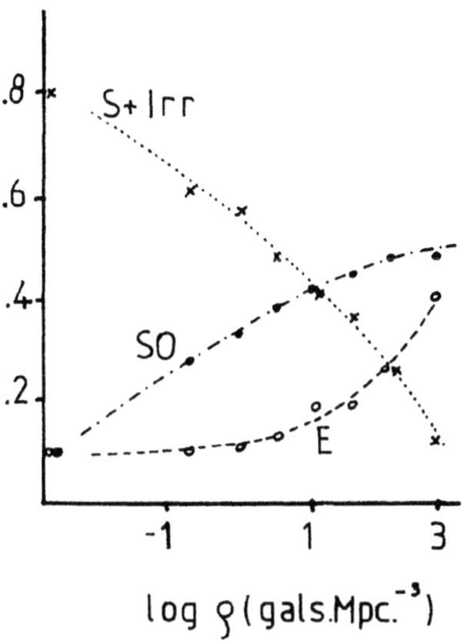

Fig. 4. The change of galaxy mix with local density according to Dressler.

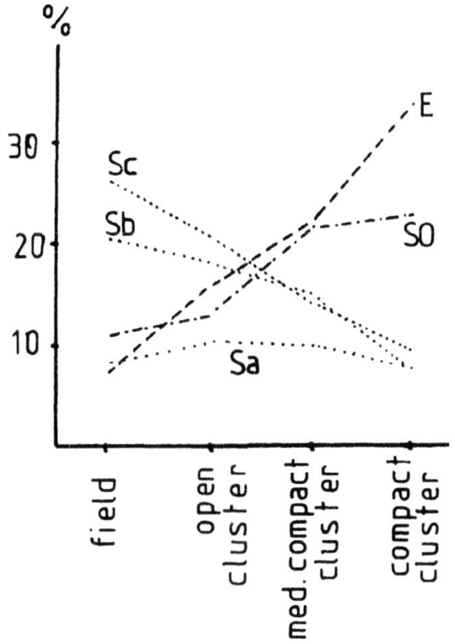

Fig. 5. The dependence of galaxy mix on environment deduced by Gisler.

A detailed study of cluster content on the basis of a large sample of clusters was carried out by Dressler (1978, 1980) Dressler chose to parametrise the cluster sample by the projected density, i.e. the number of galaxies per square Mpc. This solves the difficulties connected with the determination of cluster centre, or cluster size (which in the case of irregular clusters is a serious problem), and ensures a fair treatment of 'field' and cluster galaxies. The surface densities were calculated for the ellipticals, lenticulars and spirals in each of the clusters, moving all clusters to a distance corresponding to a redshift z = 0.04 in order to remove distance-dependent effects. The results for the whole sample are shown in Figure 4.

The change of content with projected density is quite simple. This run is observed over a range of projected density from 1/10 galaxy per square Mpc to 100 galaxies per square Mpc. In space this corresponds to a density ranging from 1/100 to 1000 galaxies per cubic Mpc. This is a very important diagram showing clearly how galaxy type depends directly on the environment. The high and low concentration clusters and also the strong X-ray

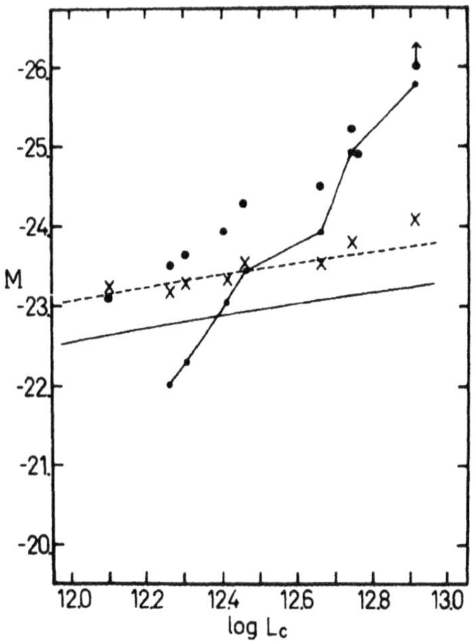

Figure 6. Oemler's (1976) relationship between the magnitude of the brightest cluster member and the cluster luminosity. ● denotes the whole cD galaxy, × the central part of the cD and • its envelope.

emitting clusters were studied separately. The differences between the two concentration groups are not significant and both are similar to the overall picture. The average content of galaxies is thus independent of the distribution of galaxies within the cluster. The strong X-ray emitting clusters have different composition: there is a deficit of spirals and an excess of lenticulars.

Gisler (1980) studied the distribution of the galaxy content of clusters taking account of different spiral types. He analysed 2004 galaxies with known redshifts and morphological types taken from the UGC. The association of galaxies with near Zwicky clusters was taken also from UGC and the cluster type was taken as an indicator of galaxy density. His diagram (Figure 5) is quite similar to Dressler's. However the fraction of Sa galaxies is almost constant, whereas the fraction of Sc galaxies decreases toward higher densities.

In summary, the average content of a cluster depends on its density. It is of further interest to ask whether cluster members have the same properties in different types of clusters. Strom (1978a,b,c) investigated 6 clusters with this in mind: one cD cluster (Coma), two spiral-poor clusters (Perseus and A2199) and three spiral-rich clusters (A1367, A1228, Hercules). The effective radii, as well as r_{26} (the radius at the isophote 26^m per square arcsec) radii were measured for 96, 72, 70, 48, 33 and 59 galaxies in each cluster respectively. The results obtained using these two different measures of galaxy size are in qualitative agreement. Their main results can be summarized as follows:

1) Galaxies with greater radii are more luminous.

2) For the same absolute magnitude galaxies belonging to dense (cD and spiral-poor) clusters have smaller effective radii than galaxies in low density (spiral-rich) clusters.

3) For the same absolute magnitude galaxies lying in the inner, more dense part of clusters have smaller radii than galaxies lying in the outer parts of the cluster.

This relationship is particularly striking in the special case of cD galaxies (Oemler 1976). The distribution of light in the central part of a cD galaxy is similar to that in normal giant ellipticals. However, cD's have in addition an 'envelope' with a very long flat light profile extending typically from 100 kpc to 3 Mpc in radius. The core radii of the brightest cluster members tend to be greater than those of the non-brightest cluster galaxies of the same absolute magnitude. Oemler's data is summarised in Figure 6. The large dots represent the absolute magnitude of the brightest galaxy in the cluster and the relationship is

$$L_{gal} \propto L_{cl}^{1.25}.$$

The crosses are the data for the central parts of the galaxies (for cD galaxies this is just the absolute magnitude of the galaxy without its envelope), and the best fit line to these points is 0^m5 above the theoretical prediction (solid line): the brightest galaxies are overluminous, even if we exclude their envelopes. The luminosity of the envelope is sensitively dependent on the cluster luminosity

$$L_{env} \propto L_{cl}^{2.25}$$

Bahcall (1980) and Romanishin and Thuan (1981) studied cD galaxies in small groups of galaxies and concluded that they have smaller envelopes than their counterparts in rich clusters.

5. ORIENTATION OF POSITION ANGLES

The earliest investigations of orientation of position angles of galaxies were based mainly on Brown's data (Brown 1968 and references therein: Reinhardt 1972). Reinhardt and Roberts (1972) showed, on the basis of these data that in the Local Supercluster position angles of galaxies tend to lie parallel to the supergalactic plane. (The analysis of UGC data performd by Nielson (1974) led to the same conclusion, but Jaaniste and Saar (1977) concluded that the spirals lay preferentially perpendicular to that plane'). However, Brown's measurements were made visually and there are strong selection effects which could introduce serious systematic errors (Holmberg 1946, 1975).

Hawley and Peebles (1975) investigated the UMa, Coma and Virgo clusters using measurements of the Palomar Sky Suvey Plates made by a professional plate scanner who knew nothing of the purpose of the work. By various means, such as changing the angle at which the image was presented to the scanner, they were able to study the visual selection effects. They also introduced a better statistical model in terms of which the significance of any result could be quantified. They found evidence for a radial alignment of position angles in the Coma cluster, but not at a compelling level of significance.

An investigation of eight clusters was carried out by Thompson (1976). These covered the full range of Bautz-Morgan and Rood-Sastry types and all belong to richness group not smaller than 1 (redshift $z \sim 0.05$). All galaxy-like images within the cluster radius 1.5 Mpc were studied. For each object its major and minor axis and position angle of major axis were measured. (All measurements were done twice, with the change of plate position of 112°.) Then, all diameters were corrected for Holmberg and inclination effects. After all corrections, galaxies with diameters greater than 15 kpc were chosen for detailed study. Only one cluster (A2197) exhibits significant alignment. The maximum of the distribution of position angles occurs at 100°, which is the cluster position angle too. The radial alignment of position angles of galaxies was also checked but only the Coma cluster showed any tendency for the major axes to lie radially from the cluster centre.

A2197 is an L-cluster according to Rood-Sastry classification. Seven clusters of this type were investigated by Adams, Strom and Strom (1980). All these clusters are spiral-rich, six of them are richness group 0, one of richness group 1. 4-m Mayall telescope plates were taken and measured on a PDS microdensitometer. (Automatic measurements, of course, ought not to be corrected for Holmberg and Opik effects.) They found that in two clusters (A999 and A2197) the distribution of position angles was not random. The peaks of the distribution coincide with position angles of clusters. For A179 marginally significant alignment in the position angles of axes for fainter galaxies was noticed. In this case the peak of the distribution was almost perpendicular to cluster position angle. For a detailed study of radial alignment the whole sample (285 galaxies) was divided into groups according to morphological types and absolute brightness of alaxies. They showed that the distribution of position angles for elliptical galaxies is random, whereas spiral and lenticular galaxies exhibit a strong tendency to have their position angles parallel or perpendicular to the radius vector, this tendency being stronger for brighter galaxies than for fainter ones.

The similar tendency of position angles to be either parallel or perpendicular to the radius vector in a deep sample of galaxies in the group centered at NGC 439 was noted by MacGillivray and Dodd (1977). Carter and Metcalfe (1980) investigated 21 clusters of galaxies. They demonstrated that cluster shape is elliptical rather than circular and clusters are intrinsically flatter than elliptical galaxies. The position angles of cD galaxies in cD clusters displayed distinct radial alignment. From 16 cD clusters only two have the position angles of their axes differing more than 22°5 from the position angles of cD galaxies.

6. BRIGHTNESS PROFILES OF CLUSTERS

The investigation of the distribution of galaxies within the cluster is usually done by counts in strips or concentric rings. Such investigations give two-dimensional data (i.e. surface density), from which the three-dimensional distribution (i.e. spatial density) can be obtained using methods of stellar statistics or the new methods (see e.g. Press 1976). The description usually is given in terms of one of three models, projected isothermal gas sphere model (Zwicky 1957), King model (King 1966, 1972), or de Vaucouleurs model (1960). All these models are based on spherical symmetry of clusters. The differences between them are negligible in comparison to errors introducted by this assumption and uncertainties of the observational data themselves.

It has long been thought that the radial density profiles of many clusters display a secondary maximum, or plateau, at large radii. Sharov (1959) first noticed this in a study of the Coma, Cancer and Hydra clusters, and more recently Bahcall (1977a,b and references therein) showed the existence of a secondary maximum in the profiles of a number of clusters. While in each cluster the secondary peak is significant only at the 2σ level, it might be argued that its occurrence in every cluster is evidence that it really is a general feature of galaxy clusters. On the other hand, Taff (1975) has argued on the basis of different methods of dealing with galaxy counts that the reality of this secondary maximum is dubious.

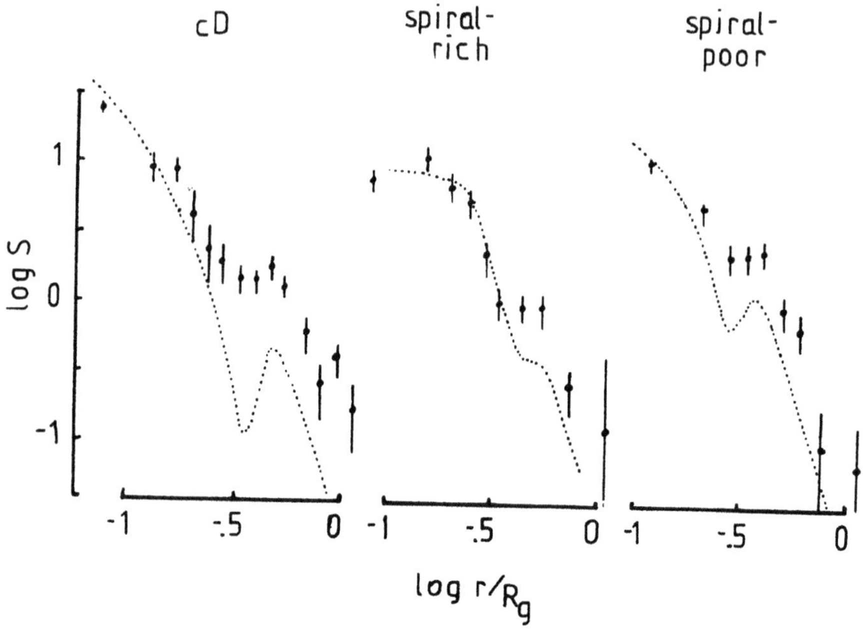

Figure 7. The composite brightness profiles of three kinds of cluster, according to Oemler (1974). The dots represent the projected profile and the dotted curves the inferred spatial profile.

Oemler's (1974) study of galaxy clusters revealed plateaux in the run of counts with projected radius, which when transformed into three-dimensional profiles corresponded to 'holes' in the density profile (see Figure 7). Moreover, he felt that the feature was most pronounced in cD clusters and least pronounced in spiral poor clusters. Hickson (1977) expressed the opinion that the strength of the feature correlated with Rood-Sastry type, being most marked in types B and C. Contrary to Oemler, he found the feature to be less strong in the cD clusters' In the most recent study Dressler (1978) found no evidence for features, but argued that deviations from a simple King-type profile were perhaps due to subclustering.

The issue is certainly not settled, but there is no lack of explanations for the phenomenon should it prove to be real. These are mostly based on the idea that the feature would reflect some aspect of the conditions at the time the cluster formed (Trevese and Vignato 1977: Occhionero, Vignato and Vittorio 1978: Gerbal and Salvadore-Sole 1980: Dekel and Shaham 1980: Dekel 1981).

Other important information that can be obtained from profiles is the cluster core radius, the cluster size and information on mass segregation. Unfortunately, even determining a core radius is not free from controversy: Bahcall (1977a,b) obtains a mean core radius of r_c = 0.25 Mpc for her sample, whereas Dressler (1978) gets r_c = 0.47 Mpc. The reason for this is not clear. Quintana (1979) showed that parameters derived from counts depend on the magnitude limit of the sample and felt this may be a consequence of mass segregation. The issue of mass segregation is important from the point of view of the dynamical evolution of the clusters. White (1976, 1977) considered this in relation to the Coma cluster.

Conclusions

In these lectures I have summarised the evidence that:

1) The variety of structures formed from galaxies is enormous.

2) There exists a close link between cluster general properties and galaxies belonging to cluster.

3) The positional angles of axes of galaxies in some clusters are not distributed in random.

4) The brightness (density) profiles of clusters may reveal interesting information about origin and/or evolution of clusters.

REFERENCES

Abadi, H.I. and Edmunds, M.G. 1978. Astr.Astrophys., 70, 189.
Abell, G.O. 1958. Ap.J.Suppl. 3, 211.
Abell, G.O. 1965. Ann.Rev.Astr.Astrophys., 3, 1.
Adams, M.T., Strom, K.M. and Strom, S.E. 1980. Ap.J., 238, 445.
Bahcall, N. 1977a. Ann.Rev.Astr.Astrophys., 15, 505.
Bahcall, N. 1977b. In Highlights of Astronomy, 4, Part 1, D. Reidel Publ. Co., Dordrecht, Holland, p. 247.
Bahcall, N. 1980. Ap.J., 238, 117.
Bautz, L. and Morgan, W.W. 1970. Ap.J., 162, L149.
Brown, F.G. 1968. MNRAS, 138, 527.
Burns, J.O. and Owen, F.N. 1977. Ap.J., 217, 34.
Carter, D. and Metcalfe, N. 1980. MNRAS, 191, 325.
de Vaucouleurs, G. 1960. Ap.J., 131, 585.
de Vaucouleurs, G. 1970. Science 167, 1203.
de Vaucouleurs, G. 1971. PASP, 83, 113.
Dekel, A. 1981. Preprint Caltech OAP-611.
Dekel, A. and Shaham, J. 1980. Astr.Astrophys., 85, 154.
Dressler, A. 1978. Ap.J., 226, 35.
Dressler, A. 1980. Ap.J., 236, 351.

Gerbal, D. and Salvador-Sole, E. 1980. Astr.Astrophys., 87, 165.
Gisler, G.R. 1980. Astr.J., 85, 623.
Hawley, D.L. and Peebles, P.J.E. 1975. Astr.J., 80, 477.
Hickson, P. 1977. Ap.J., 217, 16.
Holmberg, E. 1946. Medd.Lunds Astr.Obs. ser 2, No. 117.
Holmberg, E. 1975. In Galaxies and the Universe.
Jaaniste, J. and Saar, E.E. 1977. Tartu Obs.Preprint A-2.
King, I. 1966. Astr.J., 71, 64.
King, I. 1972. Ap.J., 174, L123.
Leir, A. and van den Bergh, S. 1977. Ap.J.Suppl. 34, 381.
MacGillivray, H.T. and Dodd, R.J. 1977. MNRAS, 186, 69.a
Melnick, J. and Sargent, W.L.W. 1977. Ap.J., 215, 401.
Nielson, P. 1974. Uppsala Rep. No. 6.
Occhinero, F., Vignato, F. and Vittorio, N. 1978. Astr. Astrophys., 70, 265.
Oemler, A. 1974. Ap.J., 194, 1.
Oemler, A. 1976. Ap.J., 209, 693.
Oemler, A. 1977. In Highlights of Astronomy, 4, part 1, D. Reidel Publ. Co., Dordrecht, Holland, p. 253.
Peebles, P.J.E. 1979. MNRAS, 189, 89.
Peebles, P.J.E. 1980. The large scale structure of the Universe, Princeton University Press.
Press, W.H. 1976. Ap.J., 203, 14.
Quintana, H. 1979. Astr.J., 84, 15.
Reinhardt, M.1972. MNRAS, 156, 151.
Reinhardt, M. and Roberts, M.S. 1972. Astrophys.Letters 12, 201.
Rood, H. and Sastry, G. 1971. PASP, 83, 313.
Rudnicki, K. 1964. Acta Astr. 14, 273.
Rudnicki, K. 1978. Irish Astr.J., 13, 241.
Rudnicki, K., Zieba, S., Snopkowski, R. and Wrona, S. 1978. Astr.Astrophys. 52, 1.
Shanks, T. 1979. MNRAS, 189, 89.
Shugarov, A.S. 1959. Astr.Zh. 36, 807.
Strom, K.M. and Strom, S.E. 1978a. Astr.J., 83, 73.
Strom, S.E. and Strom, K.M. 1978b. Astr.J., 83, 732.
Strom, K.M. and Strom, S.E. 1978c. Astr.J., 83, 1293.
Taff, L.G. 1975. Astr.Astrophys., 42, 242.
Thompson, L. 1976. Ap.J., 209, 22.
Van den Bergh, S. 1977. Vistas in Astronomy 21, 71.
White, R.A. 1978. Ap.J., 226, 591.
White, S.D.M. 1976. MNRAS, 177, 717.
White, S.D.M. 1977. MNRAS, 179, 33.
Zwicky, F. 1957. Morphological Astronomy, Springer Verlag, Berlin.
Zwicky, F. 1959. Handbuch der Physik, vol. 53, p. 390. Springer Verlag.
Zwicky, F., Herzog, E., Karpowicz, M., Kowal, C. and Wild, P. 1961-1968. Catalogue of Galaxies and Clusters of Galaxies, Vol. I-VI, Pasadena, Calif.Inst.Technology.

EPILOGUE: DO WE UNDERSTAND HOW GALAXIES FORMED?

Bernard J.T. Jones

Institute of Astronomy, Cambridge.

There are two approaches to the problem of galaxy formation: we can look at galaxies now and hope to find some fossil evidence for their early history, or we can study theoretical cosmology and discuss the evolution of various kinds of inhomogeneity. The difficulty with the first approach is that a lot of chemical and dynamical evolution has taken place since the formation of galaxies, and these processes tend to wipe out memories of time past. We are then led to studying the oldest stars in our own Galaxy, or trying to search for evolution in galaxies at great redshifts. The second approach is limited only by our ability to write down, solve and understand equations: though it cannot of course provide any answers without observational confrontation. The two approaches meet in the epochs corresponding to redshifts 10 - 100. The dynamical and chemical evolution is too complex to follow reliably into that regime, and that era is almost inaccessible to direct observation. There is in addition the great uncertainty concerning the nature and quantity of 'missing mass' in the Universe: that fraction of the Universe we can see may be irrelevant to understanding the origin and evolution of cosmic structure. It is interesting to reflect just how much our ideas would change if neutrinos turn out to have masses in excess of 10eV, and then to think that there may yet be other species of particle we know nothing about! We are therefore some way from being able to claim a proper understanding of how galaxies formed, but we nevertheless have some pretty good ideas on the subject.

The emphasis at present is on cosmological models that deviate only slightly from the homogeneous and isotropic Friedman-Lemaitre cosmology. Apart from the advantages such a

B. J. T. Jones and J. E. Jones (eds.), The Origin and Evolution of Galaxies, 347–352.

hypothesis offers from the point of view of mathematical tractability, it is a minimal assumption that does not do violence to the standard Big Bang nucleosynthesis, and moreover is adequate to explain the presence of structure at the present time. The deeper question of the origin of or the necessity for such ad hoc initial conditions is clearly an issue that we are on the verge of tackling with numerical cosmology and theories for the state of matter at the earliest instants. Even if we adopt the almost-homogeneous picture, we have the fundamental problem of what kind of perturbations were present: adiabatic or isothermal (or even both!). Because of the physical processes that take place during the fireball phase of the expansion, the adiabatic and isothermal hypotheses each lead to very different kinds of picture for the formation of galaxies. In the standard isothermal picture, cosmic structure is built hierarchically on ever increasing scales: galaxies form first and then clusters. In the adiabatic picture, cluster-scales condense out first (forming the so-called 'pancakes'), and galaxies form afterwards. It should surely be straightforward to distinguish such different hypotheses observationally, yet such tests are frustrated by the difficulty of computing reliable models in the essentially non-linear adiabatic theory.

The closest we can come to deciding this issue perhaps lies in the observation of the galaxy clustering correlation function, introduced ten years ago by Peebles (see Peebles 1981). The featureless form of the function argues that no specially distinguished scales exist in the range spanned by galaxies, galaxy clusters and super-clusters. Taken at face value this would argue against the adiabatic theory, which contains a preferred scale. However, numerical simulations show that this point of view is overly naive and that the non-linearity of the process can establish reasonable correlation functions over a wide range of scales (Shandarin, this volume). The isothermal, or hierarchical clustering, picture seems to run into difficulty here for numerical simulations of the hierarchical clustering process invariably produce correlation functions that are too steep (Efstathiou and Eastwood, 1981): they have too much structure on the smallest (galactic) scales. The problem is not insurmountable. The simple expedient of introducing a small-scale cut-off in the initial conditions, so as to inhibit the appearance of small scale structure, does the trick. However, it is necessary to introduce the cut-off on scales comparable to galaxy clusters. We then have a theory closely resembling the adiabatic theory without having any physical reason in the isothermal case for introducing that particular cut-off. There are a number of ways out of this dilemma in the standard isothermal picture which are quite acceptable. Merely changing the slope of the spectrum of initial inhomogeneities may work (no convincing numerical experiments have yet been done for non-Poisson initial conditions). The most interesting results so far, however, are those taking account of galaxy 'mergers' (Roos and Norman 1979: Jones and Efstathiou 1979: Aarseth and Fall 1980: Efstathiou and Jones 1980).

Numerical simulations, being limited in time evolution and spatial resolution, do not follow galaxies once they are 'formed'. They do not in their simplest form take account of the merging of substructure to form larger structures. Such mergers are not at all rare since relatively little orbital angular momentum is generated and objects that form are thus on orbits that bring them into contact and coalescence with their nearest neighbours. The net effect is a depletion in the number of objects in the densest regions, and a flattening of the correlation function towards the observed slope (Jones and Efstathiou 1979: Aarseth and Fall 1980).

This 'merger' picture should not, however, be viewed merely as a graceful exit from a difficult position. It is a natural consequence of the hierarchical clustering picture (the only question being at what level the members of the hierarchy preserve their identity) and provides a basis for understanding why elliptical galaxies are found preferentially in the central regions of galaxy clusters. This last point entails the reasonable identification of elliptical galaxies with the products of mergers of smaller systems: the fact that the merging rate is greatest in cluster environments explains the rest.

Elliptical galaxies have been observed to have a dynamically insignificant amount of rotation (Illingworth 1981): they evidently owe their shapes to velocity dispersion anisotropies arising out of the initial conditions of their formation (be that a consequence of anisotropic collapse or mergers). The numerical simulations of dissipationless hierarchical clustering lead to a prediction of the value of this small amount of rotation that agrees very well with the observations (Efstathiou and Jones 1979). This is surely encouraging though if we go all the way with mergers we should wonder what happens as a result of the collision between two systems containing gas: that is certainly not dissipationless! Disc galaxies, on the contrary, are rotation dominated systems and have undergone dissipative collapse. It is a straightforward problem to calculate the collapse of gas in a massive halo (spherical for simplicity (Fall and Efstathiou 1980), non-spherical to be realistic (Jones and Wyse 1982)), but it is non-trivial to understand why some galaxies end up with discs while others do not without appealing to the merging hypothesis. The merging hypothesis appears to have a good deal to commend it, though it should be remembered that the adiabatic theory is possibly less appealing only because of the difficulty of making concrete predictions.

The cosmic background radiation provides a tool for investigating the Universe at very early times: information is contained both in the spectrum of the radiation and its angular distribution. The positive detection of anisotropies on scales of 6° by Melchiorri et al. (1981) and of a quadrupole anisotropy (Boughn et al. 1981) and their implications for cosmology have been reviewed here by Partridge and Peebles. These measurements will certainly constrain models for early cosmic inhomogeneity, but we will not get a definitive picture until anisotropy is detected on sub-arcminute scales. The spectrum of the radiation has been measured by Woody and Richards (1979) and by Gush (1981), though the results are confusing and contradictory. Woody and Richards did not find the expected Compton-type distortion (Chan and Jones 1975a,b), but found instead an excess of radiation relative to a Planck law near the peak of the spectrum. If this measurement is correct (see Weiss 1980 for a discussion) it probably tells us something about the first stars to form (Negroponte et al. 1981: Puget and Heyvaerts 1980). Clearly the microwave background experiments have reached a crucial stage, and it is only a pity that we have to wait so long for new results!

Measures of the chemical properties of galaxies lead to correlations such as those found by Terlevich et al. (1981) between the luminosity, velocity dispersion, and metallicity in a sample of elliptical galaxies: this undoubtedly tells us something about the past history of these systems. Precisely what this tells us is for the moment unclear. There are two problems: firstly there is the observational problem of understanding the spectra of mixed stellar populations and secondly there is the theoretical problem that we have no idea how these stars have formed. The first problem is the less serious: we have some hope of deciding to what extent the line strength indices that are used as metallicity indicators reflect different stellar population mixes, even if the galaxies are made up of stars the like of which do not exist in the solar neighbourhood. The problem of star formation is, however, likely to be with us for some considerable time. Even if we understand the star formation going on at present in our Galaxy (Lequeux 1980), this may not be of any relevance to what happened in elliptical galaxies. Notwithstanding this slightly pessimistic point of view, it seems important to do more modelling of galaxy formation after the style of Larson's earlier work: we can at least hope that what happens is not too sensitive to the details of the star formation, and that there are enough observational constraints to tell us whether or not a model is acceptable.

Our own Galaxy evidently provides clues: these were identified twenty years ago in the classical paper of Eggen,

Lynden-Bell and Sandage (1962). During the time since then we have better metallicity calibrations and larger samples containing more extreme metal-poor stars (see Jones and Wyse 1982), but the basic picture has not really changed: the present disc was made from material that previously occupied a larger volume, and that was enriched during its collapse. What has become clearer is the dynamical and chemical distinction between the disk and so-called halo population, and the fact that stars in the halo have a minimum metallicity (the analogue of the G-dwarf problem: see Chiosi, this volume). The halo looks like a flattened elliptical galaxy in which a disk formed at a later stage (Ostriker and Thuan 1975: Jones and Wyse 1982). The absence of extreme metal poor stars ($Z < 10^{-3}\ Z_0$) implies the existence of a previous generation of massive stars that enriched the cosmic medium prior to the time of galaxy formation (see the papers of Carr and of Jones, this volume).

So, even if it is not certain how galaxies and clusters evolved, we have some good ideas and the important prospect of confronting these ideas with observation. The studies of the spectrum and isotropy of the cosmic background radiation are important in this respect, as are the ongoing studies of elliptical galaxies. If I had to point to a neglected area of potential importance I would choose the study of dwarf galaxies (both ellipticals and irregular) and particularly those of the Local Group. It is not clear where these fit into the scheme of things, but their paucity in metals identifies them as chemically unevolved relics of a time long past. They may have had complex dynamical histories: the stars we see in them now almost certainly did not form in such a low density environment. What, if anything, is their relationship with the globular clusters or moving groups of stars seen in the Galactic halo? Are they only found as satellites of larger galaxies? They are important because they are chemical fossils from the time when stars first accumulated to form galaxies, and their progenitors may even have been the building blocks for larger more impressive systems.

References

Aarseth, S.J. and Fall, S.M. 1980. Ap.J., 236, 43.
Boughn, S.P., Cheng, E.S. and Wilkinson, D.T. 1981. Ap.J.Lett. 243, L113.
Chan, K.L. and Jones, B.J.T. 1975a. Ap.J., 195, 1.
Chan, K.L. and Jones, B.J.T. 1975b. Ap.J., 200, 561.
Efstathiou, G. and Eastwood, J.W. 1981. MNRAS, 194, 503.
Efstathiou, G. and Jones, B.J.T. 1979. MNRAS, 186, 133.
Efstathiou, G. and Jones, B.J.T. 1980. Comm.Astr. and Space Phys., 8, 169.
Eggen, O.J., Lynden-Bell, D. and Sandage, A. 1962. Ap.J., 136 748.

Fall, S.M. 1980. Nature 281, 200.
Fall, S.M. and Efstathiou, G. 1980. MNRAS, 193, 189.
Gush, H.P. 1981. Phys.Rev.Lett., 47, 745.
Illingworth, G. 1981. In The Structure and Evolution of Normal Galaxies, ed. S.M. Fall and D. Lynden-Bell. Cambridge.
Jones, B.J.T. and Efstathiou, G. 1979. MNRAS, 189, 27.
Jones, B.J.T. and Wyse, R.F.G. 1982. Astr. and Astrophys, to appear.
Lequeux, J. 1980. In Star Formation (10th Saas Fee Course ed. A. Maeder and L. Martinet).
Melchiorri, F., Melchiorri, B.O., Ceccarelli, C. and Pietranera, L. 1981. Ap.J.Lett., 250, L1.
Negroponte, J., Rowan-Robinson, M. and Silk, J. 1981. Ap.J., 248, 38.
Ostriker, J.P. and Thuan, T.X. 1975. Ap.J., 202, 353.
Peebles, P.J.E. 1981. The Large Scale Structure of the Universe. Princeton.
Puget, J.L. and Heyvaerts, J. 1980. Astr.Astrophys., 83, L10.
Roos, N. and Norman, C.A. 1979. Astr.Astrophys., 76, 75.
Terlevich, R., Davies, R.L., Faber, S.M. and Burstein, D. 1981. MNRAS, 196, 381.
Woody, D.P. and Richards, P.L. 1979. Phys.Rev.Lett., 42, 925.
Weiss, R. 1980. Ann.Rev.Astr.Astrophys., 18, 489.

INDEX

9 789027 715173

GPSR Compliance

The European Union's (EU) General Product Safety Regulation (GPSR) is a set of rules that requires consumer products to be safe and our obligations to ensure this.

If you have any concerns about our products, you can contact us on ProductSafety@springernature.com

In case Publisher is established outside the EU, the EU authorized representative is:

Springer Nature Customer Service Center GmbH
Europaplatz 3
69115 Heidelberg, Germany

Batch number: 10583998

Printed by Printforce, the Netherlands